Reimagining Systems Thinking in a Post-Pandemic World

M. Elizabeth Azukas
East Stroudsburg University of Pennsylvania, USA

Minkyoung Kim
University of West Florida, USA

A volume in the Advances in Public Policy and Administration (APPA) Book Series

IGI Global
PUBLISHER of TIMELY KNOWLEDGE

Published in the United States of America by
IGI Global
Information Science Reference (an imprint of IGI Global)
701 E. Chocolate Avenue
Hershey PA, USA 17033
Tel: 717-533-8845
Fax: 717-533-8661
E-mail: cust@igi-global.com
Web site: http://www.igi-global.com

Library of Congress Cataloging-in-Publication Data

Names: Azukas, Mary Elizabeth, 1969- editor. | Kim, Minkyoung, 1977- editor.
Title: Reimagining systems thinking in a post-pandemic world / edited by M. Elizabeth Azukas, and Minkyoung Kim.
Description: Hershey, PA : Information Science Reference, an imprint of IGI Global, [2023] | Includes bibliographical references and index. | Summary: "Reimagining Systems Thinking in a Post-Pandemic World examines the role of systems thinking in a post-pandemic world. It identifies effective models of systems thinking and destems design and generates continuous knowledge building on systems thinking by addressing a multitude of industries and service communities. This book provides value in understanding the complexities of an interconnected world and in the exploration of effective approaches to systems thinking and design. Covering topics such as blended learning, local governments, and systems thinking, this premier reference source is an excellent resource for practitioners, policymakers, healthcare providers, business leaders and managers, educators of both K-12 and higher education, pre-service teachers, administrators and faculty, teacher educators, sociologists, librarians, researchers, and academicians"-- Provided by publisher.
Identifiers: LCCN 2022062299 (print) | LCCN 2022062300 (ebook) | ISBN 9781668472859 (h/c) | ISBN 9781668472897 (s/c) | ISBN 9781668472866 (ebook)
Subjects: LCSH: Systems engineering. | Distance education. | COVID-19 Pandemic, 2020---Social aspects.
Classification: LCC TA168 .R45 2023 (print) | LCC TA168 (ebook) | DDC 620.001/1710715--dc23/eng/20230124
LC record available at https://lccn.loc.gov/2022062299
LC ebook record available at https://lccn.loc.gov/2022062300

This book is published in the IGI Global book series Advances in Public Policy and Administration (APPA) (ISSN: 2475-6644; eISSN: 2475-6652)

British Cataloguing in Publication Data
A Cataloguing in Publication record for this book is available from the British Library.

For electronic access to this publication, please contact: eresources@igi-global.com.

Advances in Public Policy and Administration (APPA) Book Series

ISSN:2475-6644
EISSN:2475-6652

Mission

Proper management of the public sphere is necessary in order to maintain order in modern society. Research developments in the field of public policy and administration can assist in uncovering the latest tools, practices, and methodologies for governing societies around the world.

The **Advances in Public Policy and Administration (APPA) Book Series** aims to publish scholarly publications focused on topics pertaining to the governance of the public domain. APPA's focus on timely topics relating to government, public funding, politics, public safety, policy, and law enforcement is particularly relevant to academicians, government officials, and upper-level students seeking the most up-to-date research in their field.

Coverage

- Government
- Law Enforcement
- Political Economy
- Politics
- Public Administration
- Public Funding
- Public Policy
- Resource Allocation
- Urban Planning

IGI Global is currently accepting manuscripts for publication within this series. To submit a proposal for a volume in this series, please contact our Acquisition Editors at Acquisitions@igi-global.com or visit: http://www.igi-global.com/publish/.

The Advances in Public Policy and Administration (APPA) Book Series (ISSN 2475-6644) is published by IGI Global, 701 E. Chocolate Avenue, Hershey, PA 17033-1240, USA, www.igi-global.com. This series is composed of titles available for purchase individually; each title is edited to be contextually exclusive from any other title within the series. For pricing and ordering information please visit http://www.igi-global.com/book-series/advances-public-policy-administration/97862. Postmaster: Send all address changes to above address.

Titles in this Series

For a list of additional titles in this series, please visit: www.igi-global.com/book-series

Government Response to Disruptive Innovation Perspectives and Examinations
Sam B. Edwards, III (Quinnipiac University, USA) and James R. Masterson (Morehead State University USA)
Information Science Reference • © 2023 • 371pp • H/C (ISBN: 9781668464298) • US $215.00

Examining Colonial Wars and Their Impact on Contemporary Military History
Miguel Madueño (Rey Juan Carlos University, Spain) and Alberto Guerrero (Universidad de Granada, pain)
Information Science Reference • © 2023 • 311pp • H/C (ISBN: 9781668470404) • US $215.00

Governance as a Catalyst for Public Sector Sustainability
Neeta Baporikar (Namibia University of Science and Technology, Namibia & SP Pune University, India)
Information Science Reference • © 2023 • 392pp • H/C (ISBN: 9781668469668) • US $250.00

Acceleration of the Biopsychosocial Model in Public Health
Simon George Taukeni (University of Namibia, Namibia)
Medical Information Science Reference • © 2023 • 347pp • H/C (ISBN: 9781668464960) • US $325.00

Handbook of Research on the Regulation of the Modern Global Migration and Economic Crisis
Emilia Alaverdov (Georgian Technical University, Georgia) and Muhammad Waseem Bari (Government College University, Faisalabad, Pakistan)
Information Science Reference • © 2023 • 393pp • H/C (ISBN: 9781668463345) • US $285.00

Maintaining International Relations Through Digital Public Diplomacy Policies and Discourses
Türker Elitaş (Hatay Mustafa Kemal University, Turkey)
Information Science Reference • © 2023 • 270pp • H/C (ISBN: 9781668458228) • US $240.00

Handbook of Research on Transforming Government, Nonprofits, and Healthcare in a Post-Pandemic Era
Joanne E. Howard (Illinois Institute of Technology, USA)
Information Science Reference • © 2023 • 590pp • H/C (ISBN: 9781668423141) • US $270.00

Minding the Gap Between Restorative Justice, Therapeutic Jurisprudence, and Global Indigenous Wisdom
Marta Vides Saade (Ramapo College of New Jersey, USA) and Debarati Halder (Parul Institute of Law, Parul University, India)
Information Science Reference • © 2023 • 260pp • H/C (ISBN: 9781668441121) • US $215.00

IGI Global
PUBLISHER of TIMELY KNOWLEDGE

701 East Chocolate Avenue, Hershey, PA 17033, USA
Tel: 717-533-8845 x100 • Fax: 717-533-8661
E-Mail: cust@igi-global.com • www.igi-global.com

Editorial Advisory Board

Table of Contents

Detailed Table of Contents

The effects of COVID-19 on global societies are profound. The United Kingdom's approach to viral infection notifications and tracking in response to the pandemic can be used as a lesson. More specifically, it can assist in lessons learned in helping identify an alternative approach to improve epidemic containment management. Rather than use the UK's Coronavirus smartphone application, as implemented by the National Health Service's digital arm and Department of Health and Social Care, this study promotes a different approach and justifies the conceptual model pertaining to a distributed test, track, and trace system (DTTTS). By using soft systems methodology, this chapter discusses how the latter can provide a more meaningful and reliable disease surveillance strategy. The study purports that soft systems methodology can break down the defined problem. In addition, applying systems thinking will justify how the DTTTS model could be used in the real world to fulfill data processing and interoperability considerations for epidemic containment measures.

The coronavirus disease 2019 (COVID-19) pandemic caused issues related to public vaccination and nucleic acid testing management and mental and physical health pressures resulting from pandemic lockdowns. Gamification has played an important role in addressing these issues and the effect of COVID-19 by promoting behaviour changes, providing positive information, relieving mental stress, and improving happiness. In the COVID-19 prevention and control experience of China, gamification has been successfully applied in many effective cases, such as 1) the collection of gamified rewards to increase the public's acceptance for vaccination and nucleic acid testing; 2) the reduction of negative emotions and improvement of happiness during lockdown by gamified social interaction; 3) the encouragement of home fitness; and 4) the improvement of online education effects. This chapter focuses on the role of gamification in a pandemic, specifically as a support tool, by people in China.

This chapter will provide readers with how a systems based approach was established and implemented during and post pandemic to maintain educational and clinical services for graduate students studying to become speech-language pathologists and the clients they provide care. Achieving a master's degree is essential to becoming a licensed and certified speech-language pathologist (SLP). A board accredited program in communication sciences and disorders has specific standards that must be met to ensure all students gain comprehensive knowledge and skills across all domains within an SLP's scope of practice. Working directly with clients requires a systems approach whereby the SLP recognizes the relationship between culture, environment, language, and disability. The unprecedented nature of the pandemic had a profound impact on approaches used to meet student learning outcomes as well as provision of services for those clients receiving their therapeutic services.

The pandemic has challenged local government organizations to respond to highly dynamic environments involving health, social, educational, and economic changes. Recognizing local government organizations as systems optimizes them to respond and excel in their ability to serve the public. The purpose of this conceptual chapter is to assess the viable system model's (VSM) efficacy as a vehicle for evaluation within the context of local government organizations, and to determine what enhancements to the model might be needed for post-pandemic application. The viable system model may benefit from the integration of complementary enhancements that may humanize the VSM for use in a post-pandemic context. Ultimately, the new reality of the post-pandemic era calls for systems thinking to reveal ways to structure service delivery reliably and innovatively.

The chapter analyses the effects of income support measures implemented during the first year of the COVID-19 pandemic on poverty and income inequality in Serbia and proposes more efficient alternatives. Results indicate a decrease in poverty and inequality in 2020, compared to 2019, with much of the observed reduction due to the universal cash transfer to the adult population (UCT). However, UCT had high expenditures and contributed to the between-group inequalities increase. The second main measure: transfer to pensioners and social assistance recipients was poorly targeted, as pensioners had below-average poverty rates before the pandemic. Children and the unemployed, who had the highest pre-pandemic poverty rates, were not targeted by specific support measures. In the case of future extreme events, when formulating income support measures, the government should consider pre-crisis differences in income distribution and assess the impact of the crisis on different groups.

This chapter specifically deals with the response of the Serbian economy and society to the challenges of the crisis, focusing on research conducted from 2020 to 2022. Bearing in mind that economic systems remember and learn from both good and bad experiences, it was interesting to see if, in this relatively short time, in terms of data availability, the analyses that were done offered results not only from a short-term but also from a long-term perspective. After the analysis of the available studies and the systematization of the main findings, it was concluded that the majority of these studies provide only analyses of the short-term effects without applying rigorous evaluation methods. This is an important limit in terms of their reliability, but also with regards to usefulness of the obtained results for the policy making purposes.

The COVID-19 pandemic may constitute what Kurt Lewin refers to as an unfreezing moment for higher education institutions. When COVID-19 started to spread worldwide, traditional brick-and-mortar institutions were forced to close due to mandated lockdowns. Concerned with student success, universities rushed to move course content online through recorded lectures and synchronous class sessions via videoconferencing platforms. In transitioning back to normal, evaluating how materials developed during the pandemic can be integrated into the classroom setting is vital. Before the world returns to a post-pandemic environment, the question arises regarding what faculty should do with content created during the pandemic. Should they be archived, never to be used again? Or can they look for ways to integrate them into coursework to enhance learning? One-way schools can help to foster quality online education is by preserving the materials created and lessons learned during the pandemic, especially video content.

This chapter about the systemic changes that took place at one small university during the pandemic tells the story of what happened in March 2020 and how the department was positioned to address the needs of the students. The Corona-19 virus was the event that closed schools and forced the college's remote response for teaching and communication. A short survey was issued to students about their experience. The survey yielded a positive response about the department's readiness and response to the students. Yet, when looked at holistically and systemically, another paradigm emerges that resembles the mass customization concept evolving in the manufacturing and service industries for greater than 60 years. Mass customization as a systems thinking paradigm is the theoretical frame of this chapter. The mass customization paradigm is adapted from the manufacturing industry. This paradigm addresses the decision to customize products based upon customer needs.

Preface

A new coronavirus, SARS-COV-2, appeared in Wuhan, China in December of 2019. Outbreaks quickly escalated during a period of uncertainty about how the virus was transmitted. This respiratory syndrome, COVID-19, spread from China to 114 countries in three months (van Bavel, et al., 2020). In March of 2020, the World Health Organization declared COVID-19 a global pandemic (World Health Organization, 2020). To thwart the spread of the virus, governments introduced regulations and restrictions, such as mask mandates, social distancing requirements, contact tracing, curfews, and lockdowns, impacting many sectors of society, and with varying degrees of success (Moon, 2020; Sibley, et al., 2020). Governments faced numerous challenges posed by the COVID-19 pandemic including a lack of preparedness (Coggi & Regazzoni, 2022), managing communication, misinformation, and the public trust (Moon, 2020; Sibley et al., 2020), strain on the healthcare systems (Myers & Liu, 2022), policy responses and containment measures (Chung et al., 2021; OECD, 2020), divergent national approaches (Council on Foreign Relations, 2020), international cooperation (Coggi & Regazzoni, 2022), and economic impacts ((Donthu & Gustafsson, 2020; van Bevel et al., 2020; World Economic Forum, 2020; Zieba, 2021).

Although the pandemic affected all sectors of society directly or indirectly, the crisis initially took the biggest toll on the healthcare systems, particularly those in developing countries where the systems were already overburdened (Haileamlak, 2021). The virus presented challenges in terms of human resource allocation, facility utilization, and medical supply management. In addition to the direct disease impact of COVID-19, the situation resulted in indirect morbidity and mortality from other preventable and treatable diseases as a result of the disruptions in essential health services (Department of Health and Social Care, Office for National Statistics, Government Actuary's Department and Home Office, 2021). Critical gaps in service were caused by the need to shift healthcare workers to provide COVID-19 related services, decreases in public transportation, loss of income to pay for services, and cancellations of planned services due to fear of contracting the virus (Haileamlak, 2021). There were also shortages of medicines, diagnostics, and other technologies during the pandemic (World Health Organization, 2021). Finally, there were higher rates of morbidity and mortality among healthcare workers caring for people with the virus, further impacting shortages of medical personnel (Haileamlak, 2021).

The COVID-19 pandemic also caused a significant disruption in the economic sector. Although the global markets have faced other challenges, the world has not seen such wide-spread government interventions since World War II (Verhoef, et al., 2022). All major stock exchanges saw massive declines as the number of COVID-19 cases grew in the first few months of the crisis (Jones, et al., 2021) and gross domestic product (GDP) reflected unprecedented global decreases in 2020 (Verhoef, et al., 2021). The lockdowns introduced by governments to contain the virus also significantly decreased consumption impacting businesses, particularly small-scale services and medium sized enterprises (Zieba, 2021).

The pandemic took a devastating toll on the hospitality, tourism, transportation, art, entertainment, and retail industries, while other industries, such as those that produced internet communication tools, digital entertainment suppliers, and online retail operations experienced increases in demand and consumption (Jones et al., 2021; Verhoef, et al., 2021; Zieba, 2021). In order for businesses to survive, they had to shift from the strategies used in "normal times" to emergent knowledge strategies, shifting to digital operations almost overnight (Bratianu & Benjinaru, 2021).

Global supply chains were disrupted causing chaos in business operations in almost all countries (Alsmairat, 2021; van Hoek, 2020). This affected the availability of goods and services as companies attempted to strategically manage their inventories to respond to demand fluctuations (Mollenkopf, et al., 2021). Improving the resilience of the supply chain has become a strategic goal for business organizations (Alsmairat, 2021). In response to these economic problems, central banks kept their interest rates very low to support economic recovery (Bloomberg, 2021). Additionally, governments implemented large scale stimulus packages to support both businesses and individuals economically impacted by the pandemic. These programs, coupled with pent up demand, have resulted in a wave of post-pandemic global inflation (Binici et al., 2022).

The COVID-19 pandemic also disrupted the education sector, creating the largest disruption of education services in history, impacting 1.6 billion learners in more than 190 countries on all continents (United Nations, 2020). The length of school closures varied widely across the world. Schools in south Asia were full or partially closed the longest, at an average of 84 weeks (Bryant, et al., 2022). This was followed by Latin American and the Caribbean schools which were closed for an average of 77 weeks (Bryant, et al., 2022). The length of school closures was frequently impacted by the relative wealth of the society, with schools in Europe being closed for only an average of 29 weeks (Bryant et al., 2022). Access to quality remote and hybrid learning also varied both across and within countries. Much of this depended on the school's ability to quickly shift to digital operations as well as access to the needed software, hardware, and infrastructure to facilitate online instruction. Thus, significant concerns have been raised regarding "learning loss." It is estimated that south Asia learners are experiencing an average delay of 12.4 months, with Latin American and the Caribbean learners exhibiting, on average, an 11.7 month learning delay (Bryant et al., 2022). Learning delays are less in the areas where schools reopened sooner. For example the estimated learning loss in Europe and central Asia is estimated to be 3.6 months per learner and 4.3 months per learner in North America (Bryant et al., 2022). On average, the global learner is estimated to be 8.2 months behind (Bryant, et al., 2022). The World Economic Forum (2022) estimated that students now risk losing $17 trillion in lifetime earnings in present value, equivalent to 14% of the global GDP, because of school closures and economic shocks.

The pandemic also exacerbated inequality within systems. For example, in the United States, gaps increased in academic performance between white students and students of color and in Ethiopia gaps between urban and rural education were compounded (Bryant et al., 2022). Inequities were also observed in higher education including differential treatment of students based on their backgrounds, limited access to knowledge, unevenness in global patterns of research collaboration, and a lack of the basic requirements for digital education, including devices, Internet access, and electricity (Abrasheva, et al., (2022). There was also a major impact on physical mobility. Students faced challenges related to visa issues, health, and well-being, and were often the victims of xenophobia and discrimination (Abrasheva, et al., (2022). Concerns about the pandemics' impact on the mental health of students have been raised across all sectors of education (Abresheva et al., 2022; Bryant et al., 2022; United Nations, 2020).

Preface

Although people are still being infected with COVID-19, with thousands dying every week and millions of people suffering debilitating long-term effects of the virus, the World Health Organization declared in May of 2023 that COVID-19 was no longer a global health emergency (World Health Organization, 2023). As of June 2023, there have been 767, 984, 989 confirmed cases of COVID-19 and 6, 943, 390 confirmed deaths (World Health Organization, 2023). The pandemic has proven to be a multi-faceted phenomenon impacting multiple sectors, including government and public administration, media, transportation, education, and the business sector and the global economy (Donthu & Gustafsson, 2020; van Bevel et al., 2020; Zieba, 2021). It has highlighted the world's complex interdependent structures, and it will require a multidisciplinary systems thinking approach for post-pandemic recovery and future pandemic prevention.

Systems thinking is an approach that focuses on understanding complex systems as interconnected and interdependent entities (Arnold & Wade, 2015; Checkland, 1981; Senge, 1990). It recognizes that a system is more than the sum of its parts and emphasizes the relationships, feedback loops, and dynamics within the system (Arnold & Wade, 2015; Senge, 1990). Rather than analyzing individual components in isolation, systems thinking examines the interactions and interdependencies among these components to gain a holistic understanding of the system as a whole. It recognizes that changes in one part of the system can have ripple effects throughout the entire system, identifying the underlying structures and patterns that shape system behavior over time.

The idea of systems thinking has garnered intuitive recognition due to the growing complexity that pervades the natural, economic, social, and political spheres (Cabrera, 2006; Cabrera & Cabrera, 2015). Systems thinking approach enables a deeper comprehension of the complexity and dynamics of the system allowing for better decision-making, problem-solving, and intervention strategies (Nelson, 2020), and it is a truly interdisciplinary endeavor that encompasses both scholarly and theoretical explorations as well as practical applications.

Given the complexities inherent in a pandemic and the subsequent post-pandemic world, the adoption of a systems thinking approach becomes crucial. It facilitates a deeper understanding of the intricate dynamics involved in pandemics and enables the development of effective strategies for post-pandemic recovery and future pandemic prevention efforts. By embracing systems thinking, we can unravel the interconnections and interdependencies within our global systems, enabling us to develop sustainable solutions to address the challenges posed by pandemics.

The systems thinking approach encourages a shift from linear cause-and-effect thinking to a more nuanced understanding of the dynamic relationships and feedback loops that influence system behavior. It acknowledges that changes in one part of the system can have unintended consequences in other parts. By considering the broader context and long-term implications, systems thinking offers a framework for addressing complex problems, fostering innovation, and promoting sustainable solutions (Checkland, 1981).

As we navigate a post-pandemic world, the relevance of systems thinking extends beyond specific industries or sectors. It is a tool for fostering collaboration, addressing interconnected challenges, and envisioning systemic change. This book acknowledges the inherent complexity of the issues we face and encourages readers to embrace a systems thinking mindset. By doing so, we can effectively navigate uncertainties, identify leverage points, and design innovative solutions that account for the interconnectedness of our social, economic, and environmental systems.

This book examines the role of systems thinking in a post-pandemic world. It identifies effective models of systems thinking and systems design and generates continuous knowledge building on systems thinking by addressing a multitude of industries and service communities. The book is organized thematically and explores the impact of the pandemic on the health sector, governments and economies, and education both in the higher education and K-12 contexts.

In Chapter 1, David Bird examines the United Kingdom's approach to viral infection notifications and tracking during the pandemic. He proposes an alternative approach utilizing a distributed test, track, and trace system (DTTTS) model. It discusses the ways in which DTTTS can enhance disease surveillance strategies and address data processing and interoperability considerations.

In Chapter 2, Huang Miao and Mohamad Saifudin Mohamad Saleh explore the role of gamification in pandemic response. It examines the successful application of gamification in China, highlighting its effectiveness in promoting behavior changes, disseminating positive information, reducing mental stress, and enhancing happiness. It further explores a range of gamified applications related to vaccination, social interaction, home fitness, and online education.

In Chapter 3, Rachel Wolf and Akila Rajappa highlights the challenges encountered by graduate students pursuing speech-language pathology (SLP) studies amid the pandemic. It delves into the adoption of a systems-based approach to ensure the uninterrupted provision of educational and clinical services. It emphasizes the interplay between culture, environment, language, and disability within the practice of SLP.

In Chapter 4, Sheila Murphy and Tracey Reginold explore the local government response to the pandemic and propose the Viable Systems Model (VSM) to improve the delivery of social services by local government in a post-pandemic context.

In Chapter 5, Marko Vladisavljević analyzes the impact of government income support measures implemented in Serbia during the pandemic on poverty and income inequality. He discusses the need for policy-makers to consider important social and economic factors, such as pre-existing inequalities and risks faced by different groups, so as not to inadvertently increase inequality among groups.

In Chapter 6, Kosovka Ognjenović and Mihajlo Đukić evaluate government policies implemented during the pandemic focusing on both long and short term impacts. They emphasize the need for rigorous systemic evaluation methods for reliable and useful evidenced-based policy making.

In Chapter 7, Audrey Lombardo and Timothy Lombardo examine how materials developed during the pandemic can be effectively integrated into the classroom setting to enhance learning during the transition from the pandemic back to normalcy. Adopting a systems theory approach, they explore strategies to optimize student engagement with reused video content. The chapter also emphasizes the significance of preserving these materials and highlights their potential to enhance learning when integrated into coursework.

In Chapter 8, Al Galloway and Daniel Gutmore discuss the systemic changes that unfolded in a small university during the pandemic. It draws insightful parallels between the concepts of mass customization in manufacturing and the evolving demands of education in a post-pandemic context, emphasizing the significance of tailoring products to meet customer needs as a key aspect of adapting to the changing educational landscape.

In Chapter 9, Dewi Wahyu Mustikasari and Keith Heggart explore the external factors that impacted teacher design processes during the implementation of blended and online learning focusing on higher education settings. They employ a systems perspective to comprehend the intricate relationship between university leaders, educators, and other stakeholders. It highlights the crucial role of institutional support services and tools in facilitating successful implementation.

In Chapter 10, Monalisa Dash, Ericka Eppler, and Matthew Meador explore the People, Process, and Environment (PPE) approach within post-pandemic online education systems, with a focus on foundational systems thinking concepts. It compares multiple systems thinking models and presents the PPE model as a straightforward optimization tool. It analyzes case studies from different educational settings to highlight the need for technological awareness, adaptability, and increased utilization of learning management systems.

In Chapter 11, Sarah Beach and Helen Min apply Critical Systems Heuristics to explore the concept of success in the U.S. educational system, particularly during the pandemic. They explore how a reimagined system could move away from standardized methods of measurement, expand agency and collaboration, and prioritize growth for students and teachers.

In Chapter 12, Sarah O'Dell and Maureen Burgess discuss how the pandemic prompted the need to reexamine their mission at the Hewitt School, a K-12 girl's school in New York City. They explore the need to use systems thinking and evidence-based research to revise their mission and improve the education of girls.

In Chapter 13, Jennifer Bailey and Gary Miller Chapter examine the national teacher crisis in the U.S., exacerbated by the pandemic and explore measures introduced to address teacher recruitment and retention challenges. They incorporate practitioner voices to provide insights into the daily challenges faced by school and teacher leaders. Bailey and Miller propose a potential solution through an open systems lens to build collective capacity in mitigating the post-pandemic teacher shortage crisis.

This book provides value in understanding the complexities of an interconnected world and in the exploration of effective approaches to systems thinking and design. Addressing themes such as health, government and economics, and education in the context of systems thinking, this premier reference source is an excellent resource for practitioners, policymakers, healthcare providers, business leaders and managers, educators of both K-12 and higher education, pre-service teachers, administrators and faculty, teacher educators, sociologists, librarians, researchers, and academics.

M. Elizabeth Azukas
East Stroudsburg University of Pennsylvania, USA

Minkyoung Kim
University of West Florida, USA

REFERENCES

Abrashera, D., Escribens, M., Sabzalieva, E., Viero do Nascimento, D., & Yerovi, C. (2022). *Resuming or reforming: Tracking the global impact of higher education after two years of disruption.* UNESCO International Institute for Higher Education in Latin America and the Caribbean. https://unesdoc.unesco.org/ark:/48223/pf0000381749

Alsmairat, A. K. M. (2021). Transformative supply chain drivers during COVID-19: A customer perspective. *Polish Journal of Management Studies*, *24*(2), 9–23. doi:10.17512/pjms.2021.24.2.01

Arnold, R. D., & Wade, J. P. (2015). A definition of systems thinking: A systems approach. *Procedia Computer Science*, *44*, 669–678. doi:10.1016/j.procs.2015.03.050

Binizi, M., Centorinao, S., Cevik, S., & Gwon, G. (2022). *Here comes the change: The role of global and domestic factors in post-pandemic inflation in Europe.* International Monetary Fund. https://www.imf.org/en/Publications/WP/Issues/2022/12/09/Here-Comes-the-Change-The-Role-of-Global-and-Domestic-Factors-in-Post-Pandemic-Inflation-in-525754#:~:text=Global%20inflation%20has%20surged%20to,to%20become%20an%20entrenched%20phenomenon

Britanu, C., & Bejinaru, R. (2021). COVID-19 induced emergent knowledge strategies. *Knowledge and Process Management*, *28*(1), 11–17. doi:10.1002/kpm.1656

Bryant, J., Child, F., Espinosa, J., Hall, S., Kola-Oyeneyin, T., Lim, C., Panier, F., Sarakatsannis, J., Schmautzer, D., Unger, S., & Woord, B. (2022). *How COVID-19 caused a global learning crisis*. McKinsey & Company. https://www.mckinsey.com/industries/education/our-insights/how-covid-19-caused-a-global-learning-crisis

Cabrera, D. (2006). *Systems thinking* [Doctoral dissertation]. Cornell University.

Cabrera, D., & Cabrera, L. (2015). *Systems thinking made simple: New hope for solving wicked problems in a complex world*. Odyssean Press.

Challenges and opportunities in the post COVID-19 world. (2020). World Economic Forum. https://www.weforum.org/reports/post-covid-19-challenges-and-opportunities

Challenges of global governance amid the COVID-19 pandemic. (2020). Council on Foreign Relations, International Institutions and Global Governance Program. https://www.cfr.org/report/challenges-global-governance-amid-covid-19-pandemic

Checkland, P. (1981). *Systems thinking, systems practice*. Wiley.

Chung, H. W., Apio, C., Goo, T., Heo, G., Han, K., Kim, T., Kim, H., Yeonghyeon, K., Doeun, L., Lim, J., Lee, S., & Park, T. (2021). Effects of government policies on the spread of COVID-19 worldwide. *Scientific Reports*, *11*(1), 20495. doi:10.103841598-021-99368-9 PMID:34650119

Coggi, P. T., & Coggi, C. J. (2022). *COVID-19 after two years: The failure of pandemic governance*. Council of Councils. https://www.cfr.org/councilofcouncils/global-memos/covid-19-after-two-years-failure-pandemic-governance

Department of Health and Social Care, Office for National Statistics, Government Actuary's Department and Home Office, (UK). (2021). *Direct and Indirect Impacts of COVID-19 on Excess Deaths and Morbidity.* https://assets.publishing.service.gov.uk/government/uploads/system/uploads/attachment_data/file/907616/s0650-direct-indirect-impacts-covid-19-excess-deaths-morbidity-sage-48.pdf

Donthu, N., & Gustafsson, A. (2020). Effects of COVID-19 on business and research. *Journal of Business Research*, *117*, 284–289. doi:10.1016/j.jbusres.2020.06.008 PMID:32536736

Haileamlak, A. (2021). The impact of COVID-19 on health and health systems. *Ethiopian Journal of Health*, *31*(6), 1073–1074. doi:10.4314/ejhs.v31i6.1 PMID:35392335

Jones, L., Palumbo, D., & Brown, D. (2021). *Coronavirus: How the pandemic has changed the world economy*. BBC News. https://www.bbc.com/news/business-51706225

Mollenkopf, D. A., Ozanne, L. K., & Stolze, H. W. (2021). A transformative supply chain response to COVID-19. *Journal of Service Management*, *32*(2), 190–202. doi:10.1108/JOSM-05-2020-0143

Myers, L. C., & Liu, V. X. (2022). The COVID-19 pandemic strikes again and again and again. *JAMA Network Open*, *5*(3), e221760. Advance online publication. doi:10.1001/jamanetworkopen.2022.1760 PMID:35262720

Nelson, H. G. (2020). The promise of systemic designing: Giving form to water. *Learning, design, and technology: An international compendium of theory, research, practice, and policy*, 1-49.

Policy brief: Education during COVID-19 and beyond. (2020). United Nations. https://www.un.org/development/desa/dspd/wp-content/uploads/sites/22/2020/08/sg_policy_brief_covid-19_and_education_august_2020.pdf

Senge, P. (1990). *The Fifth Discipline: The Art and Practice of The Learning Organization*. Currency Doubleday.

Sibley, C. G., Greaves, L. M., Satherley, N., Wilson, M. S., Overall, N. C., Lee, C. H. J., Milojev, P., Bulbulia, J., Osborne, D., Milfont, T. L., Houkamau, C. A., Duck, I. M., Vickers-Jones, R., & Barlow, F. K. (2020). Effects of the COVID-19 pandemic and nationwide lockdown on trust, attitudes toward government, and well-being. *The American Psychologist*, *75*(5), 618–630. doi:10.1037/amp0000662 PMID:32496074

The global education crisis is even worse than we thought. Here's what we need to do about it. (2022). World Economic Forum. https://www.weforum.org/agenda/2022/01/global-education-crisis-children-students-covid19/

The territorial impact of COVID-19: Managing the crisis across levels of government. (n.d.). OECD Policy Responses to Coronavirus (COVID-19). https://www.oecd.org/coronavirus/policy-responses/the-territorial-impact-of-COVID-19-managing-the-crisis-across-levels-of-government-d3e314e1/

Ultra low interest rates are here to stay: 2021 central bank guide. (2021). Bloomberg News. https://www.bloomberg.com/news/articles/2021-01-05/ultra-low-interest-rates-here-to-stay-2021-central-bank-guide#xj4y7vzkg

van Bavel, J. J., Baiker, K., Boggio, P. S., Capraro, V., Cichocka, A., Cikara, M., Crockett, M. J., Crum, A. J., Douglas, K. M., Druckman, J. N., Drury, J., Oeindrila, D., Ellemers, N., Finkel, E. J., Fowler, J. H., Gelfand, M., Han, S., Haslam, S. A., Jetten, J., & Willer, R. (2020). Using social and behavioral science to support COVID-19 pandemic response. *Nature Human Behaviour*, *4*(5), 460–471. doi:10.103841562-020-0884-z PMID:32355299

Verhoef, P. C., Noordhoff, C. S., & Sloot, L. (2022). Reflections and predictions on effects of COVID-19 pandemic on retailing. *Journal of Service Management*, *34*(2), 274–293. doi:10.1108/JOSM-09-2021-0343

World Health Organization. (2020). *COVID-19 disease situational reports*. https://www.who.int/emergencies/diseases/novel-coronavirus-2019/situation-reports/

World Health Organization. (2021). *COVID-19 significantly impacts health services for noncommunicable diseases.* https://www.who.int/newsroom/detail/01-06-2020-covid-19-significantly-impacts-health-services-for-noncommunicable-diseases

World Health Organization. (2023). *Coronavirus disease (COVID-10) pandemic.* https://www.who.int/emergencies/diseases/novel-coronavirus-2019

Zieba, K. (2021). How can systems thinking help us in the COVID-19 crisis? *Knowledge and Process Management*, *29*(3), 221–230. doi:10.1002/kpm.1680

Chapter 1
Using Soft Systems Methodology to Prove the Principles for a Distributed Test, Track, and Trace System to Support Epidemic Containment Measures

David Anthony Bird
https://orcid.org/0000-0002-5953-7107
British Computer Society, UK

ABSTRACT

The effects of COVID-19 on global societies are profound. The United Kingdom's approach to viral infection notifications and tracking in response to the pandemic can be used as a lesson. More specifically, it can assist in lessons learned in helping identify an alternative approach to improve epidemic containment management. Rather than use the UK's Coronavirus smartphone application, as implemented by the National Health Service's digital arm and Department of Health and Social Care, this study promotes a different approach and justifies the conceptual model pertaining to a distributed test, track, and trace system (DTTTS). By using soft systems methodology, this chapter discusses how the latter can provide a more meaningful and reliable disease surveillance strategy. The study purports that soft systems methodology can break down the defined problem. In addition, applying systems thinking will justify how the DTTTS model could be used in the real world to fulfill data processing and interoperability considerations for epidemic containment measures.

DOI: 10.4018/978-1-6684-7285-9.ch001

INTRODUCTION

The aim of this chapter is to provide logical and nuanced evidence regarding the United Kingdom's (UK) response to the coronavirus disease (COVID-19) outbreak. More specifically, review the UK's adoption of e-health and telehealth methods that were directly influenced by the pandemic. Building on publications by Bird (2020a, 2021), this chapter explores an alternative approach to the COVID-19 smartphone application used by the UK National Health Service (NHS) Test and Trace service. This alternative approach can be used to discern how far viral infections have spread across the country, as well as how widespread COVID-19 has infected society itself.

Research Aim

This chapter analyzes the past use-case of the NHS Digital (NHSx) smartphone application (herein referred to as the NHSx app), by applying soft systems methodology (SSM). This research paints a picture of an alternative technology option in which a distributed test, track, and trace system (DTTTS) model can be used instead.

Using evidence sourced from a literature review from many reputable sources this chapter employs the SSM seven-stage model as a means to discern, and make sense, of the existing NHSx app. The seven-stage model is used to scrutinize the real-world problem to identify gaps within the NHSx app ecosystem and uses systems-thinking for problem-solving purposes. The author explores the alternative DTTTS approach proposed in Bird (2020a, 2021) and undertakes a compare-and-contrast analysis to highlight the pros and cons of both the NHSx app and the DTTTS.

The author is able to establish that the DTTTS concept could be a key enabler to help the UK Department of Health and Social Care (UKDHSC) respond more effectively and efficiently to an evolving severe acute respiratory syndrome (SARS) epidemic. Thereby, the DTTTS concept could enable the UK Health Security Agency (UKHSA) rapidly deploy epidemiological surveillance (ES) teams to viral hotspots, and their proximities. Enabling the UKHSA ES teams to select further test subjects for additional testing in order to monitor the spread of infection and by doing so inform UKHSA containment decisions and requisite actions by the UKDHSC. As a consequence, the author is able to justify how the DTTTS conceptual model could provide additional benefits hitherto not realized through the NHSx app.

BACKGROUND

On July 19, 2021, the prime minister of the UK relaxed England's legal restrictions used to counter the COVID-19 pandemic (Shearing & Lee, 2021). Contact tracing efforts were conducted in the most part via the NHSx app implemented by the NHSx service organization. The NHSx app formed part of a wider ecosystem for the NHS Test and Trace service. The wider contact-tracing ecosystem also included more person-centric efforts such as calling registered phone numbers assigned to individuals across cellular and public switched telephone networks. Citizens of England, and the devolved nations, were encouraged to download the NHSx app, or variants thereof developed for each of the devolved nations and these were still part of the wider NHSx app system. The NHSx app system was then used to detect whether people had been in contact with or in the proximity of someone who was suspected or had subsequently tested positive for the virus.

Amid the wave of infections in 2021, there was a considerable rise in alerts from the app. This was known as the "pingdemic" (Lynch, 2021). After making several concessions to avoid highly dependent areas of industry from grinding to a halt, the Secretary of State for Health and Social Care, on behalf of the UKDHSC, considered adjusting the sensitivity of the associated NHSx app system (Wearmouth, 2021). While infection rates from the Delta variant at the time were on the decline, hospitalizations and death rates were on the rise due to the first strain of Omicron.

An Alternative Approach

The UK's borders are less porous than many other countries; therefore, it is plausible to implement a different approach to deal with an epidemic that may escalate to pandemic status. The main goal is to deploy UKHSA test sites at key ingress and egress choke points on UK borders. With a change in legislation, travelers who enter the country would be compelled to take a Polymerase Chain Reaction (PCR) swab test at key border transit points. The samples would then be dispatched to UKHSA test laboratories. Each person would be assigned a unique identifier linked to their enrollment to the system via a front-end thick client operated by staff of the UKHSA test team. The thick clients would securely interface with an application programming interface resident on cloud back-end infrastructure. The back-end system would be used to process and store the test subject's personal data, as well as the resultant medical data comprising test outcomes from the test laboratories.

A sequence of anonymized data could then be correlated to track infectious people within the country who had been allowed to return to their places of residence for a defined quarantine period. Their respective locales could be used by the Joint Biosecurity Centre (JBC) on behalf of the UKHSA to assess positive test hotspots (groupings of anonymized people represented in a heat-map format). This reliable and specific data would help direct epidemiological teams to hotspot areas as they monitored the spread of infection into the wider communities (Bird, 2020a, 2021).

Ultimately, this approach is aligned with a recently rediscovered 16-year-old draft governmental paper. This document suggests better control of borders as one of many listed measures for a UK government response to a SARS disease outbreak. Notably, COVID-19 is triggered by the viral load known as SARS Coronavirus 2 (SARS-COV-2). Ironically, these identified measures from the noughties are similar to those considered for implementation (far too late) in 2020 as part of the UK government's response to COVID-19 (Lovett, 2021). Bird (2020a, 2021) proposed a DTTTS approach, which comprises a front-end client/back-end server model hosted with cloud-based technology. This method's Concept of Operations (CONOPS) would enable the UKHSA JBC to operate as an Epidemiological Operations Center (EOC).

The DTTTS concept, as defined by Bird (2020a), would enable the UKDHSA, on behalf of the UK government, to:

- Respond to an epidemic by using UKHSA test staff at the borders to register test subjects into the back-end system and issue PCR test identifiers associated with tested individuals.
- Enable PCR swab test results processed by the UKHSA test laboratories to be assimilated in the back-end system based on their unique associations with registered test subjects.
- Enable the UKHSA JBC to process and store data that can be used "definitively" for COVID-19 infection monitoring, tracking and real-time trending to identify potential extended viral spread originating from hotspots.

- Provide ES teams with the ability to target hotspots and their environs for further testing, to then collate evidence of intra-populace infection outbreaks and thereby inform potential UKHSA containment recommendations.
- Provide UKHSA with the ability to trace COVID-19 transmission in the community and make decisions on regional containment options.

This concept is reflected at a high level in Figure 1.

Figure 1. DTTTS operational concept defined

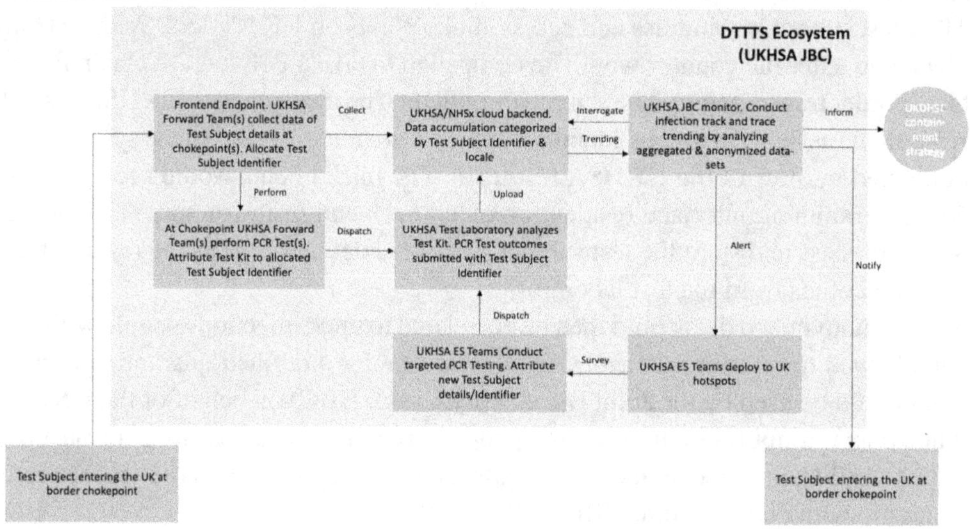

SOFT SYSTEMS APPROACH

The UK government has previously advocated the use of soft systems approaches in order to: (1) understand a system's view and differing perspectives in response to stakeholder need; (2) recognize influencers such as the environment as well as the comprehend sub-systems that all contribute to an efficient systemic solution; and (3) appreciate component interactions and wider system dynamics which can be used to breakdown technical complexity (Government Operational Research Service, 2012). In fact, the World Economic Forum (2021) used systems thinking to understand the impact of COVID-19 and used causal relationships to comprehend its effect in relation to innovation and decision making as part of society's response to the pandemic.

Checkland (1981) re-envisaged the concept of systems thinking in the 1970s and 1980s. He considered it to be a human endeavor to solve complex problems using declared perspectives and purposeful actions (The Open University, 2022). In effect there are different paradigms with the soft systems approach and these include formal system, control, communications, engineering reliability, and human factors such as cultural and organizational contexts (The Open University, 1993b). Checkland (1981) described the meaning of a system to be:

The central concept 'system' embodies the idea of a set of elements connected together which form a whole, this showing properties which are properties of the whole, rather than properties of its component parts (p, 3).

Consequently, the author utilized Checkland's SSM in this chapter and used his experience of applying this methodology. SSM techniques were used to work out problems associated with NHSx's app and compare it to the DTTTS concept detailed in Bird (2020a, 2021).

Method

SSM will detach and scrutinize a real-world problem into a systems-thinking world for problem-solving purposes (The Open University, 2007a). The SSM approach can be used to identify the problems and produce recommendations for improvements using a range of techniques, and other supporting methods. This approach is particularly versatile for this problem setting because 'other systems thinking' methods can be considered as part of the analysis. SSM can be used to compare conceptual models and the real-world and be used to focus on areas for improvement within a system of interest. A system of interest is an assembly of connected components that are structured in an organized manner. In addition, the system is dependent upon these components and the system's behavior would be affected if one or more components leave the system (The Open University, 2023).

SSM provides a versatile seven-stage model (The Open University, 1993a). It is compatible with the define, review, investigate, verify, and execute (DRIVE) principles for problem solving and improvements (The Open University, 2007b):

- **Define:** Real-word aspects relating to:
    - The problem situation where the problem definition is assessed in an open and unstructured manner to prevent bias.
    - The problem is normally expressed in diagrammatic form in order to summarize the problem and find out more about the problem situation.
- **Review:** Apply systems-thinking about the real world for:
    - Consideration of other systems thinking techniques in order to provide insight into relevant systems (RS) pertinent to the problem situation.
    - A description of the RSes in narrative form and then precisely identify a fundamental definition for each of these purposeful systems called root definitions.
    - Elaboration of what each system does expressing the activities, components or processes using a conceptual model of RSes that are defined by the root definitions.
- **Investigate:** Systems thinking about the real world by:
    - Testing how adequate or complete the relationship is between the conceptual model and root definitions for the elaborated RSes. This then forms a consensus system containing the entirety of previously defined RSes.
- **Verify:** Relate back to the real world in order to:
    - Compare the abstract nature of the systems thinking world with the problem defined for the real-world situation. Undertake a compare and contrast exercise to discern similarities and differences.
    - Identify admissible, desirable and feasible changes that will provide benefit in the real world.

- **Execute:** Relate back to the real world by:
    - Communicating changes to key stakeholders who are participants within the problem situation. Thereby, improve the problem situation by taking action, which is the intended and desirable outcome of using soft systems approaches.

(The Open University, 1993a, 1993b).

This chapter will use DRIVE up to the verification stage. The execute stage would be an actual implementation of identified changes in the real-world context and is only explored in the Further Research Directions section. Beyond this, the execute stage extends beyond the conceptual model proposed in this chapter and therefore an implementation of this model is outside the scope of this chapter.

SSM utilizes several constructs to derive each system model under analysis. The root definition is the foundation for defining each sub-system, also known as RS (Wilson, 2008). The system of interest referred to in this chapter is known as the wider system of systems (WSoS) within the analysis. SSM uses predefined criteria to which a system must operate that includes: (1) the customer of the system; (2) people who are stakeholders of the system; (3) transformation measured in terms of efficacy, efficiency and effectiveness; (4) worldview or the purpose of the system from the customer organizational perspective; (5) system owner; and (6) environment in which the system operates. These facets are accumulated into the acronym CATWOE (University of Cambridge, 2016).

Define

Problem Definition

The NHSx app uses combined Google and Apple technology to allow the front-end application client to interface with the NHSx back end. The back-end cloud technology ingests data for analysis by the NHS Test and Trace service (National Cyber Security Centre, 2021). Fundamentally the NHSx app is installed on a smartphone client. It uses Bluetooth low energy (BLE). BLE is used to detect other smartphones within the proximity of the user that have the NHSx app installed (UK Health Security Agency, 2023). The application client interfaces with a web-based back-end system, allowing the NHSx app to act as a broker. Peoples' identities within the NHSx app system are protected by an anonymization technique using NHSx app pseudonyms.

In 2020, the UK government encouraged smartphone users to voluntarily download the NHSx app. Based on a literature review, as well as the author's personal experience of participating in PCR and Rapid Lateral Flow (RLF) tests, the relationship between the NHSx app and other NHS Test and Trace services is shown in Figure 2. Citizens were also entrusted to request a PCR test kit at specific times during the first year of the pandemic, but this required the test kit to be dispatched to a nominated laboratory for it to be processed and the resulting test outcome provided. Citizens in the UK were encouraged to use RLF tests as a self-test method from 2021 onwards. During the height of the pandemic, RLF test kits were distributed by the NHS through pharmacies or by mail (via an NHS supply chain method that is outside the scope of this chapter). Initially, PCR tests were the most reliable and only testing option; however, by 2021, RLF tests were developed and favored over PCR tests due to their rapidity.

Figure 2. NHS test and trace ecosystem diagram containing the defined problem system flow

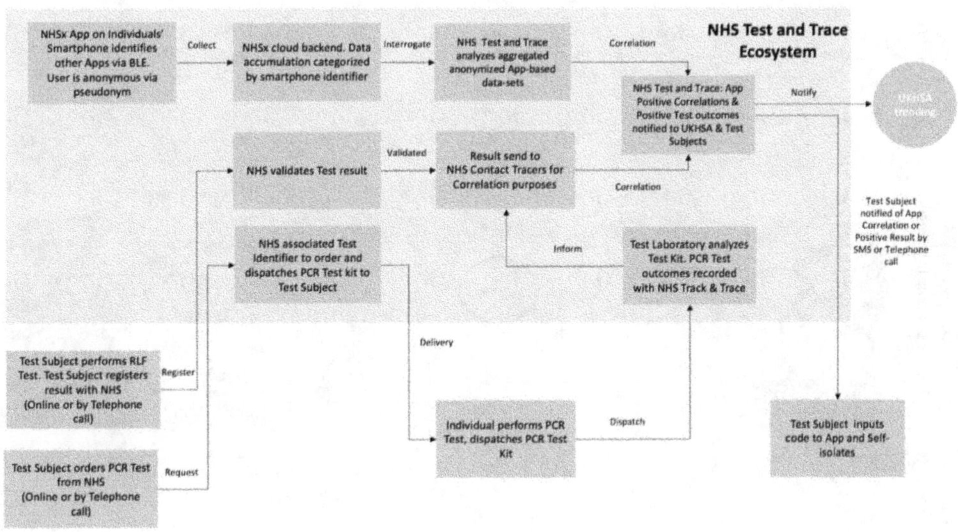

The NHS service contacted test subjects by telephone or short message service (SMS) mobile phone notifications. The NHS provided citizens with a code to enter into the NHSx app if they had been contacted through the NHSx app or a positive test result had been received. The NHSx app system was then used to trace the pseudonymized identifiers of detected NHSx app clients to notify other people if they had been near a suspected infectious person. Subsequently, test subjects would be required to self-isolate for a defined period or be subject to a fine of £1000 (UK Health Security Agency, 2020, 2022a, 2022b). The self-isolation requirement was also extended to people who had been notified via the NHSx app.

Problem Situation

Based on research by Bird (2020a, 2021), a cause-and-effect diagram was created (see Figure 3) to illustrate external influencers that drove the UK government, under their sphere of influence, to develop the NHSx app. The technology was originally based on a UK concept; however, unreliability during testing (Burgess, 2020) forced the NHSx to adopt the Google/Apple consortium algorithm; additionally, the initial NHSx findings were shared with Google and Apple to help enhance their own algorithm. Academia and the National Cyber Security Centre (NCSC) assessed the final algorithm used to process front-end data anonymously by means of a pseudonym and the sharing of anonymized data with a cloud back-end system (National Cyber Security Centre, 2021).

To be effective, upon its release the application required buy-in and conformity from the British public to achieve widespread adoption. At its peak, the app was downloaded at least 23.3 million times. However, only about 16 million people enabled the BLE radio element (Trendell, 2021), a key constituent of the application's data derivation functionality. As a direct consequence of the removal of lockdown restrictions in 2021, there was a drop in the use of the NHSx app. Midway through the pandemic the mainstream press reported that at least one in three users may have deleted the application (Lynch, 2021) rather than continue with self-isolation measures. Occurrences such as these were problematic to the efficiency of the NHSx app system.

Figure 3. Cause-and-effect diagram of the problem defined

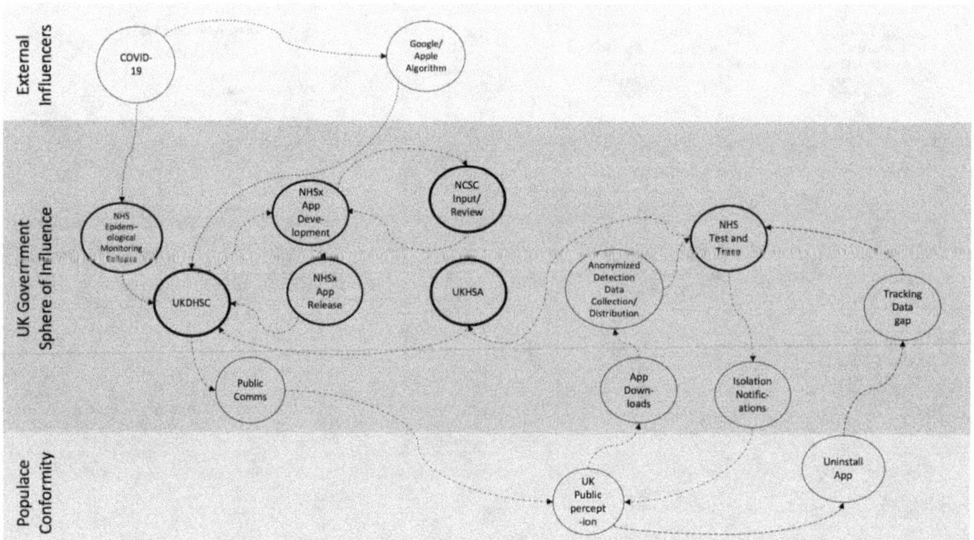

Review

The DTTTS WSoS is a system of sub-systems that are connected to the global environment. The global environment's internal and external forces may influence or disrupt the DTTTS. Based on data assimilated from Bird (2020a, 2021), as well as the approach in Bird (2019, 2020b), a formal system model (FSM) was adapted to define the WSoS as a systems model. The FSM effectively includes: (1) a decision-making sub-system; (2) a performance-monitoring sub-system, and (3) other subsystems and components that carry out tasks to effect transformation (The Open University, 1993b). This analysis is shown in Table 1. A diagrammatical version is shown in systems terms in Figure 4. The cloud is a pivotal, crucial element of the DTTTS. Therefore, the "shared security responsibility" cascades across all elements of cloud use where the responsibilities of the cloud security provider (CSP) and the UKDHSC and UKHSA are defined.

The DTTTS concept (Bird, 2020a) presents several capabilities, controls, and cultural considerations and these are shown in Table 2 illustrating each sub-system or RS in the DTTTS WSoS paradigm.

Root Definitions and Conceptual Model Components

This section contains the systems-thinking elements that describe the root definitions associated with RSes and the resulting CATWOEs for each context taken from Table 2.

There are associations between RSes in the DTTTS, which are complex end-to-end. Inter-dependencies between RSes as applicable subsystems of the DTTTS can be distilled agnostically at a systems-thinking level using root definitions and conceptual models of the RSes using the CATWOE approach. These are comprised of: (1) the root definition that applies a perspective; (2) the CATWOE analyses that inform the applicability of each RS; and (3) the transformation element relevant to the functionality of each RS and requisite considerations/controls contained therein. The RSes have been reordered during the SSM approach and shown in Tables 3 to 12. Transformations and environment definitions were transposed from the FSM outcomes in Table 1 during this analysis.

Figure 4. FSM diagram of the DTTTS concept

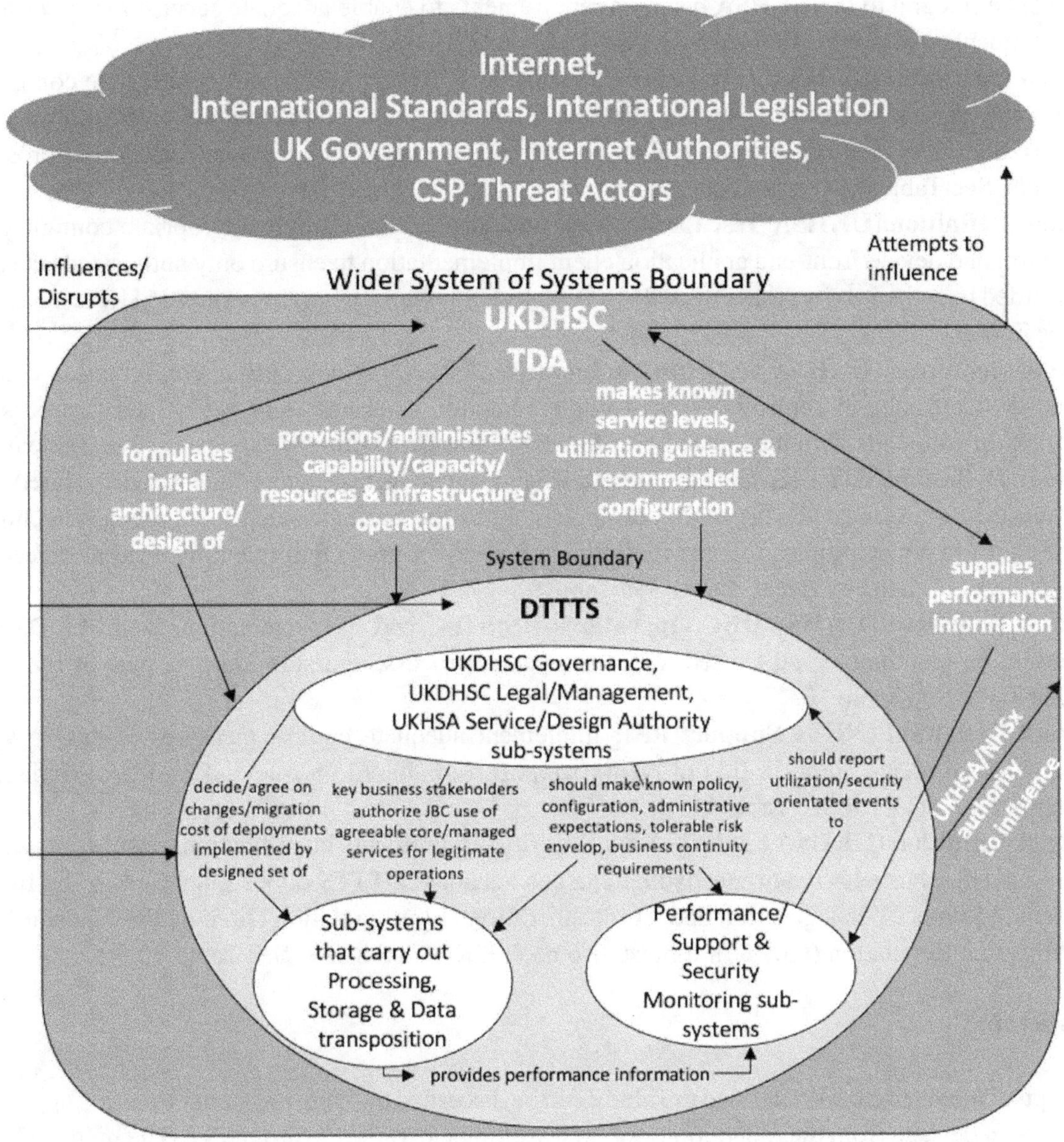

- **Root Definition (UKDHSC Governance RS):** WSoS-level consideration for the functionality of the DTTTS must be integral to the UKHSA business requirements to enable adequate measures to be implemented within the UKDHSC defined service level agreements (SLA). See Table 3.
- **Root Definition (UKDHSC Legal RS):** WSoS consideration of the functionality and ecosystem for the DTTTS must be integral to the UKHSA business requirements to enable the DTTTS to be implemented within legal frameworks and UKDHSC defined SLAs. See Table 4.

- **Root Definition (UKHSA Service RS):** Consideration of the service requirements for the DTTTS must be integral to the UKHSA business requirements to enable adequate service measures to be implemented within the defined SLAs by the UKDHSC. See Table 5.
- **Root Definition (UKHSA Test Team Endpoint(s) RS):** Applicable and appropriate configuration control and device/front-end application client implementations applied to ensure only authorized access is afforded to endpoints by authenticated UKHSA test teams at border crossing points. See Table 6.
- **Root Definition (UKHSA Test Laboratory Endpoint(s) RS):** Apply appropriate configuration control and device/front-end application client implementation to ensure only authorized access is afforded to endpoints by authenticated UKHSA test laboratory teams at permitted UK test centers. See Table 7.
- **Root Definition (UKHSA ES Team Endpoint(s) RS):** Apply appropriate configuration control and device/front-end application client implementation to ensure only authorized access is afforded to endpoints of authenticated UKHSA ES teams conducting survey tasks. See Table 8.
- **Root Definition (DTTTS Cloud System RS):** Design Infrastructure-as-a-Service (IaaS) and Platform-as-a-Service (PaaS) instances applicable to the business-need to implement technical measures in a noncomplex manner that offers diversity of protection recommended to defeat cyberattacks initiated by threat actors. See Table 9.
- **Root Definition (UKHSA JBC Operations RS):** JBC end user interactions with the DTTTS should be identifiable and trusted to access, parse, process, and store data as part of the EOC CONOPS. See Table 10.
- **Root Definition (NHSx Support RS):** Implement adequate control measures to ensure valid authentication and authorization of NHSx support users based on role and by least privilege to support the DTTTS. See Table 11.
- **Root Definition (UKHSA Security RS):** Security requirements identification must be integral to the UKHSA business requirements that enables adequate DTTTS control measures to be implemented to meet UK legislation and European Union (EU) regulation (such as the General Data Protection Regulation (GDPR)). It must also meet UKDHSC SLAs. See Table 12.

Investigate

A conceptual model of each RS defined in Table 2 can be discerned by using a systems-thinking approach. This has been devised from the root definitions and workings in Tables 3 through 12. The model, shown in Figure 5, also shows the control function within which RSes of UKDHSC governance, UKDHSC legal, and UKHSA service reside. The control aspect is a vital part of the SSM model.

The SSM conceptual model provides relevant systems-thinking aspects applicable to the DTTTS as it relates to its operational function articulated in previous works (Bird, 2020a, 2021):

1. UKHSA test teams to register test subjects
2. DTTTS to ingest data through exposed APIs
3. UKHSA JBC to identify regional hotspots where infected/socially isolated people reside
4. UKHSA JBC to monitor and track expanded outbreaks from hotspot locales

Figure 5. SSM conceptual model

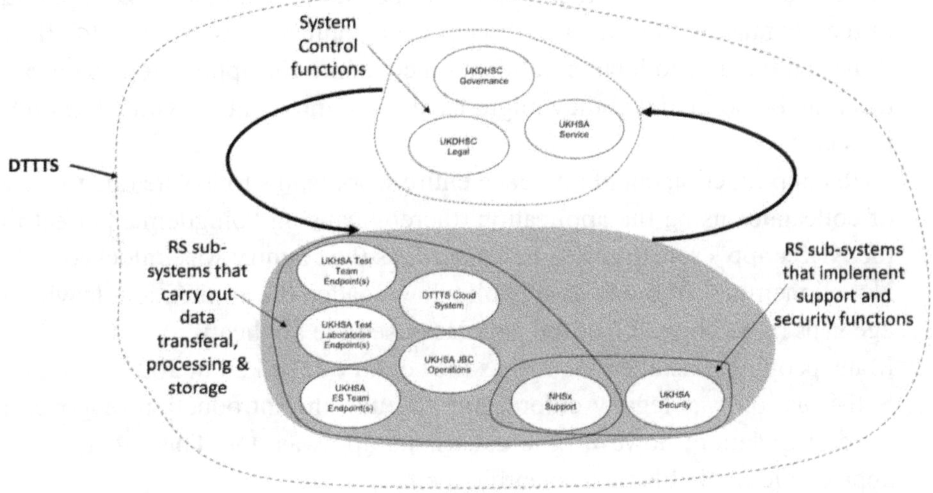

5. UKHSA JBC to instigate UKHSA ES team deployments to hotspot areas to conduct further PCR testing/registration and trace the expansion of COVID-19 infections
6. Provide definitive data for the UKDHSC to consider in an informed manner and implement epidemiological containment measures to mitigate the associated viral risk

Verify

By analyzing the NHSx app implementation compared to the DTTTS concept and associated outline CONOPS by Bird (2020a, 2021), the following pros and cons are identified pertaining to each approach:

- **NHSx App**
  - Pros:
    - User data anonymized by the application at source, thereby not requiring special consideration as sensitive personally identifiable information (PII).
    - Available for mass download by British public.
    - Application not operated in isolation and traditional methods remained available in conjunction with the application, such as telephone contact-tracing and SMS notifications.
    - NHS Test and Trace service coordinated trending with the JBC for UKDHSC statistics and generic reporting at a regional level.
  - Cons:
    - Long lead-time to implement and test the NHSx app approach from scratch and rework.
    - Relies on good will of users in relation to proper use of the application and leave BLE beaconing out.
    - PCR and RLF results processed outside the NHSx app contain PII by virtue of the test subject's name and UK national insurance number. Therefore, the test subject is

requested to manually input NHS notification codes into the NHSx app so a relationship between the test subject's pseudonym and contacts detected by the BLE can be discerned. This enables other contacts to be alerted via the NHSx app, notifying them of their contact with someone who has been confirmed as having COVID-19.

- Relies on the user to input a notification code into the application so onward BLE contacts can be notified that they might have come into contact with a COVID-19 infected person.
- NHSx app function could influence entire work teams to isolate due to close proximity of colleagues using the application, thereby causing "pingdemic" effect that required the NHSx app's sensitivity to be adjusted as the country was unlocked.
- Not all members of the British public downloaded the application, leaving large coverage gaps (e.g., only 23% of the populace used the application).
- Many people uninstalled the app as the country unlocked in 2021.
- Self-isolation was legally enforceable through the introduction of government emergency legislation; however, the use of the app was not. This left the crowd-sourced approach less reliable than scientific methods.
- This application approach is an active system. It incurs direct operating costs. Therefore, it required UK government investments to kick-start development. Additionally, there is a continuing need to provide back-end in-service maintenance activities and updates to the application as a live system. There is also a need to provide in-app updates to ensure its functions operate as designed and any security weaknesses are also patched. But the latter is dependent on the public to routinely install or permit the installation of these updates on their smartphones.

The data collection and utilization model surmised from the NHSx app approach in Figure 6 highlights that useable data from this source was more generic. The data was used for statistical purposes by the UK government; thus, potentially influencing in-country regional wholesale lockdowns because the approach was based on a less focused regional perspective.

Figure 6. NHSx app data and UKHSA operational views

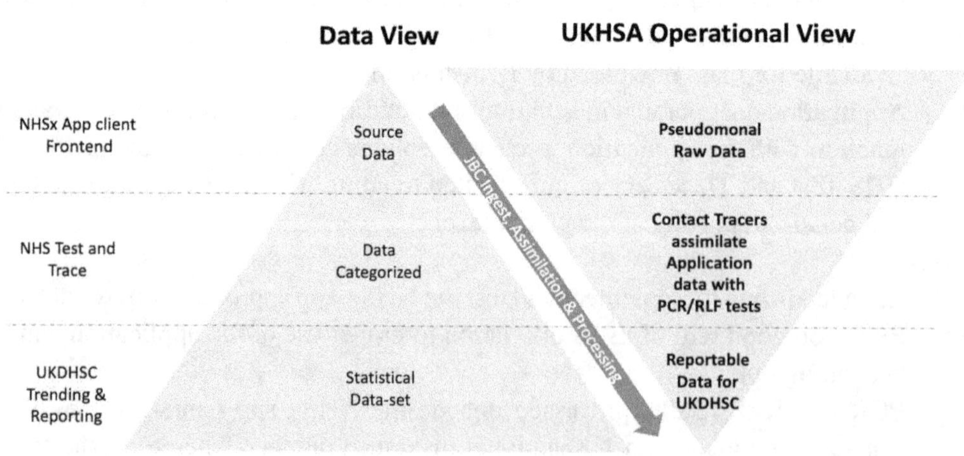

- **DTTTS**
    - Pros:
        - Test subject data is not anonymized by the system until it reaches the actionable stage where locational data is used to direct ES teams to target infection hotspots.
        - The scientific approach is dependent on more reliable PCR result outcomes.
        - PCR test outcomes are deemed to be "definitive" data. It can be used to corroborate infection trends as the UKHSA JBC tracks people from the border inwards. Thereby, it enables the identification of a conglomeration of infection hotspots when test subjects are sent home after being processed at the border and the requisite PCR test results are associated with their test identifier.
        - The EOC process of disease intelligence, and command and control of incident response activities (Georgia Department of Public Health, n.d.) provides informative data that assists the UKHSA to monitor, contain, and respond to infectious spread. The EOC approach can be used to track people ingress from the border. It can also monitor the spread of infections during an epidemic emergency. An EOC would provide the UKHSA with the ability to coordinate more effectively with local and regional authorities.
        - Definitive data can steer the UKHSA to choose containment options based on positive test results and assimilated hotspots. It enables ES activities to be vectored to specific locales for survey purposes and additional testing and viral evidence collation requirements.
        - It enables ES teams to be coordinated so the capability is not overwhelmed or mismanaged. It prevents guess work during deployment estimations and more informative information prevents poor decision making.
        - This concept could be built following lessons learned from the COVID-19 pandemic and a cold-standby system approach can be used to avoid unnecessary costs when not utilized.
    - Cons:
        - Requires the rapid deployment of UKHSA test team staff to border chokepoints to collect test subject details and oversee PCR testing.
        - Requires change of legislation to compel people to become test subjects by virtue of entering the country.
        - The DTTTS use-case processes PII. Therefore, it requires measures to safeguard PII. Its use must be proportionate, appropriate, and used for the purpose for which it was collected. It is subject to legislation and regulation and therefore subject to checks and balances by the UKDHSC.
        - Screening test subjects at border chokepoints will cause queues and inconveniences.
        - It potentially increases demand on the UKHSA laboratory infrastructure to process a higher volume of PCR tests.
        - Test subjects will be subject to new laws that force them to self-isolate or be subject to a fine.

- A cold-standby approach requires regular drills/exercises to test system standups in the cloud. Background maintenance is required on cloud images/templates to operate effectively and ensure that it is not vulnerable to performance issues or security vulnerabilities.

The data collection and utilization model proposed by the DTTTS concept is shown in Figure 7. Actionable data is an outcome from using the DTTTS. It can enable the UK government to track infections from the border and throughout the country. Thus, government authorities can direct ES teams to hotspot areas, undertake targeted trace surveys and requisite containment actions to preclude widespread lockdowns. The DTTTS would need to operate in line with emergency legislation and using PII within the bounds of the UK Data Protection Act (DPA) 2018 and the country's commitment to comply with the EU GDPR.

Figure 7. DTTTS data and UKHSA operational views

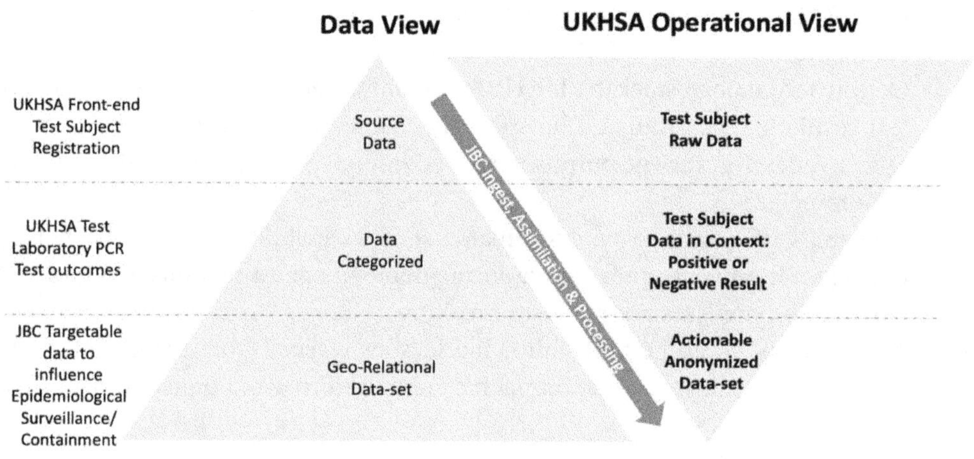

SOLUTIONS AND RECOMMENDATIONS

While the NHSx app has its place, the provenance of the collected data should be considered "lead" data rather than a wholly trustworthy data source to steer wider epidemiological containment efforts. Bill Gates, in his latest book, highlighted that the world is woefully underinvested in disease surveillance (Gates, 2022) systems. This means that trusted, "definitive" data from a more reliable source is needed to consolidate accurate monitoring of a country-wide viral spread. The latter would have been advantageous to track Coronavirus hotspots at the start of the pandemic. Consequently, in preparedness for a future event with a link to the SARS family, the author built on previous published works (Bird, 2020a, 2021) to propose a DTTTS comprising a front-end client/back-end server model based on cloud-based technology complete with an outline CONOPS related to EOC operations. The DTTTS can provide reliable, definitive data that can be used for a greater statistical purpose and provides the ability to conduct infection trending analysis by the UKHSA JBC.

FUTURE RESEARCH

Both the DTTTS concept and real-world NHSx app technologies have their place. For instance, the UKHSA has publicly announced that the NHSx app helped avert 1 million infections, avoided 44,000 hospitalizations and prevented 9,600 deaths within the first year of operation (UK Health Security Agency, 2023). That said, the UK government placed significant investments into developing, testing, deploying, and updating the NHSx app. Perhaps the "lead" data from this crowd sourcing application could be corroborated with the "definitive" data proposed for the DTTTS concept. This would provide advantages to monitor, predict, and steer responsive ES activities from a multi-dimensional perspective.

Regarding these two data sources, analysis and assessments would need to be undertaken to establish the benefits to ES on-the-ground resources used to target hotspots and trace the spread of infection across the populace. A method would need to be developed to validate both types of data sources. More work would be required to assess the credibility of each technology and investigate integrating both these capabilities into a single, co-existing system. Undoubtedly, considerations and changes would need to be made for data compatibility purposes. Properly assessing the benefits of both approaches would avoid discrepancies and confusion with regards to epidemiological response actions.

However, as already stated by mid-2021 at least one in five people had uninstalled the NHSx app, putting its future in doubt when assessing it on a cost versus reward basis. It was also regarded as a giant experiment (Cellan-Jones, 2021). Resultingly, the UK government retired the NHSx app on 27 April 2023 (UK Health Security Agency, 2023). It therefore follows that the approach taken with the NHSx app was more of a 'bespoke' solution in response to COVID-19 and therein heightens the case for a DTTTS type of approach that facilitates a better longer-term use-case.

By decommissioning the NHSx app, an opportunity has been lost to take the anonymized data-sets already collated for use by the UKHSA. However, if such data could be made available for further research, if afforded by UK law, then these data-sets could be useful for comparative purposes where a DTTTS environment is ever trialed. Therefore, if deemed proportionate and relevant in line with UK legislation, there would certainly be a valid argument to consider retaining extant anonymized trending data processed by the NHSx app cloud-backend for future research purposes. This sort of research could be conducted as part of the execution stage of the DRIVE model.

CONCLUSION

Mackenzie (2020) stated that we must improve disease surveillance and plan for epidemiological events. This is particularly poignant when a paradigm shift is expected by the World Health Organization (WHO) pandemic foresight initiative. The mantra of this initiative is the change of approach from reactive to proactive thinking as part of future pandemic preparedness (World Health Organization, 2023). The Times also recently reminded us that the next pandemic is just around the corner (Spencer, 2023).

Global COVID-19 Situation

Most of the UK population has now received at least four vaccinations. Still, the population is mixing more often and even in early 2022 at least 150 people per day were still dying of COVID-19 (Levy, 2022) and the UKHSA has announced it will no longer publicize the average infection rate figures or R number (Kenber, 2022). Worryingly, around the same timeframe the Chinese government announced a grand unlocking of COVID-19 restrictions across the country, which resulted in widespread infections. Following Italy's experience of testing passengers who had entered the country from the Orient, many countries like the United States and Germany in late 2022 were still insisting on pre-departure testing (Sky News, 2022a). During this timeframe the Secretary of State for Health and Social Care was particularly focused on the emergence of new variants of COVID-19 from the Far East (Channel 4 News, 2022; Davies & Mao, 2022).

Concerns have also been expressed by the mainstream press regarding the transparency of reporting of new variants by Chinese authorities and concerns about the timeliness of warnings issued by China (Hymas & Lovett, 2022; Whipple, 2022). Learning the lessons from the first wave of COVID-19 in 2020, the UKHSA by 2022 stipulated that people flying from China must provide a negative test result prior to their flight (Sky News, 2022b).

In 2023, published academic research identified that COVID-19 can be spread through droplets acting as a micro-aerosol and also from surfaces touched by people contaminated with SARS-COV-2 on their hands. Thus, proving how versatile the COVID-19 virus can be in propagating within the community (Keane, 2023). By the first quarter of 2023 Omicron was already mutating into sub-variants. The Arcturus sub-variant has been declared as the most transmissible to date (Haase, 2023). It has been reported that Arcturus has been found in the UK in the Northeast of England but also traced to the Northwest and Scotland (Gregory, 2023). It has also been found in 22 other countries to include the United States, India, Singapore and Australia (Bawden, 2023; Ferris, 2023). Meanwhile in the UK a spring vaccine booster campaign has been put into place for children aged five, people with a weakened immune system and those who are over 75. Yet one in 40 people still tested positive for COVID-19 in mid-March 2023 (Davies, 2023).

According to the WHO at least seven million people died during the COVID-19 pandemic. However, while COVID-19 infections seem to be reasonably prolific the actual monthly death rate has dropped. So, the Emergency Committee of the WHO called an end to the declared public health emergency of international concern, but recommended that countries should maintain their guard (Mundasad & Roxby, 2023). Subsequently, leading countries have deemed the COVID-19 public health emergency to be at an end. Yet, in the US for instance hospitalizations are still occurring and in May 2023 the disease was still causing at least 1,000 deaths per week (Rodriguez & Alltucker, 2023). Therefore, considering this latest COVID-19 information it still appears there may still be a need for a targeted surveillance approach from the border into the heart of the UK.

NHSx App Approach

The NHSx app was the result of fast-track initiative to implement a test and trace system after UK epidemiological capabilities were overwhelmed by COVID-19. The main purpose of the NHSx app was to help detect whether people had been in contact or in the proximity of someone who was suspected or had subsequently tested positive for COVID-19. The NHSx app model was based on a front-end smartphone client that interfaced with a web-based back-end system. The UK government was heavily invested in this solution and encouraged smartphone users to download the app as part of their UK-wide test and trace strategy. Participation was voluntary and due to the app's functionality, the approach as a whole was principally based on a crowd-sourced technology. The NHSx app was the mainstay technological implementation in response to COVID-19 and a key contributor to the UK government's central initiative on behalf of the UKDHSC.

While the NHSx app was the central initiative by the UK government to improve disease surveillance for COVID-19, it was a reactive endeavor. The conditions in which it was created implies a degree of group think conformity was involved (Hart et al., n.d.). Such a phenomenon could have influenced the requirement rather than pursuing a more defined scientific approach with a more useful purpose. The time and monetary investment for the NHSx app was substantial - at least £56.5 million up to 27 April 2023 (Downey, 2020, 2021; Hoeskma, 2023). Yet, with all this effort the UK government has decommissioned the NHSx app just as the Arcturus sub-variant is starting to take hold within the UK populace. The UK government justified their withdrawal of the NHSx app due to the success of the vaccination program and the fact that public use of the app was on the decline (UK Health Security Agency, 2023). If financial outlay to maintain the NHSx app is such a significant influencing factor, then should this money have been invested in developing a system like a DTTTS instead?

DTTTS Approach

The DTTTS would provide a key contribution to the ethos of WHO's pandemic foresight initiative by being proactive by design. DTTTS promotes the following: (1) the use of front-end endpoints at border chokepoints to register test subjects into the system; (2) external UKHSA test laboratories are provided with the ability to update test subject data sets with PCR test results; (3) the distributed system back end operated jointly by NHSx and the UKHSA, including a robust implementation of public cloud platform technology; (4) the DTTTS back end that can be used by the UKHSA JBC under the principles of an EOC; and (5) if appropriately maintained and updated in the background, the DTTTS could be held in reserve as a cold-standby system at significantly less operational cost until it is truly required during the next SARS viral epidemic.

The DTTTS could be used by the JBC to assess positive test hotspots by correlating the addresses of travelers from overseas but represented in less distinct groupings in the form of a heat-map. In turn, this more reliable and specifically collected data could be used to help direct epidemiological teams to hotspot areas and also further monitor infection breakouts in our communities. The DTTTS conceptual model could provide more longevity by design, with the ability to wind-up and wind-down the DTTTS cloud back-end with associated scalable supportability measures. This would make it more cost effective. In which the focus would be concentrated upon epidemiological data collection and processing as necessary and as required. This approach would provide the UKHSA with a better through-life intelligence collation capability for epidemiological containment purposes.

A key component of the DTTTS concept is the test subject's anonymity in certain data analysis settings coupled with the reliability of each test subject's residential area. This approach would be an absolute necessity for the data to be transparently used as a tactical tool in epidemic containment management efforts. The DTTTS concept provides "definitive" data to monitor, predict, and track hotspots, and steer responsive ES team activities. Assessments could be made to coordinate the finite resources of on-the-ground ES teams to target hotspots and trace the spread of virus infection across the populace.

Applied Methodology

Systems thinking has been used in this chapter as an analytical tool to highlight the problems related to the NHSx app use-case as part of the UK's response to COVID-19. SSM was deemed a prudent approach because it can be used to define, review, investigate and verify the problem situation in the real world. Thereby, the use of SSM has been instrumental in realizing the transformational benefits that the DTTTS can bring to epidemic containment management. The methodology used by the author to analyze the problem situation considered a number of aspects such as cultural paradigm, organizational structures, technological implementations and relevant processes. From a theoretical standpoint and transformational perspective, this chapter has demonstrated that a DTTTS capability would be an effective enabler to the UK's epidemic response capability. Even though the NHSx app approach was previously promoted and adopted by the UK government, reasoning has been provided indicating that the proposed DTTTS would be more efficient and effective. Therefore, a DTTTS may avert unnecessary deaths and unneeded self-isolation situations during SARS epidemics that originate from overseas. This particular point is still relevant today and was not holistically achieved using the NHSx app during the pandemic. Thus, this aspect in itself rationalizes and justifies the purpose of this study.

REFERENCES

Bawden, T. (2023, April 12). New Covid variant 'Arcturus' is even more infectious than Kraken and could push UK cases higher. *The Independent.* https://inews.co.uk/news/health/new-arcturus-covid-variant-kraken-uk-cases-higher-2270721

Bird, D. (2019). Derivation of a conceptual framework to assess and mitigate identified customer cyber-security risks by utilizing the public cloud. In *Proceedings Fourth International Congress on Information and Communications Technology* (vol. 2). Springer.

Bird, D. (2020a). Distributed test, track and trace system as a contributor to epidemic containment management. In *Proceedings of the Fourth World Conference on Smart Trends in Systems, Security and Sustainability.* Institute of Electrical and Electronic Engineers. 10.1109/WorldS450073.2020.9210295

Bird, D. (2020b). *Real-time and retrospective analyses of cyber security.* IGI Global.

Bird, D. (2021). A concept of operations to embody the utilization of a distributed test, track and trace system for epidemic containment management. In A. Joshi, A. Nagar, & G. Marin-Raventos (Eds.), *Advances in sustainability, science and technology* (pp. 253–266). Springer.

Burgess, M. (2020). *Why the NHS Covid-19 contact tracing app failed.* Wired UK. https://www.wired.co.uk/article/nhs-tracing-app-scrapped-apple-google-uk

Cellan-Jones, R. (2021, July 14). *NHS Covid app: Should it stay or should it go?* BBC News. https://www.bbc.co.uk/news/technology-57836325

Channel 4 News. (2022, December 29). *Channel 4 News.* Channel 4. https://www.channel4.com/news/

Checkland, P. (1981). *Systems Thinking, Systems Practice.* John Wiley.

Davies, A., & Mao, F. (2022, December 29). *Covid in China: US imposes Covid testing for visitors from China*. The BBC. https://www.bbc.co.uk/news/world-asia-china-64111492

Davies, R. (2023, April 17). Spring Covid booster 2023: Who is eligible and can you book them now? *The Evening Standard.* https://www.standard.co.uk/news/health/spring-covid-booster-vaccine-eligibility-booking-b1074507.html

Department of Health and Social Care. (n.d.). *Our governance*. GOV.UK. https://www.gov.uk/government/organisations/department-of-health-and-social-care/about/our-governance

Downey, A. (2020, September 22). *Total cost of NHS contact-tracing app set to top £35 million.* DigitalHealth. https://www.digitalhealth.net/2020/09/total-cost-of-nhs-contact-tracing-app-set-to-top-35-million/

Downey, A. (2021, June 2). *Developer of NHS Covid-19 app awarded £10m contract extension.* DigitalHealth. https://www.digitalhealth.net/2021/06/developer-of-nhs-covid-19-app-awarded-10m-contract-extension/

Ferris, N. (2023, April 16). Why the world is unprepared if new Covid variant Arcturus proves deadly – in graphs. *The Independent.* https://www.independent.co.uk/news/health/arcturus-variant-nhs-covid-vaccine-b2321575.html

Gates, B. (2022). *How to prevent the next pandemic*. Allen Lane.

Georgia Department of Public Health. (n.d.). *Function and Form of the Command Center: Advanced Healthcare Emergency Management Course*. Department of Public Health.

Government Operational Research Service. (2012). *Introduction to systems thinking*. Government Office for Science.

Gregory, A. (2023, April 14). Mapped: Worst areas in UK for Covid cases amid fears over new Arcturus variant. *The Independent.* https://www.independent.co.uk/news/health/covid-cases-uk-map-arcturus-b2323268.html

Haase, M. (2023, April 15). *What Experts Want You to Know About Arcturus, the New COVID Variant.* Prevention. https://www.prevention.com/health/health-conditions/a43592554/arcturus-xbb16-new-covid-variant-what-to-know/

Hart, P., Stern, E., & Sundelius, B. (Eds.). (n.d.). Beyond Groupthink: Political group dynamics and foreign policy-making. Michigan. doi:10.3998/mpub.11178

Hoeskma, J. (2023, January 6). *Palantir gets £11.5m six-month NHS contract extension.* DigitalHealth. https://www.digitalhealth.net/2023/01/palantir-gets-11-5m-six-month-nhs-contract-extension/

Hymas, C., & Lovett, S. (2022, December 30). China accused of covering up Covid data amid new variant fears. *The Telegraph.* https://www.telegraph.co.uk/politics/2022/12/30/china-withholding-covid-data-new-variant-fears/

Keane, D. (2023, April 7). London study finds first evidence Covid spreads on household surfaces. *The Evening Standard*. https://www.standard.co.uk/news/health/covid-spreads-household-surfaces-imperial-college-london-study-b1072709.html

Kenber, B. (2022, December 27). Covid data dumped as publishing R value 'no longer necessary'. *The Times*. https://www.thetimes.co.uk/article/covid-data-dumped-as-publishing-r-rate-no-longer-necessary-gzhclctnf

Levy, O. (2022, August 15). Covid booster jab approved. *BBC News*.

Lovett, S. (2021, August 4). Revealed: Britain's 'lost' blueprint for fighting a coronavirus outbreak – written 16 years ago. *The Independent*. https://apple.news/AswZ6ocEwSnaZJ0NoNv3zlQ

Lynch, R. (2021). Third of people delete NHS app as 'pingdemic' spreads. *The Telegraph*. https://www.telegraph.co.uk/business/2021/07/16/third-people-delete-nhs-app-pingdemic-spreads/

Mackenzie, D. (2020). *COVID-19: The pandemic that never should have happened, and how to stop the next one*. The Bridge Street Press.

Mundasad, S., & Roxby, P. (2023, May 5). *Covid global health emergency is over, WHO says*. BBC. https://www.bbc.co.uk/news/health-65499929

National Cyber Security Centre. (2021). *High level privacy and security design for NHS Covid-19 Contract Tracing App*. Crown Copyright.

National Health Service. (2023, March 21). *How to use an NHS COVID-19 rapid lateral flow test*. NHS UK. https://www.nhs.uk/nhs-services/covid-19-services/testing-for-covid-19/how-to-use-an-nhs-covid-19-rapid-lateral-flow-test/

National Health Service Test and Trace. (n.d.). *How to do a PCR test for coronavirus (COVID-19) at home*. NHS UK.

Rodriguez, A., & Alltucker, K. (2023, May 05). The COVID public health emergency is over in the US. Here's what that means for you. *USA Today*. https://eu.usatoday.com/story/news/health/2023/05/11/end-covid-public-health-emergency-explained/70182594007/

Shearing, H., & Lee, J. (2021, July 19). England Covid restrictions ease as PM urges caution. *The BBC*. https://www.bbc.co.uk/news/uk-57882029

Sky News. (2022a, December 29). *Half of passengers on China flight to Italy have COVID - as US becomes fifth country to tighten restrictions on Chinese travellers*. Sky. https://news.sky.com/story/covid-patients-no-longer-have-to-quarantine-in-hong-kong-as-restrictions-are-lifted-12775498

Sky News. (2022b, December 31). *Travellers arriving in England from China will need to show negative COVID test*. Sky. https://news.sky.com/story/uk-expected-to-require-travellers-from-china-to-provide-negative-pre-departure-covid-test-12777027

Spencer, B. (2023, March 19). Disease X is coming. Are we ready for the next pandemic? *The Times*. https://www.thetimes.co.uk/article/disease-x-is-coming-are-we-ready-for-the-next-pandemic-ck29hmxx7

The Open University. (1993a). *T301 Block4: The soft systems approach*. Open University Press.

The Open University. (1993b). *T301 Block2: A systems approach to failures.* Open University Press.

The Open University. (2007a). *T889 Block5: The soft systems approach.* Open University Press.

The Open University. (2007b). *T889 Block 1 problem solving and improvement: Quality and other approaches – Introduction.* Open University Press.

The Open University. (2022, March 23). *Systems explained by Peter Checkland.* Open Learn. https://www.open.edu/openlearn/money-business/leadership-management/systems-explained-peter-checkland

The Open University. (2023). *5.1 Definition of a system.* Open Learn. https://www.open.edu/openlearn/digital-computing/systems-thinking-and-practice/content-section-5.1

Tikkanen, A. (2023, April 11). *National Health Service.* Encyclopaedia Britannica. https://www.britannica.com/topic/National-Health-Service

Trendall, S. (2021). Sensitivity of contact-tracing app will not be adjusted. *PublicTechnology.net.* https://publictechnology.net/articles/news/sensitivity-contact-tracing-app-will-not-be-adjusted

UK Health Security Agency. (2020, May 27). *NHS test and trace: What to do if you are contacted.* GOV.UK. https://www.gov.uk/guidance/nhs-test-and-trace-how-it-works

UK Health Security Agency. (2022a, February 24). *COVID-19: People with COVID-19 and their contacts.* GOV.UK. https://www.gov.uk/government/publications/covid-19-people-with-covid-19-and-their-contacts

UK Health Security Agency. (2022b, April 1). *People with symptoms of a respiratory infection including COVID-19.* GOV.UK. https://www.gov.uk/guidance/people-with-symptoms-of-a-respiratory-infection-including-covid-19

UK Health Security Agency. (2023, April 27). *NHS COVID-19 app.* GOV.UK. https://www.gov.uk/government/collections/nhs-covid-19-app

UK Health Security Agency. (n.d.). *Featured.* GOV.UK. https://www.gov.uk/government/organisations/uk-health-security-agency

University of Cambridge. (2016). *Soft Systems Methodology.* Institute for Manufacturing. https://www.ifm.eng.cam.ac.uk/research/dstools/soft-systems-methodology/

Wearmouth, R. (2021). *Major overhaul to NHS Covid app to reduce number of contacts pinged to isolate.* The Mirror. https://apple.news/AtKXj1KuLQ7S4F3pla4vb2Q

Whipple, T. (2022, December 31). Could China's Covid 'catastrophe' be bad for Britain too? *The Times.* https://www.thetimes.co.uk/article/04648ec8-8855-11ed-bb21-8f4d97ec7b02?shareToken=3c0b21689f49dcd47784b8b6ff64d330

Wilson, B. (2008). *Soft systems methodology: Conceptual model building and its contribution.* Wiley.

World Economic Forum. (2021, January 18). *What 'systems thinking' actually means - and why it matters for innovation today.* World Economic Forum. https://www.weforum.org/agenda/2021/01/what-systems-thinking-actually-means-and-why-it-matters-today/

World Health Organization. (2023). *Epidemic and Pandemic Foresight Initiative.* WHO. https://pandemic-foresight.who.int/

ADDITIONAL READING

Gates, B. (2022). *How to prevent the next pandemic.* Allen Lane.

Mackenzie, D. (2020). *COVID-19: The pandemic that never should have happened, and how to stop the next one.* The Bridge Street Press.

KEY TERMS AND DEFINITIONS

Concept of Operations (CONOPS): Is a documentation set that diagrammatically and descriptively represents the operational use-case for a system. It also contains configuration and support aspects, work flows and modus operandi for its operation by authorized personnel.

Epidemiological Operations Center (EOC): Is a concept based on security operations center precepts in which the dedicated Joint Biosecurity Centre (JBC) team can monitor data using machine learning algorithms to discern clusters of infection hotspots. Trending approaches can be used to notify regional authorities of greater congregations of people with positive test results so more focused and targeted ES activities can be undertaken to detect infection breakouts. Thereby, enabling more informed and effective containment decision-making (Bird, 2020a, 2021).

Epidemiological Surveillance (ES): Is also known as shoe leather epidemiology in which teams from UKHSA can be deployed into society in order to survey the environs of an infection and utilize several swab testing methods (defined below) to collect samples for analysis. These tests can be used to determine viral spread in the case of the SARS Coronavirus 2 (SARS-COV-2) strains that underly the COVID-19 disease (Bird, 2020a, 2021).

National Cyber Security Centre (NCSC): This organization is the UK's National Technical Authority responsible for cyber security matters in line with the National Security Strategy. It provides influence, advice and guidance to stakeholders and practitioners within both public and private sectors.

National Health Service (NHS): Is a national healthcare system that is sponsored by the UKDHSC under government administration and is publicly funded by citizens, and people with a right to work in the UK, through national insurance deductions taken from their wages (Tikkanen, 2023).

Polymerase Chain Reaction (PCR) Test Kit: Specialist test kits that are used to take a swab sample from test subjects, which are then stored in a liquid vile and uniquely barcoded for transportation to medical laboratories for processing. The PCR results are considered the most reliable method of testing for SARS-COV-2 because analysis of each test sample is conducted by specialized staff within registered medical laboratories (National Health Service Test and Trace, n.d.).

Rapid Lateral Flow (RLF) Test Kit: This type of test is also known as an antigen test which can be self-administered by test subjects who take a swab sample from the back of their throat and mix it with a specially designed solution. The solution is then placed onto a test strip and within 30 minutes the RLF test strip provides a visual indication of whether the test sample is positive, negative or invalid. RLF tests are far quicker, but potentially less accurate, than PCR tests (National Health Service, 2023).

UK Health Security Agency (UKHSA): An organization responsible for the coordination of the UK government's response to the onset of an epidemic or pandemic. The UKHSA coordinates and receives epidemiological intelligence from the JBC and samples taken by ES teams in the field. The UKHSA primarily replaces the UK's previous Public Health organizations within each of the four nations across the UK. Based on lessons that had been learned by the UK government, the Public Health organization was restructured into the UKHSA so it can be more effective in its response to epidemics and pandemics (Bird, 2020a, 2021).

APPENDIX

Table 1. WSoS formal system model defined

FSM Aspect	Analysis
Environment	The environment comprises the Internet-wide area network, international standards and legislative bodies, organizations that influence the use of the Internet, and governments like the UK government. Public CSPs have defined the shared security responsibility for cloud customers to understand their role in cloud security. Customers and CSP may be targeted by hackers, becoming victims of inadvertent data center or platform outages under the responsibility of the CSP. The complexity of the public cloud is simplified into guidance by the CSP - for ease of customer use and CSP declarations of conformity/compliance to certifying bodies, standards organizations, and legislation. UKHSA/NHSx organizations promote other frameworks to achieve compliance, indicating relevant controls of accessing the cloud online.
Formulates initial design/decides upon transformations	The WSoS considers the UKDHSC Technical Design Authority (TDA) influence upon the DTTTS, UKDHSC, and UKHSA governance functions that are responsible for deciding the functioning of the DTTTS system. This includes the use of front-end computing such as endpoints. The UKHSA Cloud Environment depends on both CSP and UKHSA/NHSx supporting services. CSP are responsible for the platform architectural design, declaring compliance and certifications within their area of responsibility and the shared security responsibility. UKHSA/NHSx operate within the shared security responsibility model by applying salient security controls to safeguard data. The DTTTS is to permit authorized access and avoid accidentally or negligently exposing data online.
Provides resources and legitimates area of operation	The UKDHSC governance function is responsible for engaging with their own stakeholders. It is authorized (e.g., business, commercial, and security sign-off) to use the DTTTS. The UKHSA management and governance function should make known their policy for the use of DTTTS. This includes services to be used, the access control policy, change management and configuration control, and administrative requirements to support the processing, storage (performance and security monitoring), and use of DTTTS sub-system components for operations.
Makes known expectations	The UKDHSC management and governance function must disseminate their policy for the use of cloud services. The UKHSA/NHSx stipulate management aspects to include access control, change and configuration control, administrative support, performance, and security monitoring. This enables the processing and storage of data by UKHSA JBC. CSP provide the cloud platform and data center resourcing. It defines service level agreements for operation and guidance related to core services, managed services, and configuration.
Supplies performance information	UKHSA/NHSx administrators can conduct an analysis of logs to analyze ingress and egress data flows pertaining to the DTTTS. UKHSA JBC administrators analyze and monitor data using authorized monitoring tools or logging servers to conduct performance and data processing of the DTTTS. CSP conduct security monitoring and threat intelligence related to connections and data flows that transit their cloud platforms. CSP also provide data rate and volumetric service performance monitoring. They inherently provide data that can be viewed using various dashboards by the NHSx team.
Decision-making subsystem	Key stakeholders (i.e., UKDHSC executive, project/program managers, finance and information security managers) must have a business operations view and mitigate risks associated with the use of cloud technologies. The department's risk appetite should drive policy and procedures for the use of the DTTTS, defining authorization to use services for a particular purpose, financial authority to purchase capacity/capability in the cloud, and expectations for using the cloud platform, including championing the tolerable security risk envelope to operate to and a disaster recovery regime. In addition, an organizational account and associated structure should be explored with the CSP. Thus, privilege can be applied at the project level and considered by the overarching business.
Subsystems that carry out transformations	NHSx administrators and UKHSA JBC users must adhere to the expected security risk envelope for their respective business area. They must follow organizational policy and procedures when using the DTTTS. DTTTS should be architected, considering front- and back-end elements. Frontend devices should be robustly designed to utilize secured and authenticated connections over fourth and fifth generation cellular networks. UKHSA back-end elements must also be robust, using models like infrastructure-as-a-service (IaaS) and/or platform-as-a-service (PaaS) and PaaS-like services to conduct legitimate operations within the cloud. Controls include protection mechanisms for IaaS instances hosted in the VMs or data in storage, access control to data, employing adequate authentication/authorization measures, accounting, and audits of their environments. CSP is responsible for the implementation of platform systems hardware and baseline soft infrastructure: (a) hypervisor for IaaS and hypervisor; (b) host and middleware, including provisioning of application capability for PaaS; and (c) data center aspects. The CSP provides guidance and configuration recommendations like access control, networking, management of keys, security monitoring, and the protection of data in transit and at rest.
Boundaries	UKHSA Design Authority need to design in their own boundary protections for the DTTTS. It should include network architecture for interactions with IaaS and PaaS hosts. API interactions need to be protected and authenticated at the front-end device level and user level. Thus, the UKHSA test team users are authorized to pass data to the DTTTS cloud element. CSP provide detective and reactive boundary monitoring controls, privileged access workstations for administrative access, virtualization, and sandboxing technologies. CSP also manage data center governance, physical security, equipment maintenance, and equipment support.
Others	UKHSA/NHSx technical communities can exert influence on the UKDHSC regarding the technicalities of the system design and implementation. CSP can influence the publishing of controls to customers. Trust is established through being externally certified and audited by trusted third parties.

Table 2. WSoS defined

RS Sub-System Category	Description
UKHSA Test Team Endpoint(s)	Front-end endpoints are used as data input nodes with teams on border chokepoints. They are also used to register test subject details and unique PCR test identifiers into the cloud's back-end system. Utilizing fourth or fifth generation cellular bearers, the endpoints will connect to a centralized cloud back end via secure end-to-end communications to an exposed API. The front-end application is sandboxed on the endpoint, providing encryption to protect the data-at-rest.
UKHSA Test Laboratories Endpoint(s)	PCR tests are processed. The results are added to the cloud's back-end via Internet bearers and an exposed API in accordance with the unique PCR test identifiers by UKHSA staff at test laboratories. PCR results are associated with test subjects in the back-end data set for analysis by UKHSA JBC analysts.
UKHSA ES Team Endpoint(s)	Front ends are used by ES teams to investigate anonymized data that is stored on the cloud's back-end. This targets COVID-19 hotspots in different areas. They undertake further targeted epidemiological surveys in regional locales as a means of measuring the spread of COVID-19. Fourth and fifth generation cellular bearers are used. Endpoints connect to a centralized cloud back end via secure end-to-end communication to an exposed API. The front-end application is sandboxed on the endpoint, providing encryption to protect the data-at-rest.
DTTTS Cloud System	This is based on a schema data. The cloud's back-end will ingest the data through a hardened (but secure) API with end-to-end encryption applied from endpoint to ingress interface. Data ingress is from endpoints and egress as data is pulled to test laboratories. The updated data is appended. This can be controlled via secure identity and access management, coupled with multi-factor authentication in the cloud environment.
NHSx Support	JBC user profiles and credentials are managed using robust configuration and change management processes. Templates and host builds need to be updated, verified, and validated through functional and security testing. Back-end application schemas and data validation processes need to be implemented. Virtual hosts are configured and patched in line with defined processes and procedures. Performance monitoring will ensure the system is operating optimally with an ability to scale horizontally as required based on loading.
UKHSA Security	The end-to-end system will need to be architected in accordance with security-by-design principles. Defense mechanisms to cyberattacks need to be in place, such as web application firewalls, counter-distributed denial of service attacks by threat actors, properly configured network routing and protocol filtering, process user authentications measures for the endpoints to authenticate to the back-end, and security protective monitoring.
UKHSA Service	NHSx standard service offerings and policies are required as defined in International Standards Organization (ISO) 27001:2022. They need to consider ISO27017:2015 and ISO27018:2019 due to cloud native aspects and incorporation of personal data in line with the UK Information Commissioner's Office (ICO) directives. Provide Design Authority oversight over the DTTTS on behalf of the UKDHSC TDA.
UKHSA JBC Operations	Operating as an EOC, as described in Bird (2021), will receive ingested data sets presented to the DTTTS in aggregated form, comprising: • Test subject personally identifiable information (PII) and home address data are collected by the teams at the border chokepoints using the endpoints. • Test subject PCR test results data provided by the UKHSA test laboratories to UKHSA JBC and then collated to regional hotspots by locale. Both aspects will enable the UKHSA JBC to conduct monitoring in accordance with the CONOPS, resulting in a more informed tracking and monitoring effort. This will allow for trending in outbreak analysis and enable coordination of UKHSA epidemiological teams to target hotspot areas to conduct more directed, focused tests as part of trace efforts.
UKDHSC Legal	The UKHSA can use cloud computing; however, it must comply with the UK Data Protection Act (DPA) 2018 and the UK's commitments to the General Data Protection Regulation (GDPR) of the European Union. Pseudo anonymization and full anonymization techniques for PII data is permissible in the cloud per the UKDHSC. It is envisaged that Parliament would need to pass emergency laws to enforce the DTTTS approach in line with the DPA 2018 and GDPR for the purposes of PII collection and medical epidemic testing purposes, infection tracking, and tracing the chains of infectious transmission.
UKDHSC Governance	Oversight is required by the UKDHSC and the executive of the UKHSA in line with service requirements defined by polices (that are delegated into processes and procedures) and budget constraints. To remove over-reliance on the cloud platform supplier by leveraging existing SaaS or PaaS technologies, IaaS would enable greater government control but demands in-service governance. This, thereby, enables oversight over the installation, operation, templating, and tear-down of the cloud's back-end when no longer needed. PII can only be retained for its primary use under the DPA 2018 as explained to test subjects (providing their details and medical samples for testing at border chokepoints).

Table 3. UKDHSC governance RS CATWOE analysis

Criteria	Test
Customer	UK government
Actors	Proposed stakeholders: • Secretary of State for Health and Social Care • Permanent Secretary of UKDHSC • Second Permanent Secretary of UKDHSC • UKDHSC Director General Finance • Chief Medical Officer • UKDHSC Senior Information Risk Officer (SIRO) • UKDHSC Senior Responsible Officer (SRO) • UKDHSC Change Manager (UK Department of Health and Social Care, n.d.)
Transformation	Take a balanced strategic view for the UK government's use of DTTTS under the control of the UKHSA as part of the country's epidemiological test, track, and trace requirements.
Worldview	Stand up a capability to support epidemiology and the identified need to conduct effective track and trace operations to facilitate and enable infection containment measures within the UK.
Owner	Permanent Secretary
Environment	Develop and deploy capability in line with UK government policy.

Table 4. UKDHSC legal RS CATWOE analysis

Criteria	Test
Customer	UK government
Actors	Proposed stakeholders: • Secretary of State for Health and Social Care • Permanent Secretary of UKDHSC • Second Permanent Secretary of UKDHSC • UKDHSC Director General Finance • UKDHSC Legal Council
Transformation	Provide oversight for compliance and adherence to ICO stipulations for the UKHSA to process PII in accordance with DPA 2018 and GDPR. Ensure there are effective control measures to anonymize PII appropriately and effectively by the DTTTS system to the UKHSA JBC. This approach will track hotspots and alert ES teams, allowing them to target hotspots and survey regional locales to trace infection spread within the country.
Worldview	Process PII in accordance with the operational need and context of use.
Owner	Permanent Secretary
Environment	Consider legal implications and responsibilities to anonymize data due in compliance to legislation and ICO guidelines.

Table 5. UKHSA service RS CATWOE analysis

Criteria	Test
Customer	Permanent Secretary of UKDHSC
Actors	Proposed stakeholders: • UKHSA Chief Executive • Chief Medical Officer • UKHSA Director General of Finance • UKHSA Director General Data & Analytics • UKHSA SIRO •UKHSA SRO • UKHSA JBC Operations Manager (UK Health Security Agency, n.d.)
Transformation	Provide oversight of UKHSA strategic service requirements, ensuring they are embedded for DTTTS implementation. Take a balanced view of the context for using DTTTS reliant on the public cloud, an understanding of the precepts of the shared security responsibility based on CSP guidelines and recommendations. The executive shall promote the adoption of best practices and direct development of organizational cloud-centric policies, influence UKHSA processes and procedures for change/configuration management, expect robust security practices, and support measures to be applied within defined budgetary constraints in accordance with UKDHSC tempo.
Worldview	Stand up DTTTS in line with UKDHSC TDA architecture to meet the UKDHSC need and context of use.
Owner	UKHSA Chief Executive
Environment	Must consider due diligence to establish compliance to ISO standards and recognized certificating frameworks.

Table 6. UKHSA test team endpoint(s) RS CATWOE analysis

Criteria	Test
Customer	UKHSA JBC Manager
Actors	Potential customer stakeholders: • NCSC Specialists • UKHSA SIRO • UKHSA Security Architect • UKHSA Technical Architect • UKHSA Information Security Team • UKHSA Change Control Board • NHSx Administrators
Transformation	Must be configured on a locked down endpoint device. Front-end application client sandboxed on the endpoint, providing encryption to protect the transitory data at-rest on the device. Front-end application must interface with the DTTTS cloud back-end across fourth and fifth generation cellular networks. The device and UKHSA test team user must authenticate with DTTTS using public key infrastructure (PKI) and pass PII data securely to the exposed API in-transit. UKHSA test teams must conform to UKHSA organizational policies, processes, and procedures.
Worldview	To implement appropriate front-end interface compatible with back-end API schema requirements, encryption to register details, test identifiers assigned, and PII data correctly distributed. This allows it to be processed and stored. Thereby, the UKHSA JBC operational users can access the right data processing/storage functions at the right time.
Owner	UKHSA SRO
Environment	Must ensure identifiable connections/services are verified prior to transmission of data applicable by law, identified technical controls to protect data in-transit and at-rest, and NCSC guidance and recommendations.

Table 7. UKHSA test laboratory endpoint(s) RS CATWOE analysis

Criteria	Test
Customer	UKHSA JBC Manager
Actors	Potential customer stakeholders: • NCSC Specialists • UKHSA SIRO • UKHSA Security Architect • UKHSA Technical Architect • UKHSA Information Security Team • UKHSA Change Control Board • NHSx Administrators
Transformation	Must be configured on a locked down endpoint device. Front-end application client sandboxed on the endpoint, providing encryption to protect the transitory data at-rest on the device. Front-end application must interface with the DTTTS cloud back-end across Internet bearers. The device and user must authenticate with the DTTTS via PKI and pass PII and medically confidential data via secure means to the exposed API in-transit. UKHSA test laboratory teams must conform to UKHSA organizational policies, processes, and procedures.
Worldview	To implement appropriate front-end interface in accordance with the API schema, encryption so the data is correctly interrogated using test identifiers and appending medical data that is added, and thereby processed and stored appropriately by UKHSA JBC operational users that have access to the right data processing/storage functions at the right time.
Owner	UKHSA SRO
Environment	Must ensure identifiable connections/services are verified prior to transmission of data applicable by law, identified technical controls to protect data in-transit and at-rest, and NCSC guidance and recommendations.

Table 8. UKHSA ES team endpoint(s) RS CATWOE analysis

Criteria	Test
Customer	UKHSA JBC Manager
Actors	Potential customer stakeholders: • NCSC Specialists • UKHSA SIRO • UKHSA Security Architect • UKHSA Technical Architect • UKHSA Information Security Team • UKHSA Change Control Board • NHSx Administrators
Transformation	Must be configured on a locked-down endpoint device. Front-end application client sandboxed on the endpoint, providing encryption to protect the transitory data at-rest on the device. Front-end application must interface with the DTTTS cloud back-end across fourth and fifth generation cellular networks. The device and user must authenticate with DTTTS via PKI and pass PII data securely to the exposed API in-transit. UKHSA test teams must conform to UKHSA organizational policies, processes, and procedures.
Worldview	To implement appropriate front-end interface in accordance with the API schema, encryption so anonymized data is correctly interrogated and new test identifiers registered and PII details appended, processed, and stored appropriately. Thus, enabling UKHSA JBC operational users to have access to the right data processing/storage functions at the right time.
Owner	UKHSA SRO
Environment	Must ensure identifiable connections/services are verified prior to transmission of data applicable by law, identified technical controls to protect data in-transit and at-rest, and NCSC guidance and recommendations.

Table 9. DTTTS cloud system RS CATWOE analysis

Criteria	Test
Customer	UKHSA Director General Data & Analytics
Actors	Potential customer stakeholders: • UKHSA SIRO • UKHSA JBC Operations Manager • UKHSA Technical Architect • UKHSA Security Architect • UKHSA Information Security Team • UKHSA Change Control Board • UKHSA Programme/project manager(s)
Transformation	Taking guidance from CSP guidance and shared security responsibility recommendations design fit-for-purpose cloud solution required for business usage. Cloud instances must implement various security and dependency control measures to protect the IaaS instance(s), IaaS/PaaS application(s) and APIs, data processed, stored, and associated data flows. Adequate UKHSA JBC EOC tooling must be installed and be effectively functional to perform informed monitoring and tracking of data sets held by the system. Must conform to organizational cloud-centric policies, processes, and procedures, as well as budgetary constraints.
Worldview	To design cloud back end that serves the UKHSA business requirement and fulfils the UKDHSC strategic approach. Implement appropriate controls for adequate protection of both management and operations perspectives throughout the life of operation in line with organizational policy, processes, and procedures. Perform baseline architectural change management through life.
Owner	UKHSA SRO
Environment	Maintain awareness of Internet attack vectors, hacker/cyber-research modus operandi, compliance to legislation, and use of design standards.

Table 10. UKHSA JBC operations RS CATWOE analysis

Criteria	Test
Customer	UKHSA Chief Executive
Actors	Potential customer stakeholders: • UKHSA JBC Operations Manager • UKHSA Security Architect • UKHSA Technical Architect • UKHSA Information Security Team • UKHSA JBC Privileged Users
Transformation	Implementations must verify the identity and provide validation processes to ensure UKHSA JBC trusted users can access, parse, process, and store data to monitor and track infections across the UK. Must conform to organizational policies, processes, and procedures for escalation of actionable data to dispatch ES teams to hotspot areas to trace infections within the UK.
Worldview	Only verifiably authorized UKHSA JBC operations users can interact with the data processing/storage functions in the DTTTS. UKHSA JBC to use the DTTTS cloud back-end for EOC monitoring and alerting purposes.
Owner	UKHSA Director General Data and Analytics
Environment	Must ensure UKHSA JBC users are identifiable and authorized prior to accessing the DTTTS in a manner applicable by law, UKDHSC authority, and UKHSA processes and procedures to act on anonymized data in a proportionate and justifiable manner for epidemiological containment purposes.

Table 11. NHSx support RS CATWOE analysis

Criteria	Test
Customer	UKHSA Director General Data & Analytics
Actors	Potential customer stakeholders: • UKHSA Security Architect • UKHSA Technical Architect • UKHSA Information Security Team • UKHSA Change Manager • NHSx Administrators
Transformation	Taking guidance from CSP guidance and shared security responsibility recommendations to implement access control, authorization, and authentication services to access IaaS instance(s) and PaaS application(s) and data sets. Taking guidance from CSP guidance and shared security responsibility recommendations to implement mechanisms to verify identity and prove trustworthiness of internal/external connections/systems that interface with IaaS instance(s) and PaaS application(s). Must conform to organizational cloud-centric policies, processes, and procedures.
Worldview	To implement appropriate authorized levels of privilege for defined roles, management of cloud security groups, boundary and internal access control lists, network segmentation, and isolation measures like encryption and business continuity options so the correct operational UKHSA JBC human and system-level users can access the right data processing/ storage functions at the right time.
Owner	UKHSA SRO
Environment	Must ensure admission into instances is controlled to users (human and system-level) in a sufficient manner applicable by law, UK government enforcement, alignment to CSP contract, and in line with UKDHSC authoritative audit and resulting recommendations.

Table 12. UKHSA security RS CATWOE analysis

Criteria	Test
Customer	UKHSA SIRO
Actors	Proposed customer stakeholders: • NCSC Specialists • UKHSA SRO • UKHSA Change Manager • NHSx Support Manager • NHSx Administrators
Transformation	Taking a balanced view of the context for using the public cloud, an understanding of the precepts of the shared security responsibility based on CSP guidelines and recommendations. Promote the integration of security requirements into the business requirements capture process through best practice and due diligence. The board shall promote adoption of best practice and direct development of organizational cloud-centric policies, processes, and procedures for change/ configuration management, as well as support and define budgetary constraints in accordance with business tempo.
Worldview	To stand-up DTTTS to meet the UKHSA need and context of use. Verify front-end application client functions correctly and is appropriately hardened, ensuring sandboxing works effectively on the endpoint and encryption is configured correctly to protect the transitory data at-rest on the device. Ensure PKI cryptography is configured to facilitate protections and authentication measures as a key enabler so only verifiable operational connections/systems interface with the correct data processing/storage functions in IaaS/PaaS elements of the cloud's back-end. Review schema to ensure the exposed API is robustly set up and data is correctly ingested to the system. Verify encryption is effectively instigated for data in-transit and at-rest. Validate data anonymization processes are appropriate and effective. UKHSA security must conform to UKHSA organizational policies, processes, and procedures.
Owner	UKHSA SIRO
Environment	Must consider end-to-end protection of data in-transit while transiting over fourth and fifth generation cellular networks/Internet and authorized access to data within the DTTTS, including the protection of data at-rest. Must consider selection of CSP by undertaking due diligence to establish compliance to legislation, standards, and recognized certificating frameworks. Must ensure identifiable connections/services are verified prior to accessing IaaS/PaaS to instances in a manner applicable by law, government authorities' requirements, aligned with CSP contract and in line with UKDHSC authority audit, and requisite recommendations. Must consider NCSC guidelines and recommendations applicable to the end-to-end design of DTTTS.

Chapter 2
The Role of Gamification in a Post-Pandemic World: The Experience of China

Miao Huang
Communication University of China, Nanjing, China

Mohamad Saifudin Mohamad Saleh
Universiti Sains Malaysia, Malaysia

ABSTRACT

The coronavirus disease 2019 (COVID-19) pandemic caused issues related to public vaccination and nucleic acid testing management and mental and physical health pressures resulting from pandemic lockdowns. Gamification has played an important role in addressing these issues and the effect of COVID-19 by promoting behaviour changes, providing positive information, relieving mental stress, and improving happiness. In the COVID-19 prevention and control experience of China, gamification has been successfully applied in many effective cases, such as 1) the collection of gamified rewards to increase the public's acceptance for vaccination and nucleic acid testing; 2) the reduction of negative emotions and improvement of happiness during lockdown by gamified social interaction; 3) the encouragement of home fitness; and 4) the improvement of online education effects. This chapter focuses on the role of gamification in a pandemic, specifically as a support tool, by people in China.

INTRODUCTION

In most countries, the COVID-19 outbreak was rapid and unexpected. The first known case occurred in late December 2019. Subsequently, the World Health Organization (WHO) declared COVID-19 a pandemic on March 11, 2020 (Cucinotta & Vanelli, 2020), at which time many countries imposed strict movement restrictions through early 2022 (Ioannidis, 2022). By the second half of 2021, more than 70% of the global population was expected to have been vaccinated or infected (Ioannidis, 2022). Although the current COVID-19 case burden is decreasing, virus transmission continues to demonstrate high in-

DOI: 10.4018/978-1-6684-7285-9.ch002

fectivity and seasonal characteristics (Ioannidis, 2022). Most of the population has reduced COVID-19 contraction risks; however, the perceived individual risk remains highly overestimated.

Following the 2019 identification of COVID-19 and its initial characterization in early January 2020, the containment goal of China was to interrupt local transmission of the novel coronavirus (Li et al., 2020). The Chinese government actively conducts nucleic acid testing on the public to produce an accurate illustration of COVID-19 spread. These efforts also aim to control its spread in an efficient timeframe. The routine nucleic acid testing is enhanced by testing all close contacts of an infected person (Li et al., 2021).

Since September 2020, all close contacts of infected people are required to undergo three tests throughout a 14-day intensive quarantine period (Li et al., 2021). Given the vast and dense distribution of the Chinese population, the government has developed a detailed and highly effective nucleic acid testing strategy (Han et al., 2022). Ensuring that the public completes nucleic acid testing promptly has become a challenge for the government (Han et al., 2022).

In response to the rapid COVID-19 spread in early 2020, the Chinese government implemented the highest-level (first level) emergency response in most provinces and cities. This was accompanied by the strictest restrictions on public life nationwide (Chen et al., 2020). The most important Chinese New Year celebrations were cancelled and people had to abide by home quarantine requirements. Schools, factories, government offices, and most public facilities were closed until the outbreak was effectively controlled (Chen et al., 2020).

The rapid COVID-19 outbreak highly influenced Chinese university operations, specifically the teaching and learning systems (Qiu et al., 2021). Under the Chinese Ministry of Education guiding principle of "interrupting courses without interrupting learning," Chinese universities ensured that teaching continued through various means (Huang et al., 2020). The motivation to ensure social distancing led to remote teaching becoming the primary approach to reduce virus spread during the pandemic (Kwon et al., 2020).

During online teaching, educators were required to resolve the challenges of sustaining student engagement and motivation. Simultaneously, long teaching hours, social isolation, and technological interventions distracted students from the classroom, which led to online fatigue among both teachers and students (Nair, 2022). Students were mentally fatigued from attending too many online classes. In addition, teachers struggled to teach via digital platforms, asserting that the new teaching models increased their workloads (Lei & So, 2021). Specifically, tourism educators designed customized virtual practicums (virtual lectures, discussion boards, virtual group projects, augmented or virtual reality [VR]-based learning, simulations, and gamifications) and students welcomed the new teaching model (Seo & Kim, 2021).

COVID-19 presents known risk factors for mental health (Moreno et al., 2020). In addition to their unpredictability and uncertainty, social distancing and lockdowns can lead to the following: social isolation; reduced income; loneliness; reduced physical activity; difficulty accessing social services; increased access to food; alcohol and online gambling; and reduced family and social support (Moreno et al., 2020). During the first-level epidemic emergency response period, the proportion of the Chinese public with moderate or higher psychological symptoms increased, specifically people with obsessive-compulsive disorder, fear, anxiety, interpersonal sensitivity, and mental illness (Tian et al., 2020). It is widely recognized that social isolation exerts a substantial effect on physical and mental health (Edwards, 2013). Edwards (2013) demonstrated that loneliness and social isolation may be as harmful to health as smoking 15 cigarettes per day.

The current study will refer to several classic cases of China's epidemic prevention and control to form empirical research. The cases cited in this study come from the reports of formal Chinese media and the verification of researchers, which is very representative. Through these cases, the researchers hope to provide references for gamification solutions for the prevention and control of possible epidemic diseases in the future. The study's aims to provide creative suggestions for developers and researchers in the fields of games and mobile applications.

GAMIFICATION AS A USEFUL TOOL

Gamification generally refers to game mechanics being used in a non-game context (Deterding et al., 2011). Gamification can also refer to user experience and behavioral consequences through competition, involvement, and participation (Aguiar-Castillo et al., 2020). The increasing gamification use context has transformed it into a broader concept (Montagni et al., 2020), where serious games are not purely for entertainment purposes. Instead, they are being used in training and education contexts (Alvarez & Djaouti, 2011). Digital tools, such as mobile applications, do not encompass the full gaming experience. They do, however, contain gamification elements like scores, rewards, and tasks, which are common gamified use contexts (Fleming et al., 2017). Gamification-based applications are effective in enhancing user motivation and engagement (Hamari et al., 2014).

With the popularity of games in daily life, gamification has formed an information system that can be designed to influence user behavior (Koivisto & Hamari, 2019). Although the research on gamification lacks coherence and consistency in the theoretical system, it is still valued by the fields of education, crowdsourcing, and health (Koivisto & Hamari, 2019). As a mature human-computer interaction system, gamification research will be strengthened on the basis of the development of human behavior theory (Rapp et al., 2019). While improving the theoretical quality and practicality of the gamification field, researchers must consider the moral reflections brought about by gamification elements in serious fields (Rapp et al., 2019). Gamification was gradually applied as an important element of software, systems, and services to attract and motivate users. This, in turn, inspired subsequent behaviors (Xi & Hamari, 2019). Gamification enhances system or service effectiveness in meeting user needs, thus increasing the probability that users will consciously use the system or service and its resulting conversion rate of beneficial behaviors (Xi & Hamari, 2019). Gamification is considered an effective personal health support tool in which positive inducement has higher potential than negative punishment (Schmidt-Kraepelin et al., 2020).

Gamification-induced immersion is related to satisfaction with autonomy needs. In gamification, the achievement elements are related to satisfaction with autonomy and satisfaction with autonomy and ability needs. The social elements are related to satisfaction with autonomy, competence, and relatedness needs (Xi & Hamari, 2019). Goal-, social-, and reward-based gamification enables behavioral change modalities, such as goal setting and self-monitoring, to change user behavior (Cho et al., 2021).

In game-based learning, the game rules present clear limits for the participants, which are relatively different from course presentations, assignments, and examinations (Buckley & Doyle, 2016). Gamification can promote culturally inclusive behavior. Participants' motivation to use game-based learning includes the pursuit of enjoyment or rewards, satisfying their enthusiasm for learning where gamification includes leader boards, badges, monitoring response, feedback, and bonus points (Nair, 2022).

Among various gamification user groups, Generation Z (or Zoomers), born between 1995 and 2010, is more active in the online digital environment (Nair & Sadasivan, 2019). Having been exposed to digital technology and devices from an early age, Zoomers are "digital natives" proficient in mobile technology and smartphones (Topic & Mitchell, 2019). Moreover, Zoomers demonstrate stronger needs for gamified health-conscious campaigns (Alahäivälä & Oinas-Kukkonen, 2016). Compared to previous generations, Generation Z exhibits unique characteristics and learning and communication styles that are more connected to technology. Gamified sports mechanics are most relevant to Zoomers, which provide new strategies for addressing Zoomer health issues (Zain et al., 2021).

Generation Z is highly accepting of gamification features, which include leader boards, levels, challenges, points, and rewards (Deterding et al., 2011). Simultaneously, this generation prefers to learn while playing games (Nair & Sadasivan, 2019). Higher-quality and more enjoyable health awareness games can, thus, equip Zoomers with important health information (Zain et al., 2021). Therefore, it is possible to encourage Zoomers to use gamified health awareness games to fight the COVID-19 pandemic (Zain et al., 2021).

GAMIFICATION IN ASSESSMENT OF SYSTEMS THINKING

In view of the outstanding effect of gamification in the field of education, researchers began to use it as an effective means of evaluating systems thinking (Tabnak & Hosseini, 2021). The dynamic simulation of systems thinking by gamification users realizes their transformation from external users to internal users (Kotsopoulos et al., 2019). This shows high applicability in the fields of automation and interaction design, in which choices and decisions need representation (Kotsopoulos et al., 2019).

In addition, the elements of role-play, virtual currency, data collection, and cooperative collaboration in gamification have played active roles in socio-economic and urban ecosystems (Davidová et al., 2022). The interactive content brought by these gamification elements can improve the urban environment and quality of life through bottom-up multiple values (Davidová et al., 2022). The practice of VR in environmental science proves that gamification simulation at the social level can realize environmental transformation and change, as well as promote communication and systems thinking in the macrostructure of environment, society, and politics (Cole et al., 2021).

The rapid iteration of information and communication technology has formed a new social system, which has given birth to learning needs and brought about changes in learning thinking (Mindeguia et al., 2019). Learners use gamification to explore and solve problems more deeply from the perspective of systems thinking (Mindeguia et al., 2019). At the technical level, the intelligent collaboration and simulation systems involved in gamification already have the capability of sustainable knowledge transfer (Nitsche et al., 2022). Gamification can effectively understand the complexity and potential dependencies of network knowledge transfer based on the system dynamics model (Nitsche et al., 2022). Gamified systems thinking through new technologies like immersion, enhanced interaction, and VR strengthen learner participation, academic rigor, and ability development. It also improves learners' confidence in problem solving and emotional stability in the face of setbacks (Rivas et al., 2022). Especially during the COVID-19 pandemic, teaching systems thinking concepts that used gamification to communicate disease and healthcare knowledge enabled the dissemination of knowledge while enhancing teamwork and a sense of community among learners (Tan et al., 2020).

GAMIFICATION PROMOTED VACCINATION AND NUCLEIC ACID TESTING

Vaccine hesitancy poses a serious threat to public health. The reasons for the lack of motivation to be vaccinated are diverse and complex and include complacency, lack of vaccine access, distrust of health or medical authorities, misinformation spread, risk underestimation, and limited understanding of the benefits of vaccination and how it works (Paterson et al., 2018). During the COVID-19 pandemic, vaccine hesitancy was the main reason for low vaccination rates (Troiano & Nardi, 2021). The most common reasons for vaccine refusal were the fear that rush-produced vaccines were unsafe, distrust, considering vaccines useless, lack of knowledge about the novel coronavirus, uncertainty about vaccine efficiency, the belief that one was immune, and doubt regarding the vaccine provenance (Troiano & Nardi, 2021). Gamification can be used as a new and effective intervention to reduce and address vaccine hesitancy frequency (Montagni et al., 2020). Gamified learning can promote users' knowledge accumulation in the vaccination context, provide vaccination-related information reserves for the parents of the vaccinated, and increase their willingness to be vaccinated (Fadda et al., 2017).

Game modes and gamification elements demonstrate varied presentation effects. Most researchers recognised the role of reward systems and role-playing, which were mainly aimed to increase public awareness of the benefits of vaccination and promote vaccination behaviour (Montagni et al., 2020). Other gamification elements include appealing interfaces, instant success, continuous feedback, challenges and competitions, teamwork, and custom avatars, which can also be used for effective immune communication (Willis et al., 2013). Serious gamification feature-based games and mobile apps are increasingly being used to train health professionals and provide preventive and health care information to the public (Mandryk & Birk, 2019; Raiff et al., 2018).

The WHO supported the development of PIPDeploy, an interactive face-to-face tabletop simulation exercise game to facilitate emergency preparedness learning to promote the development and implementation of an influenza vaccine deployment action plan to strengthen COVID-19 preparedness. The PIPDeploy uses gamification elements (game interface, time pressure, leader boards, and teamwork) to increase participant motivation (Ghiga et al., 2021). The participants received the gamified simulation effect well, which caused them to reflect on the gap between the vaccine deployment plan and the vaccine product import regulatory framework (Ghiga et al., 2021).

In the early stage of COVID-19 vaccination, local Chinese governments introduced various policies to obtain in-kind incentives for vaccination to increase the public vaccination rate (Zhong, 2021). The government or community incentives ranged from daily necessities, vouchers equivalent to a small amount of cash, to occasional lottery draws, which exerted an enhancing effect on people who were undecided regarding vaccination and increased the proportion of people vaccinated (Zhong, 2021).

When encouraging residents to actively undergo nucleic acid testing, the city of Suzhou, China, borrowed the gamification element of reward collection and converted nucleic acid testing certificates into continuous collectible cards with illustration themes, which encouraged residents to collect and save the cards after completing the test (Ju, 2022). Thus, the gamification accelerated nucleic acid testing and circulation efficiency and increased the residents' enthusiasm for nucleic acid testing (Ju, 2022).

GAMIFICATION REDUCED THE PSYCHOLOGICAL PRESSURE CAUSED BY LOCKDOWN

Gaming is a psychological mechanism to pursue the ideal experience (Csikszentmihalyi, 1991). Video games promote players' mental health, which include stress relief (Reinecke, 2009), cognitive stimulation (Barr, 2017), combating loneliness (Kaye et al., 2017), coping with trauma, and enhancing well-being (Colder Carras et al., 2018). Many people were forced to stay at home during the COVID-19 pandemic and turned to video games due to negative emotions, such as anxiety and fear. The vast majority of such people agreed that playing games positively affected happiness (Barr & Copeland-Stewart, 2022).

Vangelisti and Perlman (2006) suggested that socially-isolated people may be at greater risk for loneliness and being surrounded by a social network of meaningful personal relationships is critical to feel socially embedded and alleviate loneliness. Cole and Griffiths (2007) highlighted the benefits of playing games online with others. Up to 75% of participants fostered the appropriate friendships by participating in online games and 30% met romantic partners through gaming. These friendships were frequently extremely high quality with over one-third of participants stating that they discussed issues with their online friends that they would not share with their offline friends (Petrosino et al., 2021).

Multi-player online gaming experiences can accumulate and improve well-being (Reer & Quandt, 2019) as game-based social interactions can reduce users' loneliness (Kaye et al., 2017). In the early days of the COVID-19 outbreak in 2020, people worldwide were forced to stay at home. The home life precipitated by the COVID-19 lockdowns resulted in the need for cognitive stimulation and social opportunities (Barr & Copeland-Stewart, 2022). To sustain mental health during the home quarantine period, increasing social opportunities through television, the Internet, and telephone were recommended to provide targeted psychological interventions. Teaching young people to develop an optimistic and positive attitude to relieve panic and anxiety was also recommended (Tian et al., 2020).

During the lockdowns of 2020, many young people turned to video games to deflect anxiety and fear (Chen, 2020). At that time, the newly launched game 'Animal Crossing' received much attention, frequently breaking sales records and becoming a popular event beyond the game field in China. Animal Crossing features a simple core function and low operation difficulty but integrates numerous social functions, such as creation, dialoguing, collaboration, visiting, travelling, exhibiting, and exploring (Chen, 2020). Many players developed advanced gameplay, such as puzzle-solving, fairs, parties, weddings, and concerts, which highly aided the provision of psychological comfort, reduced loneliness, and encouraged players to overcome anxiety and panic during the epidemic (Constantine, 2020).

Most people described the Animal Crossing game experience as 'healing', which demonstrated the role of games in mediating emotions (Barr & Copeland-Stewart, 2022). The WHO advocated video game-playing as a healthy social pastime during the COVID-19 pandemic. Playing video games during the pandemic not only kept people at a safe social distance, but also diverted attention, relieved stress and anxiety caused by the pandemic, and improved well-being (Barr & Copeland-Stewart, 2022).

The success of Animal Crossing in China was largely due to its release at the right time when users were confined to their homes. Animal Crossing users discovered escapism during their self-isolation where they communicated with the outside world in a comfortable virtual world shaped by games (Zhu, 2021). The game satisfied the desire to escape reality and the longing for a better life and fulfilled the desire for social interaction and to alleviate loneliness (Zhu, 2021).

GAMIFICATION STRENGTHENED HEALTH AWARENESS AND IMPROVED PHYSICAL FITNESS

Gamification is used for healthcare-related purposes in areas that include physical activity, weight loss programmes, personal hygiene, and health care worker hygiene (Ramírez Elizondo, 2016). Zain et al. (2020) proposed a GAMEBC model to drive behavioural change in health awareness campaigns. The GAMEBC model consisted of elements, such as autonomy, competence, relevance, and engagement. Gamification was effective in different health-related contexts, such as exercise, weight control, and depression control (Egashira et al., 2022). Gamification positively influenced health awareness in an enjoyable, engaging, and accommodating manner where it yielded encouraging results for health awareness and service cost (Egashira et al., 2022).

While many types of fitness-related applications exist, many users gradually lose interest (Li et al., 2021). As a key element of gamification mechanics, effective gamified narrative design has the potential to engage users and encourage them to develop stronger motivations to achieve their real-world goals and usage intent with higher persistence for mobile fitness applications (Li et al., 2021). Gamified applications can provide users with an interesting and immersive storyline narrative mechanism that inspires them to progress steadily toward their main goal (Sailer et al., 2017). In health-related contexts, gamification elements (challenges, rules, interactions, feedback and rewards, emotional settings, and socialisation) have significant advantages. These advantages include promoting positive emotional experiences (Mullins & Sabherwal, 2020), social orientation (Majuri et al., 2018), enhancing leadership and collaboration skills through communication, judgment, and social skills, and developing psychomotor skills (Hope et al., 2021).

Fitness applications use various gamification mechanisms (points, levels, and leader boards) to encourage users to actively use them (Sardi et al., 2017). The narrative mechanism increases the attractiveness of fitness applications through other mechanisms. Interesting narratives encourage users to continue exercising and unlock subsequent storylines (Li, 2021). Therefore, adding a proper narrative mechanism to such applications can present users with higher-level immersion, which persuades them to use it regularly (Li, 2021).

The COVID-19 outbreak changed lifestyles (Cucinotta & Vanelli, 2020), which included self-isolation-induced inability to exercise outside the home and sedentary behaviour at home. Such lifestyle changes resulted in reduced physical activity and increased the demand for at-home exercise (Hall et al., 2021; Ozemek et al., 2019). With technological developments, technology-driven physical activities (exergames) emerged as a new fitness training and health maintenance option (Sato et al., 2021). Integrating different digital game modes into physical activity enabled physical exercise at home without space constraints (Wu et al., 2022).

'Ring Fit Adventure' was released in China in the early stage of the COVID-19 epidemic. A Nintendo Switch role-playing fitness game, Ring Fit Adventure uses high-precision sensors to detect player movements to provide a gaming experience. Players use the game sensors and resistance training equipment to control game characters to run, jump, stretch, and squeeze (Wu et al., 2022). Furthermore, infrared sensors on the gamepad monitor changes in the player's heart rate to ensure that the player achieves the optimal exercise intensity and modifies it based on physiological feedback (Sato et al., 2021). Ring Fit Adventure was an opportunity for indoor exercise during the COVID-19 pandemic lockdowns (Ambrosino et al., 2020). Subsequently, Chinese users sought it immediately and tripled its price (Duan, 2020).

Gamification elements encourage consistent user engagement in training and activities and are helpful for improving sleep quality and mood disturbances (Wu et al., 2022). Represented by Ring Fit Adventure, exergames are not only an alternative for at-home exercise but demonstrate the potential to evolve as a form of anti-epidemic changes (Wu et al., 2022).

GAMIFICATION IMPROVED ONLINE TEACHING EFFICIENCY

The popularity of games renders gamification feasible to motivate learners. Game features and experiential interactions can facilitate active learning (Dichev & Dicheva, 2017). Addressing specific factors, such as motivation, behaviour, and practice can facilitate self-directed learning, self-direction, assignment completion, and assessment feedback in various professional disciplines and teaching environments more easily and effectively (Dichev & Dicheva, 2017).

Gamifying education involves implementing educational games and digital simulations based on the development of modern visualisation technology (Tsytsiura & Romanova, 2020). Gamified education technology increases interaction with teaching materials in the classroom, making learning more engaging and enjoyable and therefore more effective (Aguiar-Castillo et al., 2020). Nevertheless, while gamification may motivate and encourage some students, it can demotivate others (Sanchez et al., 2020). Despite its many limitations, gamified learning is not unworkable (Nair, 2022). Instead, gamified learning must be used together with other learning interventions in specific applications, which would allow learners to make choices based on their own circumstances (Nair, 2022).

Gamification-based learning activities demonstrated the following five key benefits amidst COVID-19 disruptions: 1) increased student engagement and motivation, 2) added value to the learning experience, 3) increased inclusion and diversity, 4) simplified feedback and assessment, and 5) fostered employability (Nair, 2022). In specific gamification systems, reward-based strategies are considered an effective scheme to motivate students to actively participate and improve their performance (Rincon-Flores & Santos-Guevara, 2021). The reward-based gamified teaching system identified and improved the participants' moods, reduced their anxiety and isolation caused by the pandemic, and improved teacher-student interaction (Rincon-Flores & Santos-Guevara, 2021).

Nieto-Escamez and Roldán-Tapia (2021) reviewed academic research on gamified teaching strategies during the COVID-19 pandemic. Gamification improved students' learning motivation by creating higher learning interest among students than traditional teaching strategies. Thus, gamification exerted an enhanced effect when combined with traditional tools (Nieto-Escamez & Roldán-Tapia, 2021). The remote learning implemented to avoid the epidemic provided a solution to reduce infection risk (Kwon et al., 2020). Simultaneously, remote learning enabled participants to increase communication opportunities, reduce anxiety and distractions, and highlighted their interests (Kwon et al., 2020). Undoubtedly, these features allowed participants to better participate in the flow experience advocated by gamification (Wang et al., 2020; Zhao & Li, 2020).

Fontana (2020) examined the chemistry teaching game ChemDraw and demonstrated that using videoconferencing software as a medium for serious educational games enhanced the participants' social connection. Strengthening stories, creativity, and competition in games effectively improved learning, enhanced classroom communication, and students' confidence (Fontana, 2020). Oe et al. (2020) noted that once key themes had been communicated using traditional learning methods, gamification might help to reinforce those themes and learners could better integrate key knowledge into learning outcomes. Nonetheless, the disadvantage of gamification was that it made note-taking difficult for the participants, which hindered their learning outcomes (Oe et al., 2020).

In the early days of the COVID-19 lockdowns, many Chinese teachers passively began using social media to conduct online lessons to continue teaching because they couldn't get to school. Initially, many teachers were resistant to this unfamiliar teaching method. Nevertheless, teachers gradually began to accept the method when they experienced the entertainment function of promoting participation and the teacher–student closeness precipitated by live streaming technology (Chen et al., 2021). During the rapid transition from offline to online education, the entertainment function of online platforms played a positive role in maintaining classroom activity and good teaching effects (Chen et al., 2021).

In addition to live-streaming on social media, more teachers in China adopted online conference software as mainstream distance education platforms (Tsytsiura & Romanova, 2020). A comprehensive analysis of the distance education platforms used by Chinese higher education institutions determined that using game methods to improve students' learning motivation was an important factor in improving distance education efficiency (Tsytsiura & Romanova, 2020). Game models designed for undergraduate courses also aided the improvement of students' activity levels, engagement, and motivation in classroom learning (Wang et al., 2017).

SYSTEMATIC APPLICATION OF GAMIFICATION IN CHINA

In early 2022, the simulation game China's War Epidemic, which combined a China epidemic prevention experience and gamification application, was released on Steam, a dominant global game platform (Steam, 2022). Like many video games and mobile applications, this game popularized epidemic prevention knowledge by immersing players in a specific Chinese epidemic prevention environment via gamified interactions.

As depicted in Figures 1 and 2, players simulate the government to formulate policies, hire experts, allocate resources, and build hospitals while facing time, funding, food supply, and public opinion challenges (Steam, 2022). Through such simulations, players gain knowledge about COVID-19, better understand pandemic- or virus-related information, and enhance their comprehension of macro-level epidemic prevention policies. As experiencers, players can also learn about contingency plans in special epidemic prevention situations and increase their anti-risk ability in the post-epidemic era.

CONCLUSION

The COVID-19 pandemic prompted continuous government management measures, such as lockdowns, home isolation, and long-term nucleic acid testing. Gamification, a highly applicable method, was widely adopted to ensure a safer, more efficient, and more humane response to the pandemic.

This chapter focused on the role of gamification in major areas of epidemic prevention and control. First, it explored the promotion of vaccinations and nucleic acid sampling. Second, it looked at methods to reduced mental pressures caused by lockdowns. Third, it promoted home fitness. Fourth, it improved the effect of online teaching.

Gamification elements, specifically reward systems and role-playing functions, can improve public awareness of vaccination and drive vaccination behavior (Montagni et al., 2020). Gamification-based mobile applications and serious games are valuable sources of information on infection prevention and healthcare (Mandryk & Birk, 2019; Raiff et al., 2018). Moreover, the real-life example in Suzhou demonstrated that gamified reward elements can increase public enthusiasm for vaccination and nucleic acid testing.

The social isolation caused by pandemic lockdowns led to young people focusing more on games and related applications, which increased their happiness and reduced the effect of negative emotions. Gamification presented cognitive stimulation and social opportunities to provide targeted psychological interventions to the public. Games, which became a social pastime that promoted well-being, provided a safe social distance and distraction for people at home (Barr & Copeland-Stewart, 2022).

In addition to mental health issues, physical health is a major issue arising from home isolation. Gamification elements in health-related contexts (i.e., challenges, rules, interactions, feedback and rewards, emotional settings, and socialization) demonstrate significant advantages. Thus, at-home fitness through exergames exert a positive effect. In addition to the typical reward elements of points, levels, and leader boards, the gamified narrative mechanism immerses users and encourages them to continue their exercise (Li et al., 2021; Sailer et al., 2017).

University teachers and students struggled to acclimate to the challenges presented by live streaming or online conferences, a remote teaching opportunity organized in response to home isolation. Gamification-based learning enhanced student engagement and motivation, added value to their learning experience, increased knowledge inclusion and diversity, simplified feedback and assessment, and fostered student employability (Nair, 2022). Teachers used gamification to increase closeness between teachers and students through live broadcasts, enhance classroom activity, and achieve better teaching effects (Chen et al., 2021).

Systems thinking about gamification provides a new area for researchers to understand how the causal relationship between behavior and feedback affects social life (AlSaad & Durugbo, 2021). Gamification has had a systematic impact on the social science level. In fact, it has affected the behavior of many social groups. Daily public life will continue to face many of these challenges in a post-pandemic world. A positive attitude and healthy lifestyle are needed to deal with future COVID-19 spread risks. Among the coping strategies, gamification-based rule systems and applications that include gamification will provide government and enterprise managers with a multitude of feasible options.

ACKNOWLEDGMENT

This work is funded by Jiangsu Provincial Educational Science Planning Project Key Project (2022), issued by Jiangsu Provincial Education Science Planning Leading Group and Communication University of China, Nanjing. Project title: Innovative Research on Children's Logic Teaching Practice in the context of Gamification (B/2022/04/42).

REFERENCES

Aguiar-Castillo, L., Hernandez-Lopez, L., De Saá-Pérez, P., & Perez-Jimenez, R. (2020). Gamification as a motivation strategy for higher education students in tourism face-to-face learning. *Journal of Hospitality, Leisure, Sport and Tourism Education*, *27*, 100267. doi:10.1016/j.jhlste.2020.100267

Alahäivälä, T., & Oinas-Kukkonen, H. (2016). Understanding persuasion contexts in health gamification: A systematic analysis of gamified health behavior change support systems literature. *International Journal of Medical Informatics*, *96*, 62–70. doi:10.1016/j.ijmedinf.2016.02.006 PMID:26944611

AlSaad, F. M., & Durugbo, C. M. (2021). Gamification-as-innovation: A review. *International Journal on Management of Innovation & Technology*, *18*(05), 2130002. doi:10.1142/S0219877021300020

Alvarez, J., & Djaouti, D. (2011). An introduction to serious game definitions and concepts. *Serious games & simulation for risks management, 11*(1), 11-15. http//hayka-kultura.com/larsen.html

Ambrosino, P., Fuschillo, S., Papa, A., Di Minno, M. N. D., & Maniscalco, M. (2020). Exergaming as a supportive tool for home-based rehabilitation in the COVID-19 pandemic era. *Games for Health Journal*, *9*(5), 311–313. doi:10.1089/g4h.2020.0095 PMID:32876489

Barr, M. (2017). Video games can develop graduate skills in higher education students: A randomised trial. *Computers & Education*, *113*, 86–97. doi:10.1016/j.compedu.2017.05.016

Barr, M., & Copeland-Stewart, A. (2022). Playing video games during the COVID-19 pandemic and effects on players' well-being. *Games and Culture*, *17*(1), 122–139. doi:10.1177/15554120211017036

Buckley, P., & Doyle, E. (2016). Gamification and student motivation. *Interactive Learning Environments*, *24*(6), 1162–1175. doi:10.1080/10494820.2014.964263

Carras, M. C., Kalbarczyk, A., Wells, K., Banks, J., Kowert, R., Gillespie, C., & Latkin, C. (2018). Connection, meaning, and distraction: A qualitative study of video game play and mental health recovery in veterans treated for mental and/or behavioral health problems. *Social Science & Medicine*, *216*, 124–132. doi:10.1016/j.socscimed.2018.08.044 PMID:30257787

Chen, S. (2020). During the epidemic period, the phenomenon-level game "Animal Crossing" was born. *Beijing Youth Daily*. https://www.chinanews.com.cn/cul/2020/04-07/9149268.shtml

Chen, S., Yang, J., Yang, W., Wang, C., & Bärnighausen, T. (2020). COVID-19 control in China during mass population movements at New Year. *Lancet*, *395*(10226), 764–766. doi:10.1016/S0140-6736(20)30421-9 PMID:32105609

Chen, X., Chen, S., Wang, X., & Huang, Y. (2021). "I was afraid, but now I enjoy being a streamer!" Understanding the Challenges and Prospects of Using Live Streaming for Online Education. *Proceedings of the ACM on Human-Computer Interaction, 4*(CSCW3), (pp. 1-32). ACM. 10.1145/3432936

Cho, I., Kaplanidou, K., & Sato, S. (2021). Gamified wearable fitness tracker for physical activity: A comprehensive literature review. *Sustainability (Basel), 13*(13), 7017. doi:10.3390u13137017

Cole, D., Pereira, R., & Spray, J. (2021). Gamification and Virtual Reality for Communicating Ecoliteracy and Climate Science: Carbon Transport in the Essequibo River at Iwokrama Guyana. In *Transforming Society and Organizations through Gamification: From the Sustainable Development Goals to Inclusive Workplaces* (pp. 151–180). Springer International Publishing. doi:10.1007/978-3-030-68207-1_9

Cole, H., & Griffiths, M. D. (2007). Social interactions in massively multiplayer online role-playing gamers. *Cyberpsychology & Behavior, 10*(4), 575–583. doi:10.1089/cpb.2007.9988 PMID:17711367

Constantine. (2020). Behind the explosion of Animal Crossing during the epidemic, is it another escape from the real predicament? *Tencent News.* https://xw.qq.com/cmsid/20200402A07QY100

Csikszentmihalyi, M., & Csikzentmihaly, M. (1990). *Flow: The psychology of optimal experience* (Vol. 1990). Harper & Row.

Cucinotta, D., & Vanelli, M. (2020). WHO declares COVID-19 a pandemic. *Acta Biomedica, 91*(1), 157. doi:10.23750/abm.v91i1.9397 PMID:32191675

Davidová, M., Sharma, S., McMeel, D., & Loizides, F. (2022). Co-De|GT: The gamification and tokenisation of more-than-human qualities and values. *Sustainability (Basel), 14*(7), 3787. doi:10.3390u14073787

Deterding, S., Dixon, D., Khaled, R., & Nacke, L. (2011). From game design elements to gamefulness: defining "gamification". *Proceedings of the 15th international academic MindTrek conference: Envisioning future media environments*, (pp. 9-15). ACM. 10.1145/2181037.2181040

Dichev, C., & Dicheva, D. (2017). Gamifying education: what is known, what is believed and what remains uncertain: a critical review. *International Journal of Educational Technology in Higher Education, 14*(1), 1–36. doi:10.118641239-017-0042-5

Duan, Q. Q. (2020). The fitness ring adventure has become "hard currency" during the epidemic. *Sina Tech.* https://tech.sina.com.cn/it/2020-02-27/doc-iimxxstf4649888.shtml

Edwards, S. (2013). Forum focus-All the lonely people. *Nursing Older People, 25*(10), 10. doi:10.7748/nop2013.12.25.10.10.s12

Egashira, M., Son, D., & Ema, A. (2022). Serious Game for Change in Behavioral Intention Toward Lifestyle-Related Diseases: Experimental Study with Structural Equation Modeling Using the Theory of Planned Behavior. *JMIR Serious Games, 10*(1), e28982. doi:10.2196/28982 PMID:35188465

Fadda, M., Galimberti, E., Fiordelli, M., Romanò, L., Zanetti, A., & Schulz, P. J. (2017). Effectiveness of a smartphone app to increase parents' knowledge and empowerment in the MMR vaccination decision: A randomized controlled trial. *Human Vaccines & Immunotherapeutics, 13*(11), 2512–2521. doi:10.1080/21645515.2017.1360456 PMID:29125783

Fleming, T. M., Bavin, L., Stasiak, K., Hermansson-Webb, E., Merry, S. N., Cheek, C., Lucassen, M., Lau, H. M., Pollmuller, B., & Hetrick, S. (2017). Serious games and gamification for mental health: Current status and promising directions. *Frontiers in Psychiatry, 7*, 215. doi:10.3389/fpsyt.2016.00215 PMID:28119636

Fontana, M. T. (2020). Gamification of ChemDraw during the COVID-19 pandemic: Investigating how a serious, educational-game tournament (molecule madness) impacts student wellness and organic chemistry skills while distance learning. *Journal of Chemical Education*, *97*(9), 3358–3368. doi:10.1021/acs.jchemed.0c00722

Ghiga, I., Richardson, S., Álvarez, A. M. R., Kato, M., Naidoo, D., Otsu, S., Nguyen, P. T., Nguyen, P. N., & Nguyen, T. (2021). PIPDeploy: Development and implementation of a gamified table top simulation exercise to strengthen national pandemic vaccine preparedness and readiness. *Vaccine*, *39*(2), 364–371. doi:10.1016/j.vaccine.2020.11.047 PMID:33293160

Hall, G., Laddu, D. R., Phillips, S. A., Lavie, C. J., & Arena, R. (2021). A tale of two pandemics: How will COVID-19 and global trends in physical inactivity and sedentary behavior affect one another? *Progress in cardiovascular diseases, 64*, 108. 10. https://doi.org/ doi:1016/j.pcad.2020.04.005

Hamari, J., Koivisto, J., & Sarsa, H. (2014). Does gamification work? A literature review of empirical studies on gamification. *2014 47th Hawaii international conference on system sciences*, 3025-3034. 10.1109/HICSS.2014.377

Han, X., Li, J., Chen, Y., Li, Y., Xu, Y., Ying, B., & Shang, H. (2022). SARS-CoV-2 nucleic acid testing is China's key pillar of COVID-19 containment. *Lancet*, *399*(10336), 1690–1691. doi:10.1016/S0140-6736(22)00577-3 PMID:35397850

Hope, D. L., Grant, G. D., Rogers, G. D., & King, M. A. (2021). PROGRAMME DESCRIPTION: Integration of an extended, immersive, gamified pharmacy simulation as a capstone event. *Pharmacy Education*, *21*, 656–669. doi:10.46542/pe.2021.211.656669

Huang, R. H., Liu, D. J., Tlili, A., Yang, J. F., & Wang, H. H. (2020). Handbook on facilitating flexible learning during educational disruption: The Chinese experience in maintaining undisrupted learning in COVID-19 outbreak. *Beijing: Smart Learning Institute of Beijing Normal University, 46*. https://creativecommons.org/licenses/by-sa/3.0/igo/

Ioannidis, J. P. (2022). The end of the COVID-19 pandemic. *European Journal of Clinical Investigation*, e13782. doi:10.1111/eci.13782

Ju, Z. (2022). There are many styles of nucleic acid testing vouchers in Suzhou. *Netease News*. https://www.163.com/dy/article/H60QATJT0535B3HJ.html

Kaye, L. K., Kowert, R., & Quinn, S. (2017). The role of social identity and online social capital on psychosocial outcomes in MMO players. *Computers in Human Behavior*, *74*, 215–223. doi:10.1016/j.chb.2017.04.030

Koivisto, J., & Hamari, J. (2019). The rise of motivational information systems: A review of gamification research. *International Journal of Information Management*, *45*, 191–210. doi:10.1016/j.ijinfomgt.2018.10.013

Kotsopoulos, K. I., Chourdaki, P., Tsolis, D., Antoniadis, R., Pavlidis, G., & Assimakopoulos, N. (2019). An authoring platform for developing smart apps which elevate cultural heritage experiences: A system dynamics approach in gamification. *Journal of Ambient Intelligence and Humanized Computing*, 1–17. doi:10.100712652-019-01505-w

Kwon, R., Zhang, M. L., & VandenBussche, C. J. (2020). Considerations for remote learning in pathology during COVID-19 social distancing. *Cancer Cytopathology*, *128*(9), 642–647. doi:10.1002/cncy.22289 PMID:32497399

Lei, S. I., & So, A. S. I. (2021). Online teaching and learning experiences during the COVID-19 pandemic - A comparison of teacher and student perceptions. *Journal of Hospitality & Tourism Education*, *33*(3), 148–162. doi:10.1080/10963758.2021.1907196

Li, M., Wang, Y., Wu, Y., & Liu, H. (2021). Gamification narrative design as a predictor for mobile fitness app user persistent usage intentions: A goal priming perspective. *Enterprise Information Systems*, *15*(10), 1501–1545. doi:10.1080/17517575.2021.1941272

Li, Z., Chen, Q., Feng, L., Rodewald, L., Xia, Y., Yu, H., Zhang, R., An, Z., Yin, W., Chen, W., Qin, Y., Peng, Z., Zhang, T., Ni, D., Cui, J., Wang, Q., Yang, X., Zhang, M., Ren, X., & China, C. D. C. (2020). Active case finding with case management: The key to tackling the COVID-19 pandemic. *Lancet*, *396*(10243), 63–70. doi:10.1016/S0140-6736(20)31278-2 PMID:32505220

Li, Z., Liu, F., Cui, J., Peng, Z., Chang, Z., Lai, S., Chen, Q., Wang, L., Gao, G. F., & Feng, Z. (2021). Comprehensive large-scale nucleic acid-testing strategies support China's sustained containment of COVID-19. *Nature Medicine*, *27*(5), 740–742. doi:10.103841591-021-01308-7 PMID:33859409

Majuri, J., Koivisto, J., & Hamari, J. (2018). Gamification of education and learning: A review of empirical literature. *Proceedings of the 2nd international GamiFIN conference, GamiFIN 2018*. https://ceur-ws.org/Vol-2186/paper2.pdf

Mandryk, R. L., & Birk, M. V. (2019). The potential of game-based digital biomarkers for modeling mental health. *JMIR Mental Health*, *6*(4), e13485. doi:10.2196/13485 PMID:31012857

Mindeguia, M., Ruiz, M., Igartua, J. I., & Ganzarain, J. (2019). Gamification: An Approach in The Master of Business Innovation and Project Management. In *INTED2019 Proceedings* (pp. 1415-1422). IATED. 10.21125/inted.2019.0440

Montagni, I., Mabchour, I., & Tzourio, C. (2020). Digital gamification to enhance vaccine knowledge and uptake: Scoping review. *JMIR Serious Games*, *8*(2), e16983. doi:10.2196/16983 PMID:32348271

Moreno, C., Wykes, T., Galderisi, S., Nordentoft, M., Crossley, N., Jones, N., Cannon, M., Correll, C. U., Byrne, L., Carr, S., Chen, E. Y. H., Gorwood, P., Johnson, S., Kärkkäinen, H., Krystal, J. H., Lee, J., Lieberman, J., López-Jaramillo, C., Männikkö, M., ... Arango, C. (2020). How mental health care should change as a consequence of the COVID-19 pandemic. *The Lancet. Psychiatry*, *7*(9), 813–824. doi:10.1016/S2215-0366(20)30307-2 PMID:32682460

Mullins, J. K., & Sabherwal, R. (2020). Gamification: A cognitive-emotional view. *Journal of Business Research*, *106*, 304–314. doi:10.1016/j.jbusres.2018.09.023

Nair, A., & Sadasivan, R. (2019). Winning the talent game: HR gamification experience for Generation Z. *International Journal on Leadership*, *7*(1), 44–49. http://publishingindia.com/ijl/

Nair, B. B. (2022). Endorsing gamification pedagogy as a helpful strategy to offset the COVID-19 induced disruptions in tourism education. *Journal of Hospitality, Leisure, Sport and Tourism Education*, *30*, 100362. doi:10.1016/j.jhlste.2021.100362 PMID:34849104

Nieto-Escamez, F. A., & Roldán-Tapia, M. D. (2021). Gamification as online teaching strategy during COVID-19: A Mini-review. *Frontiers in Psychology*, *12*, 648552. doi:10.3389/fpsyg.2021.648552 PMID:34093334

Nitsche, A. M., Franczyk, B., & Schumann, C. A. (2022). Gamification for Smart and Collaborative Last Mile Networks Knowledge Transfer. In *2022 IEEE 28th International Conference on Engineering, Technology and Innovation (ICE/ITMC) & 31st International Association for Management of Technology (IAMOT) Joint Conference* (pp. 1-9). IEEE. 10.1109/ICE/ITMC-IAMOT55089.2022.10033189

Oe, H., Takemoto, T., & Ridwan, M. (2020). Is gamification a magic tool? Illusion, remedy, and future opportunities in enhancing learning outcomes during and beyond the COVID-19. *Budapest International Research and Critics in Linguistics and Education (BirLE). Journal*, *3*(3), 1401–1414. doi:10.33258/birle.v3i3

Ozemek, C., Lavie, C. J., & Rognmo, Ø. (2019). Global physical activity levels-Need for intervention. *Progress in Cardiovascular Diseases*, *62*(2), 102–107. doi:10.1016/j.pcad.2019.02.004 PMID:30802461

Paterson, P., Chantler, T., & Larson, H. J. (2018). Reasons for non-vaccination: Parental vaccine hesitancy and the childhood influenza vaccination school pilot programme in England. *Vaccine*, *36*(36), 5397–5401. doi:10.1016/j.vaccine.2017.08.016 PMID:28818568

Petrosino, S., Loria, E., & Pirker, J. (2021). Stay Home Playing LoL-Analyzing Players' Activity and Social Bonds in League of Legends During Covid-19 Lockdowns. *The 16th International Conference on the Foundations of Digital Games (FDG)*, 1-12. 10.1145/3472538.3472551

Qiu, H., Li, Q., & Li, C. (2021). How technology facilitates tourism education in COVID-19: Case study of nankai University. *Journal of Hospitality, Leisure, Sport and Tourism Education*, *29*, 100288. doi:10.1016/j.jhlste.2020.100288 PMID:34720752

Raiff, B. R., Fortugno, N., Scherlis, D. R., & Rapoza, D. (2018). A mobile game to support smoking cessation: Prototype assessment. *JMIR Serious Games*, *6*(2), e9599. doi:10.2196/games.9599 PMID:29880466

Ramírez Elizondo, E. (2016). Identification of design patterns for serious games in an educational videogame designed to create awareness on dengue and malaria fever. *IFIP World Information Technology Forum*, 85-95. 10.1007/978-3-319-44447-5_9

Rapp, A., Hopfgartner, F., Hamari, J., Linehan, C., & Cena, F. (2019). Strengthening gamification studies: Current trends and future opportunities of gamification research. *International Journal of Human-Computer Studies*, *127*, 1–6. doi:10.1016/j.ijhcs.2018.11.007

Reer, F., & Quandt, T. (2020). Digital games and well-being: An overview. *Video games and well-being*, 1-21. doi:10.1007/978-3-030-32770-5_1

Reinecke, L. (2009). Games and recovery: The use of video and computer games to recuperate from stress and strain. *Journal of Media Psychology*, *21*(3), 126–142. doi:10.1027/1864-1105.21.3.126

Rincon-Flores, E. G., & Santos-Guevara, B. N. (2021). Gamification during Covid-19: Promoting active learning and motivation in higher education. *Australasian Journal of Educational Technology*, *37*(5), 43–60. doi:10.14742/ajet.7157

Rivas, C. L. S., Pérez, H. K. F., Rivas, L. J. M., & Almaguer, C. A. G. (2022). Gamification for the Development of Competencies in Tec21 Based on Mixed Reality. In *European Conference on Games Based Learning* (Vol. 16, No. 1, pp. 482-490). 10.34190/ecgbl.16.1.679

Sailer, M., Hense, J. U., Mayr, S. K., & Mandl, H. (2017). How gamification motivates: An experimental study of the effects of specific game design elements on psychological need satisfaction. *Computers in Human Behavior*, *69*, 371–380. doi:10.1016/j.chb.2016.12.033

Sanchez, D. R., Langer, M., & Kaur, R. (2020). Gamification in the classroom: Examining the impact of gamified quizzes on student learning. *Computers & Education*, *144*, 103666. doi:10.1016/j.compedu.2019.103666

Sardi, L., Idri, A., & Fernández-Alemán, J. L. (2017). A systematic review of gamification in e-Health. *Journal of Biomedical Informatics*, *71*, 31–48. doi:10.1016/j.jbi.2017.05.011 PMID:28536062

Sato, T., Shimizu, K., Shiko, Y., Kawasaki, Y., Orita, S., Inage, K., Shiga, Y., Suzuki, M., Sato, M., Enomoto, K., Takaoka, H., Mizuki, N., Kim, G., Hozumi, T., Tsuchiya, R., Otagiri, T., Mukaihata, T., Furuya, T., Maki, S., ... Eguchi, Y. (2021). Effects of nintendo ring fit adventure exergame on pain and psychological factors in patients with chronic low back pain. *Games for Health Journal*, *10*(3), 158–164. doi:10.1089/g4h.2020.0180 PMID:33891508

Schmidt-Kraepelin, M., Warsinsky, S., Thiebes, S., & Sunyaev, A. (2020). The role of gamification in health behavior change: a review of theory-driven studies. *Proceedings of the 53rd Hawaii international conference on system sciences*, 1256-1265. http://hdl.handle.net/10125/63894

Seo, S., & Kim, H. J. (2021). How COVID-19 Influences Hospitality and Tourism Education: Challenges, Opportunities, and New Directions. *Journal of Hospitality & Tourism Education*, *33*(3), 147–147. doi: 10.1080/10963758.2021.1929531

Steam. (2022). China's War Epidemic. *Steam*. https://store.steampowered.com/app/1731360/_/

Tabnak, M., & Hosseini, S. (2021). The role of gamification and game-based learning in assessment of systems thinking. *Journal of Strategic Management Studies*, *12*(45), 119–138. https://www.smsjournal.ir/article_120683_en.html?lang=en

Tan, Y. L. L., Varma, N., & Balakrishnan, N. P. (2020). Teaching systems thinking concepts during a pandemic: 3 strategies forengaging learners. In *Higher Education Campus Conference (e-HECC) 2020*. ScholarBank@NUS Repository. https://scholarbank.nus.edu.sg/handle/10635/202904

Tian, F., Li, H., Tian, S., Yang, J., Shao, J., & Tian, C. (2020). Psychological symptoms of ordinary Chinese citizens based on SCL-90 during the level I emergency response to COVID-19. *Psychiatry Research*, *288*, 112992. doi:10.1016/j.psychres.2020.112992 PMID:32302816

Topic, M., & Mitchell, B. (2019). Generation Z & Consumer Trends in Environmental Packaging. Project Report. The Retail Institute, Leeds.

Troiano, G., & Nardi, A. (2021). Vaccine hesitancy in the era of COVID-19. *Public Health, 194*, 245–251. doi:10.1016/j.puhe.2021.02.025 PMID:33965796

Tsytsiura, K. V., & Romanova, G. M. (2020). Analysis of the Efficiency of China's Distance Economic Education in Force Majeure Circumstances. *International Journal of Learning. Teaching and Educational Research, 19*(9), 175–195. doi:10.26803/ijlter.19.9.10

Vangelisti, A. L., & Perlman, D. (Eds.). (2006). *The Cambridge handbook of personal relationships.* Cambridge University Press., doi:10.1017/CBO9780511606632

Wang, F., Wang, Y., & Hu, X. (2017). Gamification Teaching Reform for Higher Vocational Education in China: A Case Study on Layout and Management of Distribution Center. *International Journal of Emerging Technologies in Learning, 12*(9), 130. Advance online publication. doi:10.3991/ijet.v12i09.7493

Wang, S. M., Chen, Y. C., Hou, H. T., Hsu, H. Y., & Li, C. T. (2020). Exploring the effects of card game-based gamification instructional activity on learners' flow experience, learning anxiety, and performance-A preliminary study. *ICCE 2020-28th International Conference on Computers in Education, 2*, 190-198. https://apsce.net/icce/icce2020/proceedings/W1-13/W4/ICCE2020-Proceedings-Vol2-W4_5.pdf

Willis, N., Hill, S., Kaufman, J., Lewin, S., Kis-Rigo, J., De Castro Freire, S. B., Bosch-Capblanch, X., Glenton, C., Lin, V., Robinson, P., & Wiysonge, C. S. (2013). "Communicate to vaccinate": The development of a taxonomy of communication interventions to improve routine childhood vaccination. *BMC International Health and Human Rights, 13*(1), 1–16. doi:10.1186/1472-698X-13-23 PMID:23663327

Wu, Y. S., Wang, W. Y., Chan, T. C., Chiu, Y. L., Lin, H. C., Chang, Y. T., Wu, H. Y., Liu, T. C., Chuang, Y. C., Wu, J., Chang, W. Y., Sun, C. A., Lin, M. C., Tseng, V. S., Hu, J. M., Li, Y. K., Hsiao, P. J., Chen, C. W., Kao, H. Y., ... Chu, C. M. (2022). Effect of the Nintendo Ring Fit Adventure Exergame on Running Completion Time and Psychological Factors Among University Students Engaging in Distance Learning During the COVID-19 Pandemic: Randomized Controlled Trial. *JMIR Serious Games, 10*(1), e35040. doi:10.2196/35040 PMID:35315780

Xi, N., & Hamari, J. (2019). Does gamification satisfy needs? A study on the relationship between gamification features and intrinsic need satisfaction. *International Journal of Information Management, 46*, 210–221. doi:10.1016/j.ijinfomgt.2018.12.002

Zain, N. H. M., Johari, S. N., Aziz, S. R. A., Teo, N. H. I., Ishak, N. H., & Othman, Z. (2021). Winning the Needs of the Gen Z: Gamified Health Awareness Campaign in Defeating COVID-19 Pandemic. *Procedia Computer Science, 179*, 974–981. doi:10.1016/j.procs.2021.01.087

Zain, N. H. M., Othman, Z., Noh, N. M., Teo, N. H. I., Zulkipli, N. H. B. N., & Yasin, A. M. (2020). GAMEBC model: Gamification in health awareness campaigns to drive behaviour change in defeating COVID-19 pandemic. *International Journal of Advanced Trends in Computer Science and Engineering, 9*(4), 229–236. Advance online publication. doi:10.30534/ijatcse/2020/3491.42020

Zhao, J., & Li, Y. (2020). Experimental Study on Gamification Teaching of Mathematics Classroom Practice Based on Flow Experience. *2020 IEEE 2nd International Conference on Computer Science and Educational Informatization (CSEI)*, 36-40. 10.1109/CSEI50228.2020.9142485

Zhong, M. (2021). Vaccine rewards keep escalating. *Netease News*. https://www.163.com/dy/article/GFHCSMP105328YVR.html

Zhu, L. (2021). The psychology behind video games during COVID-19 pandemic: A case study of Animal Crossing: New Horizons. *Human Behavior and Emerging Technologies*, *3*(1), 157–159. doi:10.1002/hbe2.221

Chapter 3
Adaptation of Systems Thinking by Academic and Clinical Educators in the Field of Communication Sciences and Disorders Following the Pandemic

Rachel Wolf
East Stroudsburg University, USA

Akila T. Rajappa
East Stroudsburg University, USA

ABSTRACT

This chapter will provide readers with how a systems based approach was established and implemented during and post pandemic to maintain educational and clinical services for graduate students studying to become speech-language pathologists and the clients they provide care. Achieving a master's degree is essential to becoming a licensed and certified speech-language pathologist (SLP). A board accredited program in communication sciences and disorders has specific standards that must be met to ensure all students gain comprehensive knowledge and skills across all domains within an SLP's scope of practice. Working directly with clients requires a systems approach whereby the SLP recognizes the relationship between culture, environment, language, and disability. The unprecedented nature of the pandemic had a profound impact on approaches used to meet student learning outcomes as well as provision of services for those clients receiving their therapeutic services.

DOI: 10.4018/978-1-6684-7285-9.ch003

INTRODUCTION

This chapter will provide readers with how a systems based approach was established and implemented during and post pandemic to maintain educational and clinical services for graduate students majoring in communication sciences and disorders (CSD) who are studying to become speech-language pathologists (SLPs) and the clients they provide care. Achieving a master's degree is essential to becoming a licensed and certified speech-language pathologist (SLP) and includes educational coursework in addition to hands-on clinical experiences. The unprecedented nature of the COVID pandemic impacted approaches used to meet student learning outcomes as well as provision of clinical SLP services. Information will be provided on how educators in a board certified CSD program used systems thinking and implementation of a systems approach to transition and adapt academic and clinical education during the pandemic. The key take aways identified will inform future efforts and provide useful insights across educational and health care fields.

How can educators who teach graduate students studying for their master's degree in communication sciences and disorders fulfill all of the knowledge and skill requirements necessary to be competent professionals during a pandemic? The impact of the COVID-19 pandemic tested previously effective models of systems thinking which were implemented by service providers. It forced individuals to build an awareness of the interconnectedness of systems and identify where adaptations and change could enhance and progress current practices. The complex nature of educating graduate students in a clinical domain cannot be understated. While the pandemic imposed unforeseen challenges, it provided opportunities to enhance models of systems thinking. This is particularly true for an area of study, communication sciences and disorders, that trains students to practice primarily in a clinical capacity as speech-language pathologist.

The qualifications and areas of expertise of a speech-language pathologist, also called SLP, are often underextended. Although it is assumed that SLPs are experts in communication, this is a limited scope of its knowledge and practical realm. The field of study for which an SLP has competency is vast and led to a push over the last decade for educational programs to develop experts in the field of study, while not limiting or assuming those experts would become clinical practitioners. SLPs work with individuals across ages from infants to older adults who present with a variety of diagnoses in the areas of communication and swallowing. These areas include but are not limited to speech sound disorders, language disorders, literacy skills, social communication, voice disorders,fluency disorders, cognitive-communication skills, and feeding and swallowing disorders (Who Are Speech-Language Pathologists, and What Do They Do? (asha.org)).

In March 2020 while many institutions of higher education were on its spring break hiatus, the world *shut down* due to the pandemic. Faculty within the department of Communication Sciences and Disorders (CSD) at a small state university were faced with the task of figuring out how to provide the clinical opportunities needed by the graduate students to meet the knowledge and skill competencies set forth to program accredited by the Council on Academic Accreditation in Audiology and Speech-Language Pathology (CAA). The CAA serves as the American Speech-Language and Hearing Association's (ASHA) stamp of educational excellence for programs designed to educate individuals seeking to become clinical speech-language pathologists (SLP). As the unprecedented events and subsequent impacts unfolded, 30 second year graduate students, scheduled to graduate in August 2020, were in the midst of either a medical or school based clinical externship (i.e., off-campus) experience and 28 first year graduate students were performing their first semester of advanced clinical practicum working with clients at the university's on-campus speech-language and hearing clinic. Throughout academia, educators were faced with the challenge of maintaining high quality teaching, scholarship pursuits, and service commitments to the university, students, and community at large. The CSD department has always strived to implement a systems based approach to student learning outcomes. In this capacity, service learning is a fundamental means of providing early exposure to many of the more abstract and complex standards students need for future success as a clinician. This includes development of skills in such areas as interdisciplinary practice, behavior management, working with at-risk populations, adaptation of services to meet cultural and linguistic distinctions, continuum of care and prevention, as well as providing mentoring opportunities.

The events of the pandemic forced educators to take a deeper look at the role of systems thinking, the importance of interprofessional collaborations, practicing at the top of one's license, and how hybrid approaches can be used, including strengths, limitations, and areas of opportunity for development. Identification of specific patterns and models for teaching that helped define approaches to evaluate and understand system wide effects. To this end, systems thinking was used to identify the elements of the system that were working, could be adapted, sequenced, coordinated, and synchronized to enhance student success (Trbovich, 2014). Despite the undeniable impacts of the pandemic on society, valuable insights were gained from the imposed and necessary shift from a predominantly online format of teaching, to an interspersed mixture of hybrid demonstrations (i.e., online and in-person), and then back to predominantly in-person format. This was particularly pronounced when considering fields which have a clinical education component.

Description of Clinical Academic Requirements

Achieving a master's degree is essential to becoming a licensed and certified SLP. A board accredited program in communication sciences and disorders has specific standards that must be met to ensure all students gain comprehensive knowledge and skills across all domains within an SLP's scope of practice. This includes educational coursework in addition to hands-on clinical experiences. Students must gain 375 direct contact hours with clients, which are supervised by a licensed SLP. It is this application of skills that ultimately develops a student's ability to work dynamically in an applied manner. As part of graduate students' plan of study, two forms of clinical education placements are arranged for students to meet their clinical education skill competencies and hour requirements.

The initial clinical opportunities occur over three semesters serving clients across age ranges and disorder types at the on-campus clinic which is housed with the CDS department on the university campus. Subsequent to these on-campus practicum experiences, students complete two off campus clinical externship opportunities. One placement is typically at a school-based site, lasting 12 weeks and the second being at a non-school placement, traditionally a medically based site, lasting 10 weeks. All off campus practicum coordination is arranged and administered through the CSD department's externship coordinator. Students who do not accumulate the minimum number of hours and/or do not meet skill competencies across evaluation, treatment, and professional practice domains, are not eligible to receive a degree.

It should be noted that supervisors must directly observe, in real time, at least 25% of all clinical management sessions and 50% of evaluations for each client that a student is assigned to work with. Students must be observed 100% of sessions for a client whose services are billed to Medicare. The amount of direct supervision and observation of the student should be commensurate with the student's knowledge, skills and experience, and must be sufficient to ensure the welfare of the client, as judged by the supervisor.

At the onset of the pandemic, new guidelines for clinical education were intermittently set forth by the Council for Clinical Certification(CFCC) in Audiology and Speech-Language Pathology. The CFCC sets forth the Knowledge and Skills Acquisition (KASA) requirements for CSD programs related to student's acquisition of the knowledge and skills delineated in the Standards for the Certificate of Clinical Competence (SCCC). Acquisition of *knowledge* is generally achieved through academic courses, and the acquisition of *skills* is generally achieved through clinical experiences. The unprecedented situation halted access to direct clinical services, which had been provided to community members. These on-site and in-person experiences allowed graduate student clinicians to gain hours and competencies required under the 2020 Audiology and Speech-Language Pathology Certification Standards. This situation extended to the limited ability for graduate student clinicians to be physically present with clients/patients, and for clinical educators to provide on-site and in-person supervision.

Despite the acknowledgement by the CFCC that it was going to be a challenge for CAA accredited programs to provide opportunities for their graduate students to meet the minimum clinical practicum requirements in accordance with the standards, the number of clinical practicum hours was not reduced. It was determined that CAA and CFCC requirements needed to remain at levels that meet ASHA Certification standards, but also those required by state or federal organizations, such as state licensing boards, the U.S. Department of Education, and the Council for Higher Education Accreditation (CHEA). The CFCC has jurisdiction over what is required for the purposes of ASHA certification. However, ASHA does not have jurisdiction over these external agencies/bodies. One adjustment that was immediately implemented was allowing students to engage in service delivery through telepractice. Prior to the pandemic, telepractice with telesupervision was not permitted for gaining clinical education experience and hours towards standard requirements.

Temporary guidelines were set forth by the CFCC with regard to how students could accumulate hours through telepractice. These were adjusted and amended throughout the pandemic, as the critical nature of the situation fluctuated, and systems learned to adapt to the new atmosphere of benefits and concerns imposed during and after the pandemic. Initial accommodation included allowing multiple students to participate in the same session. Each student who was actively participating in the session earned the full hour toward the completion of their clinical practicum. Program and clinic directors had the authority to determine how many students could appropriately take part in an online teletherapy session with one client, keeping quality patient care, safety, and optimal clinical education in mind. If students were to observe other student clinicians but are not actively engaged in service delivery, they can only count the time as guided observation hours. Programs were tasked with carefully considering which clients were appropriate for telepractice. Initially, it was up to the discretion of each program to determine how many hours could be earned through telepractice and there was no cap for the purposes of ASHA certification. These hours were deemed to be separate from the hours earned through clinical simulation for which a minimum of 75 hours was historically acceptable in this capacity. Clinical educators were able to supervise more than one telepractice session concurrently. Clinical educators were required to be available 100% of the time to each session and maintain the minimum of 25% direct supervision of the total contact time with each client/patient (similar to in-person supervision requirements).

Components of Systems Thinking

Although described, focus on systems models in communication sciences and disorders is often minimized in favor of models which focus more directly on underlying impairment and disability within the individual. When considering models of causation for communication and swallowing disorders, systems models emphasize the problem not being within the individual, but rather in the relationship between speakers (Freeman & Cameron, 2008). Thus the problem is the environment and by changing the environment, the problem is reduced. Albeit true, contextual models raise concerns when considering labels such as disorder versus difference. Contextual models are crucial as they aim to improve communicative contexts and encourage dynamic assessments. Systems thinking and environmental contributions have been described and referenced when considering normal language development and disorders more from linguistic theories, but less common in CSD literature (Sabah, 2017). In addition, systems thinking perspectives have suggested benefits for educational instruction (Bethany, 199; Dunnion & O'Donnovan, 2014; Ison, 1999; Pedryez & Kwak, 2006; Pedryez & Vasilakos,1999).

When education is theoretically oriented from systems thinking, it is characterized as a holistic approach based on contextual learning (Astaizo-Martinez, et., al., 2021). Pre-pandemic, the components of systems thinking included contextual and service learning opportunities. The pandemic forced awareness and consideration of components and entities involved in the system. This in turn fostered identification of available opportunities to enhance and create new and improved means for implementing dynamic teaching approaches. The uncertainty surrounding the outcomes of the pandemic forced educators and clinical instructors to develop and implement systems without prior experience or knowledge of anticipated impacts or outcomes. This created multiple instances of having to critically think and impose systems thinking to ensure all resources and opportunities were considered and exhausted to ensure student success. Regardless of the pandemic, working directly with clients requires a systems approach whereby the SLP recognizes the relationship between culture, environment, language, and disability (Horton, R., 2020). The unprecedented nature of the pandemic had a profound impact on approaches used to meet student learning outcomes as well as provision of services to those clients receiving speech-language therapeutic services. It was necessary to understand and solve the complex problems that were arising, systems thinking allowed for consideration of the multiple components involved and their dynamic interaction.

In order to understand the effects of the pandemic on teaching and assessment, one must look directly at the systems involved in speech-language pathology education programs. Systems thinking and system-based practice has been described as ranging across diverse levels in healthcare including: patient/client, care team, organization, and external environment (Plack,et.al., 2017). Although valid, this is a limited scope of factors. It is now clear that additional components need to be considered, including technology and the establishment of processes that allow for flexibility and adaptation within a foundation designed for best practice procedure. It is the dynamic interaction of multiple factors that must come together in a holistic manner in order to solve problems and successfully implement efforts.

On March 13, 2020, classes were forced to change from a face to face format to online instruction. Faculty were given one week (March 16-20) to develop alternate teaching plans for each class. Professional development sessions (e.g., Basics of D2L and Zoom) were provided through the university's Center for Teaching with Technology. Although alternative instructional modalities were approved, the emphasis on maintaining academic rigor and the competition of course objectives were prioritized, as well as being reminded that this was a time for flexibility and innovation. While this was a new experience for most, transitioning the standard linear style, cause-effect teaching method to a remote learning experience was more of a transposition rather than a reinvention. The same could not be said for the adaptation of the clinical elements of teaching into a virtual realm. It should be noted that after the spring 2020 semester, the university's graduate programs returned to in person learning for all courses and clinical education in the fall 2020. Undergraduate teaching remained virtual for the entirety of the 2020-2021 academic year. Graduate students were given the option to attend classes on campus or virtually. Approximately half of the first year graduate class chose to attend in person and half chose to participate virtually. This required classes to be held at regularly scheduled times with synchronous online access. All class presentations were held in an auditorium that could accommodate 200 students, but the reality was, there were only about 15 students in each class. Students were designated a seat, on one side of the auditorium with five seats between each student. The professor had to present material with a mask and stand in visual perspective for those students who were attending online. Class schedules were arranged to have students from different classes sit on opposite sides of the auditorium as a means of reducing risk for exposure between individuals. All class presentations were recorded, downloaded, and posted for students. This was a time consuming endeavor, particularly to download and upload a class lasting three hours.

As complex as it was to adapt to a hybrid teaching model for knowledge based instruction, it was felt by the faculty of the CSD department that it was worth the extra effort to bolster student learning and success. These challenges were surmountable with some focused modifications. Alterations in methodology were far more robust when it came to courses that were more clinically / skill based. Clinical instruction requires multiple levels of dynamic interaction, synchronization and integration of networks and systems. At a foundational level, students are taught to implement the International Classification of Functioning, Disability and Health (ICF) model created by the World Health Organization (WHO) to ensure contexts, environmental factors, and individual differences are considered when measuring health and disability. It is through a systematic framework which considers both the intended and unintended outcomes and identifies the less obvious contributions of such things as contextual factors that we better understand what approach is truly useful (Clark & Hoffman, 2019). The field of speech-language pathology is an evidence-based practice involving the integration of clinical expertise, evidence, and client/caregiver perspectives. When all these components are considered, decisions are informed and the quality of services are improved (https://www.asha.org/research/ebp/). As a qualitative field, evidence is needed to document progress and effectiveness of interventions. This is not only essential for solidifying therapeutic gains, but from an educational perspective, it is necessary as a means of demonstrating student competencies and that learning standards were achieved.

Standards for education in the field of communication sciences and disorders refer to practicing at the full extent of education or training as practicing at the *top of the license (ToL).* As with systems thinking, ToL purports benefits across components of people, interprofessional practice, and community (McNeilly, 2018). Benefits of this practice to people (i.e., faculty, students, clients) have been described as increased motivation, satisfaction and retention, higher quality work performance, enhanced expertise, better individual clinical outcomes, and improved client satisfaction. There are also interprofessional benefits including more effective and efficient work flow, decreased costs, reduction in errors, and promotion of communication between entities. It has been noted when considering community benefits, that ToL results in stronger outcomes, enhanced service delivery, increased support of social and cultural diversity, functional effectiveness, and improved quality of life (Buck, et. al., 2018; Loversidge, et. al., 2018; Maslow, 2019; Russell-Babin & Wurmser, 2016). p of the license - practicing at textent of yourThe university's CSD department has adopted the teaching philosophy that knowledge is best generalized when attained in combination with skill based learning opportunities. In order to enhance and encourage hands-on learning, academic material needs to involve relevant service and reflection. According to the National Service Learning Clearinghouse, "service learning is a teaching and learning strategy that integrates meaningful community service with instruction and reflection to enrich the learning experience, teach civic responsibility, and strengthen communities (https://engage.loyno.edu/service-learning/courses)."

Maintaining service learning opportunities and a system based teaching and learning format has always been a priority.The pandemic brought these concepts to the forefront, primarily with regard to clinical education adaptations. Ensuring continued rigor and provision of dynamic learning opportunities for students was a priority. Flexibility, creative thinking, innovation and harnessing benefits, as opposed to focusing on limitations, was paramount as restrictions changed and regardless of the fluidity of the unknown situational and contextual factors posed during the course of the pandemic. Using systems thinking allows for identification of the problem and its consequences and continues the process of identifying opportunities for adaptation and change at different levels within the system (Figure 1).

Figure 1. Systems thinking map
Note. Using systems thinking to map the problem, consequences, and components to frame insights, takeaways, and solutions.

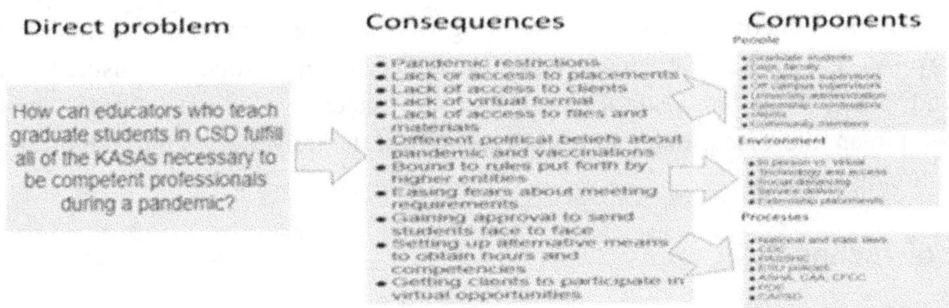

Integration of People

Coordination of on-campus and off-campus clinical placements is a multifaceted undertaking. Graduate student clinicians need access to clients with communication and swallowing disorders for whom they can diagnose and treat. Clients need to represent all age ranges and disorder types within the SLP's scope of practice including: Speech, language, cognition, social hearing, fluency, voice, swallowing, and communication modalities. For all direct contact with clients, a supervisor with ASHA certification and proper state licensure must be available and oversee practice within the stipulated standard requirements including training in supervision of students. Programs must demonstrate that all individuals providing didactic and clinical education, both on-campus and off-campus, have appropriate qualifications for the professional area in which education is provided. Arranging externship placements is a complicated and lengthy process. Formal affiliation agreements must be made between the program and the outside agency and significant follow-up is required to establish and maintain externship sites.

As a result of the pandemic modifications to clinical education were necessary. Some modifications included using simulations and telepractice for students who still needed around 50 hours of direct contact with adult clients in a medical setting. It is difficult to gain medical experience in several areas without direct contact and students need to demonstrate competencies across domains, even those where telepractice is not possible, such as diagnosis of swallowing disorders and other acute medical conditions which impact communication skills and require in-patient care. Requests had to be made to university administration for special consideration of allowing students to resume in person placement opportunities as the health crisis stabilized and facilities allowed students to return. It was clear that facilities would continue to have a hold on student placements until they had protocols and precautions in place to ensure students were not at risk beyond typical limits at a medical setting. Since SLPs were on the front line of care for many of the COVID patients who had residual swallowing, respiratory, and voice issues it seemed reasonable that training would benefit students who would soon be working in this capacity. If healthcare teaching facilities felt that students could be accommodated and were not a distraction or exhausting necessary protective resources, it was important to also educate administrators as to the importance of these opportunities.

The CSD students had additional requirements to work in hospital settings where they were subjected to specialized N-95 face mask fit testing for its safe use in health care settings. The CSD faculty underwent specialized training in Qualitative N95 fit testing based on OSHA standards (https://www.osha.gov/laws-regs/regulations/standardnumber/1910/1910.134AppA), and were actively involved in the N-95 fit testing process to ensure that the N95 face masks were individually fitted and safe for students to be placed in medical settings. The CSD faculty also participated in an interdisciplinary fit testing to support the university's nursing program to enable fit testing of nursing students.

Integration of Technology

One of the key components to adapting clinical education for student learning was establishing a technology system that would support learning objectives. Prior to the pandemic, telepractice was not a standard form of delivering services within the field of speech-language pathology in the United States of America (USA). Countries such as Australia had been using telepractice and teaching its benefits to students as part of the regular curriculum. Historically, telepractice was promoted as a means to serve those in rural areas who did not have access to more formal means of health related services. Due to the fact that insurance did not support or reimburse services coded as occurring over virtual platforms, most facilities were not equipped to implement services in this manner. In the field of speech-language pathology, the term telepractice is used to distinguish these services from those which are only used in healthcare settings. According to ASHA (ASHA, 2020), telepractice is currently considered an appropriate model of service delivery and can be the primary source or may supplement in person services (i.e., hybrid service delivery). It may also be used for supervision, mentoring, preservice, and continuing education.

It should be noted that participation in virtual therapy was dependent on a client or family member's technological access and literacy. Some clients required extensive education in order to use the Zoom platform. Other's had connectivity issues or difficulties with interactive components which allowed for ideal participation in activities. One of the major challenges to utilizing a virtual platform for providing speech-language therapy has been the code of ethics surrounding confidentiality and the Health Insurance Portability and Accountability Act of 1996 (HIPAA) which is the federal law that required the creation of national standards to protect sensitive patient health information from being disclosed without a patient's consent or knowledge. Students and supervisors are bound by the ASHA Code of Ethics. During the first semester of enrollment, students must present documentation of HIPAA training.

As a means to provide services using Zoom, a HIPAA compliant Zoom format needed to be established. Unfortunately, since no clinical materials, client charts, or information about a client can be removed from the on-campus clinic, this proposed a challenge for students transitioning to an online format. All written information about clients (i.e. evaluation reports, therapy plans, SOAP notes, test results) are stored either in the client's chart or on USB drives designated for clinical use only. Client information is not stored on any personal computers, nor in any email systems. Limited access to files during the pandemic was a minimal concern. However, students had been working with their assigned clients for several weeks prior to the unplanned shutdown of the clinic and university as a whole. Students had already established goals for their clients, but had to adapt their activities and strategies for virtual presentation.

Availability of online tests was reduced and those that were created or purchased for online use were not easily administered, particularly in a standardized manner.

Significant challenges to telepractice service delivery were encountered for different clinical populations. Speech-language pathologists treat clients with a wide array of communication and swallowing disorders. The use of telepractice presented benefits and limitations, which were distinct across type of disorder, severity of disorder, behavior and motivation factors, and age of client. Individual differences in adeptness with technology and availability of materials also impacted outcomes with using a telepractice format. The unprecedented and rapid changes brought on by the pandemic, made it necessary to modify teaching theories to better incorporate principles of interprofessional education (IPE) / interprofessional practice (IPP), collaboration and group dynamics, along with the integration of technology, provided a window into systems thinking and the role critical thinking has and was used as a means of adaptation for learning and growth (Clark & Hoffman, 2019).

Adapting assessment procedures for diagnosing clients with more organic or biological concerns was a challenge. It is clear across medical fields that objective evidence is required as opposed to those areas where diagnoses can be made based on qualitative criteria. For example, clinical swallowing examination (CSE) procedures for clients with dysphagia (i.e., swallowing disorders) required more detailed and complex adaptations through a telepractice modality. Clinical supervisors had to train the students to utilize more indirect methods of swallowing assessment versus direct assessment. Direct assessment would involve use of bolus trials administered by the clinician. However, due to risks and liability posed with a telepractice mode, caregivers had to be trained extensively to be present 100% of the time during the assessment and treatment process and facilitate the process and procedures.

Dysphagia clinical education is consistently one of the domains that students benefit from increased opportunities to work on skills. It is a complex area that requires extensive practice and precision to build confidence and independence in decision making. Telepractice has allowed for additional opportunities to target this key area of opportunity needed for the development of student clinicians' skills. Students/supervisors spent vast amount of time educating the use of safe swallow strategies at meal times, oral care provision, monitoring for respiratory signs/symptoms, seeking medical help as needed, explaining patient's swallowing physiology using their pre-recorded video fluoroscopic swallow studies, and utilizing a variety of quality of life (QOL) measures to bring awareness to their dysphagia condition. While a breakout room was advisable for clients with more behaviorally based speech and language concerns, clients with dysphagia, a more biologically based disorder, had to be seen individually for the entirety of a 50-60 min therapy session with 100% one on one online supervision.

Clients with motor speech disorders such as dysarthria and apraxia of speech (AOS) presented with a multitude of challenges when seen via telepractice mode. Clients had to be typically cued and monitored for appropriate utilization of strategies to improve their speech intelligibility. While speech parameters such as articulation and rate were monitored with fair to good accuracy, provision of feedback about client's voice (e.g., pitch, loudness and quality) and speech inflections were challenging via the Zoom modality. For example, when clients were instructed to use pitch glides, the zoom modality would not pick it up and would detect the pitch glide as noise in the sound system. Few technical adjustments had to be made on a constant basis to perceive these speech parameters accurately. Furthermore, loudness perception was also variable via zoom for clients who presented with reduced vocal loudness, since there were options to adjust loudness on the computer devices. Clients had to be frequently monitored on the loudness settings they used on the computer and their caregivers were used as surrogates in the session to provide feedback to the clinicians about the patient's actual vocal loudness. Provision of speech specific movement feedback, tactile, kinesthetic and spatio-temporal cues had to be modified and supported with a lot of visuals with consideration to speech motor dynamics on a zoom modality.

Many insights from working with clients with autism spectrum disorders (ASD) via telepractice were gleaned. One of the challenges with working with individuals with an ASD diagnosis is attentional issues. Depending on severity of diagnosis, behavioral concerns and active participation can make it a challenge to address actual speech and language goals. It can be difficult for a seasoned SLP to work one on one in a face-to-face capacity with a client with ASD, let alone in a virtual manner. Younger individuals with ASD typically respond well to approaches that have a foundation in building relationship based prelinguistic skills such as joint attention. Addressing these skills in a virtual framework proved to be a challenge. Alternatively, behavioral based approaches, such as those following an applied behavior analysis approach and have a foundation in reinforcement were more easily adapted to an online presentation. A clear benefit of using telepractice with clients with ASD was working with parents as facilitators when providing services. Having parents actively involved in sessions helped create educational and counseling opportunities. Too often in traditional face to face therapy sessions, children are left with a clinician while the parent observes virtually. This limits time for discussion of goals, what strategies are helping, and making recommendations for carry-over between sessions.

Distinctions still need to be considered when determining which populations, diagnoses, and formats (e.g., individual versus group) will be best served therapeutically by the use of or supplementation of virtual approaches. It is clear that virtual service delivery provides educational opportunities for students with regard to increased exposure, supervision, mentorship, counseling, and collaboration possibilities. To this end, its continued use has created new avenues for changing perspectives that will increase knowledge and learning.

Benefits and Limitations of Telepractice

Many insights were gained as to the benefits and limitations of using telepractice as a service delivery model as a result of the pandemic. This format had a negative connotation which would likely remain if not for its forced inception. Utilization of a virtual medium created additional opportunities for students to work collaboratively and in a mentoring capacity. Students could observe and critique each other's sessions, help collect data and gain experience with scoring and completing measures of interrater reliability. It was noted that the virtual format contributed to increases in counseling opportunities, greater availability for meetings and debriefing, and access for a higher number of students to observe clients with low incidence or unique diagnoses.In fact, many clients who would be unable to travel to the on-campus university clinic due to distance or health reasons were able to not only receive services remotely, but provide students with experiences they otherwise would not encounter. To date, many clients have opted to remain using an online format as a convenience or as a backup when an in person session cannot be attended, either consistently or on an as needed basis. The unexpected benefits of implementing a virtual format, increased graduate exposure to and experience with varied client types, which has been a strategic plan goal for the department. In addition, it has provided potential for gains towards other strategic plan goals including opportunities for research and interprofessional practice initiatives. It is a testament to the benefits of telepractice that it has continued to be retained and desired by clients post pandemic.

Initial concerns were that it would not be counted towards student hours or be a format that was discontinued in use across rehabilitative or therapeutic healthcare specialties, including speech-language pathology. While service delivery is returning to a standard in person delivery format, there is evidence that telepractice has proponents for its continued use and benefits to support its maintenance. The future of telepractice lies heavily in whether or not it will be acknowledged and covered by insurance and billing entities, as well as if longitudinal evidence supports its efficacy. There are no identified limitations where telepractice cannot be implemented. Services must comply with national, state, institutional, and professional regulations. However, post pandemic, further research stemming from widespread use of telepractice across domains will likely inform those areas for which it is most beneficial from a therapeutic perspective.

It is clear that some telepractice services cannot replace in person services for diagnosis and treatment of all communication and swallowing deficits. This is particularly true for initial formal evaluations which require the use of standardized measures, and are ideal for making accurate and precise diagnoses, recommendations, and referrals. It is also unclear if hours accrued by graduate student clinicians will continue to count towards total current contact hours required for graduation. Thus, from a clinical and academic perspective, the future use of telepractice is dependent on evidence of its effectiveness and validation and recognition by the CFCC as an educational asset.

Integration of Processes

The unprecedented and fluid nature of events during the pandemic required frequent tracking of updates and changes in policies and procedures across agencies. It also led to a need for flexibility and establishment of internal departmental processes. These changes were often determined by the interconnectedness and hierarchical governing entities which guide our society as well as our profession. In the case of the CSD department, decisions are not unilateral amongst faculty. This was particularly evident during the pandemic. Rules were being set forth by the government at a national and state level, the center for disease control (CDC), state system of higher education, administration, ASHA, CAA, CFCC, Council of Academic Programs in Communication Sciences and Disorders (CAPSD), Pennsylvania Department of Education (PDE), and a multitude of online forums and meetings to work through and develop new best practices procedures and modifications to promote safety while continuing to provide learning and development opportunities for students.

On-Campus Clinical Processes

Extensive changes were required within the university's on-campus clinic which resumed the option for face to face services in the fall of 2020 while maintaining a telepractice option though Zoom for those who preferred a remote option. Guidelines were established to ensure safety. Clients had to call their clinician upon arrival to campus and wait in their car until asked to enter the building. The waiting room within the department was closed and only one observer/caregiver was allowed in the building per client. The observation room was closed off and if an observer was present, he/she had to remain in the session room with the client. Temperatures were taken before entering the clinic and clients were asked a series of health and travel related questions before the start of each session. If any evidence of illness was noted prior to or during sessions, the session would be discontinued immediately. Fiberglass partitions were set up in each clinic room to maintain distance. Masks were required at all times as well as sanitizing hands before and after each session. All materials used during sessions had to be properly sanitized and there were limits to rooms available during the same times of day and when a subsequent session could be held in the same room.

Albeit necessary, these guidelines were not ideal for all clinical interactions. Communication is inherently an interactive practice. Service delivery models can vary from more clinician directed (i.e., structured) to more client directed (i.e., naturalistic and facilitative). Assessments and strategies often require visual observation of oral structures.

Clinical Externship Processes

In the CSD department, externship comes at the point in graduate study where students have completed their coursework and on-campus clinical practicums and are now asked to use their skills dynamically in real work practice. The pandemic significantly diminished student access to face to face clinical opportunities. Although students are permitted to gain up to 75 hours towards degree requirements through simulated experiences, these had been used sparingly prior to the pandemic. As an accommodation during the pandemic, responsible and strategic use of simulation had to be considered. This required extensive debriefing and reflection on cases which developed critical thinking skills which had the best chance of generalizability to actual clientele.

Students were given the option of opting out of face to face opportunities and receiving as an alternative a combination of adapted experiences, virtual learning opportunities, and simulated experiences. The CSD department created a Bill of Rights, which went through a university administrative approval process to opt in to face to face opportunities if and when they were made available. Due to the precarious nature of the COVID-19 virus and the uncertainty of events to come, all students were treated fairly whether or not they choose to enter into a face to face placement or not. If they opted into face to face opportunities, they had the responsibility to adhere to guidelines of their respective clinical program. The CSD department is housed in a university that is part of a state system of higher education. As a state governed entity, and thus public institution, when vaccinations for the COVID-19 virus became available, students could not be required to receive a vaccine in order to attend those classes that resumed on campus. However, students were still required to follow the regulations and requirements at their externship placement or accept an alternative to face to face opportunities. All but one student opted to forego in person clinical opportunities despite receiving the vaccinations due to concerns with still being exposed. Several students initially chose not to be vaccinated early in its dissemination, but eventually opted to comply with vaccination when they realized it would limit externship site restrictions on accepting students without vaccination. The university could not mandate vaccinations due to its state university status, but did require regular testing for unvaccinated students on campus. By the end of June 2020, restrictions had eased in particular medical settings or more virtual opportunities had been established where students could participate and complete their clinical hours either in a traditional face to face manner or through alternative opportunities. All 30 students in the CSD graduate class of 2020 graduated in August as planned.

In addition to the feedback given to students by supervisors on a weekly basis, the supervisor completes an evaluation of clinical skills at the end of the semester using the Performance Rating Scale (see Appendix A). This form is generated on CALIPSO which is a web-based application that was created by a Communication Sciences and Disorders University Clinic Director to manages key aspects of academic and clinical education designed specifically and exclusively for speech-language pathology and audiology training programs (https://calipso.com/index.html). This competency measure is completed by a supervisor as a midterm and final evaluation for each graduate student participating in a semester of on campus placement (i.e., first, second, and third semester) and off campus placements (i.e., school based and medical based). These evaluations are initiated by the supervisor but include significant student input before final submission. For off campus placements, students are required to complete the Performance Rating Scale in evaluation of their own skills which is then compared with the supervisor report for agreement or distinctions in skill analysis. Documents and tracking methods available on CALIPSO are utilized by almost every CSD graduate program accredited by the CAA. Students are able to monitor and review their progress toward meeting the standards and, ultimately, demonstrate, through a completed KASA form, that all required knowledge and skills have been met (https://www.calipsoclient.com/esu/).

Supervisors utilize a second student evaluation system to ensure thoroughness of exposure to a variety of skills. The Expected Clinical Competencies Based on Student Level checklist (see Appendix B) was developed based on specific clinical competencies that are expected following the three different levels of student clinical experience on campus, and must be completed prior to approval for externship placements. The supervisor will review these items with the student throughout the clinical practicum semester and discuss the student's progress in meeting them. Evidence is collected each year when analyzing program outcomes and it is consistently noted that there are challenges with creating opportunities for students to develop skills in the area of IPP/IPE and counseling patients and family members. It is these areas that strategic planning is focused for future initiatives.

Aims of Retrospective Analysis

Student competency assessments were completed, pre pandemic, during, and post pandemic to monitor student competencies. They were additionally used to analyze areas of strength and weakness across students. This information informed student learning outcomes. The use of existing and newly created protocols to measure student competencies provided a means for analyzing performance of students within the cohort impacted most by the pandemic, as well as compare outcomes between this group and those instructed pre pandemic.

METHOD

To assess program adaptations implemented during the pandemic, comparisons were made between cohorts who completed the graduate program pre pandemic and during the pandemic. Specific item analysis of areas of strength and weakness were carried out for students in the 2020 graduate cohort, which included those students most highly impacted by the pandemic.

Specialized rubrics were created and adapted to accompany virtual sessions to provide specific feedback on student's therapy skills as outlined by Smith (2015) (see Appendix C). These feedback forms were consistently used after each individual student therapy session and discussed individually with the students during their clinical meeting with the supervisor. Although created during the pandemic to accompany virtual sessions and subsequent meetings, their value has merited continued use post pandemic.

RESULTS

An analysis of cumulative cohort outcomes was conducted for students in the 2020 graduate cohort using their scores on the Performance Rating Scales across all clinical education opportunities. This cohort represents those students who were most displaced and impacted by the environmental/contextual and procedural adaptations during the pandemic.

Results revealed overall cohort average scores for evaluation skills = 3.84/4.00, treatment skills = 3.86/4.00, and professional practice, interaction and personal qualities = 3.90/4.00. An item analysis of specific items within each category demonstrated average scores over 3.8 with the exception of four skills in the professional practice category (see Appendix D). It should be noted that despite being skills with the lowest rating, they were still scored with a high numeric rating, in the high average range based on the Likert scale. A score of 3.00 is considered present or general effectiveness in the skill area as being good but may not have been consistent and/or independent (i.e., skill is present 60-89% of the time). A score of 4.0 is considered consistent or demonstrated abilities in the competence area that showed mastery of the skill. Performance was consistent and independent (i.e., skill is present 90=100% of the time). The areas identified are consistent with those areas identified as more complex or challenging for students in years prior to the pandemic.

1. Uses clinical reasoning and demonstrates knowledge of and ability to integrate research principles into evidence-based clinical practice = 3.77/4.00
2. Manages patient behaviors appropriately = 3.78/4.00
3. Provides counseling regarding communication and swallowing disorders to clients/patients, family, caregivers, and relevant others = 3.75/4.00
4. Collaborates with other professionals in case management = 3.78/4.00

Table 1 denotes the distribution of cumulative competency scores of student cohorts from 2014-2021. Data was averaged across students for each cohort class for further comparison of clinical outcome data between classes.

Table 1. Distribution of cumulative competency scores

N=	2014	2015	2016	2017	2018	2019	2020	2021
1	3.79	3.93	3.66	3.46	3.62	3.49	3.4	3.53
2	3.65	3.82	3.32	3.54	3.63	3.55	3.27	3.63
3	3.91	3.89	3.59	3.41	3.51	3.39	3.42	3.45
4	3.96	3.84	3.54	3.54	3.61	3.56	3.32	3.38
5	3.73	3.77	3.49	3.42	3.62	3.39	3.37	3.47
6	3.33	3.92	3.58	3.56	3.49	3.35	3.33	3.34
7	3.65	3.66	3.58	3.47	3.61	3.56	3.4	3.47
8	3.53	3.92	3.66	3.53	3.63	3.51	3.46	3.36
9	3.9	3.67	3.72	3.55	3.56	3.56	3.35	3.58
10	3.52	3.72	3.44	3.44	3.57	3.56	3.4	3.47
11	3.72	3.63	3.58	3.47	3.65	3.53	3.42	3.52
12	3.69	3.94	3.66	3.49	3.56	3.62	3.41	3.47
13	3.68	3.83	3.52	3.53	3.57	3.58	3.43	3.49
14	3.72	3.25	3.53	3.41	3.59	3.53	3.45	3.46
15	3.84	3.87	3.71	3.49	3.53	3.51	3.42	3.43
16	3.74	3.78	3.31	3.53	3.6	3.26	3.41	3.52
17	3.62	3.98	3.75	3.56	3.46	3.6	3.36	3.39
18	3.85	3.64	3.66	3.55	3.57	3.56	3.46	3.47
19	3.67	3.57	3.65	3.4	3.56	3.55	3.47	3.4
20	3.62	3.77	3.46	3.48	3.59	3.55	3.33	3.49
21	3.85	3.93	3.49	3.52	3.59	3.47	3.44	3.43
22	3.88	3.77	3.68	3.46	3.57	3.59	3.43	3.47
23	3.69	3.95	3.49	3.53	3.6	3.57	3.44	3.49
24	3.62	3.75	3.49	3.09	3.53	3.46	3.48	3.37
25	3.62	3.78	3.63	3.55	3.49	3.37	3.45	3.4
26	N/A	N/A	3.63	3.49	3.6	3.59	3.36	3.43
27	N/A	N/A	3.53	3.49	3.46	3.59	3.36	3.55
28	N/A	N/A	3.473	3.39	3.63	3.57	3.46	N/A
29	N/A	N/A	N/A	3.57	3.6	3.6	3.41	N/A
30	N/A	N/A	N/A	3.56	3.54	3.57	3.47	N/A
31	N/A	N/A	N/A	N/A	N/A	N/A	3.44	N/A
Class Average	3.711	3.783	3.565	3.4826	3.5713	3.5163	3.4071	3.4615

Note.. Data obtained from CSD Student Externship Outcome Database for student cohorts from 2014-2020.

Figure 2 shows the overall distribution of the raw scores from Table 1. As shown in Table 1 and Figure 2, overall the range of scores were similar across the years (2014-2021) with average scores ranging from 3.4071 to 3.783 on the Likert scale rating. Although some marginal differences are revealed across the years, specific comparison of pre and post-pandemic scores for the year 2017 and 2021 did not reveal any statistical difference in the parametric t-tests of independent student samples ($t(55)= 0.9815$, $p= 0.3306$). Post-pandemic outcome data thus reveals that clinical training results are strongly supportive of the adaptive systems thinking approach which yielded comparable clinical training outcomes to pre-pandemic CSD training.

Figure 2. Distribution of average clinical competency
Note. Distribution for cohort members from 2014-2021.

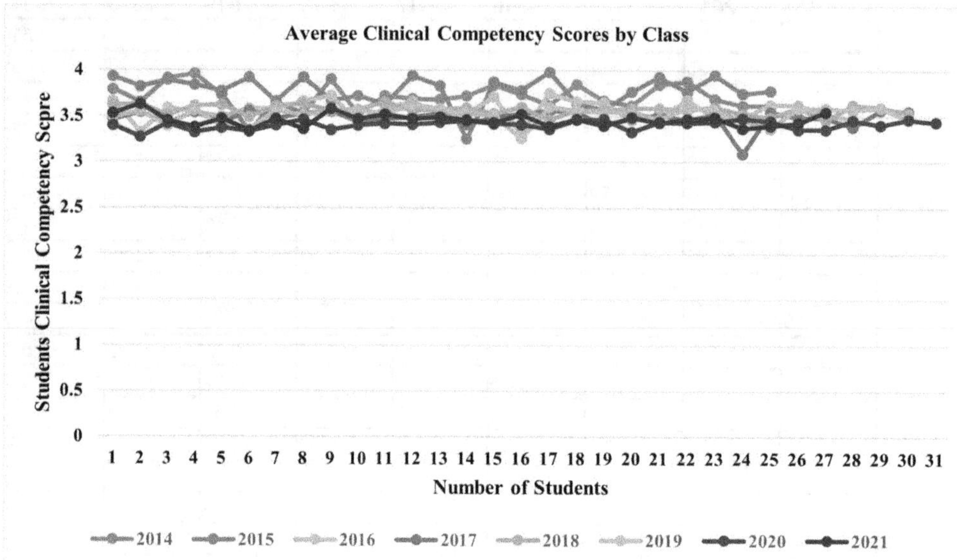

DISCUSSION

Analysis of Results

As a result of the increased systems thinking that has occurred since the pandemic, an enhanced framework for using new approaches to address these skills has been established. The addition of virtual opportunities has provided students with access to more diverse populations, diagnoses, service delivery options, prevention and client/family/community education platforms, as well as professional and interpersonal development possibilities.

All students in the 2020 cohort demonstrated competencies across skills with modifications to mode of service delivery and with restrictions to direct interaction. Skills which were mildly reduced, but still within acceptable limits, were consistent with scores for skills pre pandemic. This suggests that despite environmental and contextual changes, these skill areas are more complex for students to grasp. The realization that even with adaptations and newfound awareness of opportunities, consistent challenges persisted and will need continued focus in future strategic plans. This will be better accomplished by the continued use of benefits identified as a result of systems thinking that occurred by faculty during the pandemic. The takeaways can now be further modified and implemented to harness improved outcomes in practice.

Future Applications and Takeaways

The experience of training CSD professionals during the pandemic has created evidence as to the need to challenge traditional left brain thinking (i.e., rules, structure, order, hierarchy, etc.) and be more inclusive of a holistic right brain thinking (i.e., holistic, inclusive, integrative, interactive) approach, one that is more consistent with the tenets of system thinking.

Although the CSD department has strived to consistently create hands-on and applied learning opportunities for its students, the pandemic allowed for identification of new areas of opportunity within this capacity (Figure 3). The use of technology has created additional collaborative experiences which have been used and have built on readily available opportunities. This has expanded and exposed students to more remote content and contexts for knowledge and skill development.

Figure 3. Sytems approach and clinical education
Note. Incorporating a systems approach into clinical education through the integration of teaching, scholarship, and service learning opportunities.

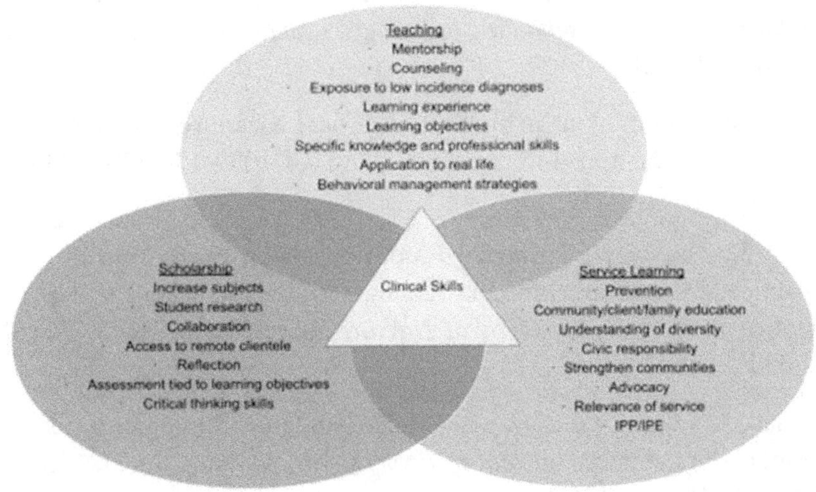

CONCLUSION

Forced realities provided the opportunity for informed reflection and critical thinking into systems thinking. This allowed for identification of strategies and components and subsequent adaptations that were more and less effective for clinically educating students in the communication sciences and disorders, most of whom are preparing to practice clinically as SLPs, The unanticipated and unprecedented global systemic disruption caused by the COVID 19 pandemic allowed for consideration and analysis of the integration of teaching, research, and service to clients and community partners, Systems thinking by department faculty ensured continued and new means for students to gain real world experience to not only enhance their clinical practice in the areas of diagnostics and treatment, but also build competencies in the area of professional practice, interaction, and personal growth. These skills are often more of a challenge to address using a traditional teaching philosophy or pedagogy.

SLPs need to be aware of socio-cultural, socio-political, and contextual factors and how they can impact speech, language, and communication behaviors necessary for learning (Horton, 2020). Through the integration of diverse entities and practical knowledge, positive outcomes emerged in a time of necessity which inform future practice and highlight the importance of system thinking on development and growth as a society.

REFERENCES

American Speech-Language-Hearing Association. (n.d.-a). *Evidence-based practice (EBP).* American Speech-Language-Hearing Association. https://www.asha.org/Research/EBP/

American Speech-Language-Hearing Association. (n.d.-b). *Telepractice.* American Speech-Language-Hearing Association. https://www.asha.org/Practice-Portal/Professional-Issues/Telepractice/

American Speech-Language-Hearing Association. (n.d.-c). *Who are speech-language pathologists, and what do they do?* American Speech-Language-Hearing Association. https://www.asha.org/public/Who-Are-Speech-Language-Pathologists/

Astaíza-Martínez, A. F., Mazorco-Salas, J. E., & Castillo-Bohórquez, M. I. (2021). Teacher-researcher training in higher education: *A systems thinking approach. Systemic Practice and Action Research, 34*(2), 187–201. doi:10.100711213-020-09532-x

Banathy, B. H. (1999). Systems thinking in higher education: Learning comes to focus. *Systems Research and Behavioral Science, 16*(2), 133–145. doi:10.1002/(SICI)1099-1743(199903/04)16:2<133::AID-SRES281>3.0.CO;2-T

Buck, J., Loversidge, J., Chipps, E., Gallagher-Frod, L., Genter, L., & Yen, P.-Y. (2018). Top of the license nursing practice: Describing common nursing activities and nurses' experiences that hinder top-of license practice Part 1. *The Journal of Nursing Administration, 48*, 266–271. doi:10.1097/NNA.0000000000000611 PMID:29672373

Clark, K., & Hoffman, A. (2019). Educating healthcare students: Strategies to teach systems thinking to prepare new healthcare graduates. *Journal of Professional Nursing, 35*(3), 195–200. doi:10.1016/j.profnurs.2018.12.006 PMID:31126396

Loyola University. (n.d.). Course list: Community Engaged Learning, Teaching and scholarship. Loyola. https://engage.loyno.edu/service-learning/courses

Department of Labor Logo United Statesdepartment of Labor. (n.d.). *1910.134 App A - Fit Testing Procedures (Mandatory). | Occupational Safety and Health Administration.* OSHA. https://www.osha.gov/laws-regs/regulations/standardnumber/1910/1910.134AppA

Dunnion, J., & O'Donovan, B. (2014). Systems thinking and higher education: The Vanguard method. *Systemic Practice and Action Research, 27*(1), 23–37. doi:10.100711213-012-9258-4

Freeman, D. L., & Cameron, L. (2008). Research methodology on language development from a complex systems perspective. *Modern Language Journal, 92*(2), 200–213. doi:10.1111/j.1540-4781.2008.00714.x

Horton, R. (2020). Systems-based approaches to speech-language pathology service delivery for school age children. In D. Scott (Ed.), *Cases on Communication Disorders in Culturally Diverse Populations* (pp. 113–136). IGI Global. doi:10.4018/978-1-7998-2261-5.ch006

Ison, R. (1999). Applying systems thinking to higher education. Systems Research and Behavioral Science. *The Official Journal of the International Federation for Systems Research, 16*(2), 107–112.

Loversidge, J., Yen, P. Y., Chipps, E., Gallagher-Ford, L., Genter, L., & Buck, J. (2018). Top-of-License Nursing Practice, Part 2: Differentiating BSN and ADN Perceptions of Top-of-License Activities. *The Journal of Nursing Administration*, *48*(6), 329–334. doi:10.1097/NNA.0000000000000623 PMID:29794597

Maslow, A. H. (2019). *A Theory of Human Motivation*. HN Publishing.

McNeilly, L. (2018). Why we need to practice at the top of the license. *ASHA Leader*, *23*(2), 10–11. doi:10.1044/leader.FMP.23022018.10

Pedryez, W., & Kwak, K. C. (2006). Linguistic models as a framework of user-centric system modeling. *IEEE Transactions on Systems, Man, and Cybernetics. Part A, Systems and Humans*, *36*(4), 727–745. doi:10.1109/TSMCA.2005.855755

Pedryez, W., & Vasilakos, A. V. (1999). Linguistic models and linguistic modeling. *IEEE Transactions on Systems, Man, and Cybernetics. Part B, Cybernetics*, *29*(6), 745–757. doi:10.1109/3477.809029 PMID:18252354

Plack, M. M., Goldman, E. F., Scott, A. R., Pintz, C., Herrmann, D., Kline, K., & Brundage, S. B. (2018). Systems thinking and systems-based practice across the health professions: An inquiry into definitions, teaching practices, and assessment. *Teaching and Learning in Medicine*, *30*(3), 242–254. doi:10.1080/10401334.2017.1398654 PMID:29283669

Russell-Babin, K., & Wurmser, T. (2016). Transforming care through top-of-license practice. *Nursing Management*, *24-28*(5), 24–28. doi:10.1097/01.NUMA.0000482527.15743.12

Sabah, S. (2017). Interactional competence and SLA: A systems-thinking perspective. [JALLR]. *Journal of Applied Linguistics and Language Research*, *4*(4), 185–191.

Smith, H. C. (2015). Student Clinicians Self-Evaluating Performance by Reviewing Session Videos and Using a Rubric. *Perspectives on Administration and Supervision*, *25*(2), 78–84. doi:10.1044/aas25.2.78

Trbovich, P. (2014). Five ways to incorporate systems thinking into healthcare organizations. *Biomedical Instrumentation & Technology*, *48*(s2, Suppl), 31–36. doi:10.2345/0899-8205-48.s2.31 PMID:25244335

KEY TERMS AND DEFINITIONS

Apraxia of Speech (AOS): Apraxia of speech can occur developementally in children or aquired in adults and is a motor planning disorder whereby the individual has limited ability coordinate speech movements and send message from the brain to the mouth.

Clinical Education: In contrast to academic (i.e., classroom based) education, clinical education involves aspect of the professional curriculum that includes the spectrum of experiential learning and clinical education settings where students practice applying knowledge, skills, and professional behaviors under the direction of a qualified clinical educator.

Communication Sciences and Disorders (CSD): Communication sciences and disorders is the term used by most professional programs to cover knowledge and skill across areas of speech, language, hearing, swallowing, and other communication abilities. Audiology and Speech-language pathology are the top two professions in the CSD field.

Clinical Competency: Standards used to ensure standardization in the amount and quality of education and training that students receive prior to beginning their practice as an independent clinician. For speech-language pathologists, the Council on Academic Accreditation (CAA) and the Council for Clinical Certification (CFCC) have a responsibility to assure stakeholders that graduates, Clinical Fellows, and certificate holders are clinically competent to practice.

Dysphagia: Dysphagia is a swallowing disorder involving the oral cavity, pharynx, esophagus, or gastroesophageal junction. Speech-language pathologists are perferred providers of dysphagia services particularly for the oral and pharyngeal phases of the swallow.

Externship: Externships are practical experiences in their fields of study. The goal of the externship in CSD programs is to provide additional insight into the field for speech-language pathology students during their educational program. Externs have not yet graduated, but placements are performed in an off-campus setting under supervision of a licensed and certified SLP.

Language: Language refers to using vocabulary or words, through any modality, to share ideas and get wants and needs met. Language encompassess both receptive (i.e., comprehension) and expressive (i.e, production) abilities.

Motor Speech: Motor-speech disorders are speech disorders that affects the motor control of speech muscles or motor programming of speech movements. The most common motor speech disorders are dysarthria and apraxia of speech. Dysarthrias are oral communication problems due to weakness, incoordination, or paralysis of the speech musculature.

Service Learning: Service learning incorporates teaching and learning strategy that integrates applied, hands-on learning often as a community service with instruction and reflection to enhance the learning experience.

Speech: Speech is how sounds and words are produced orally and includes precision of articulation, voice, and fluency characteristics.

Speech-Language Pathology (SLP): Speech-language pathologist is the title for the professional who helps individuals who have speech, language, swallowing, and other communication problems.

Top of the License: Practicing at the top of a license refers to the focus of clinial activies that require the unique knowledge, skills and attitudes within the scope of practice of the field. The goal is to facilitate functional outcomes in a cost effective manner.

APPENDIX A

Performance Rating Scale

Please refer to the Performance Rating Scale for grading criteria.
Use a score between 1 and 4, in 0.25 increments (1.25, 1.5 etc.)

1. Inadequate; skill present <30% of the time 3. Present; skill present 60-89% of the time
2. Emerging; skill present 31-59% of the time 4. Consistent; skill present 90-100% of the time

	Speech sound production	**Fluency**	**Voice**	**Lang.**	**Hearing**	**Swallowing**	**Cognition**	**Social aspects**	**Comm. modalities**
EVALUATION SKILLS									
1. Conducts screening and prevention procedures									
2. Collects case history information and integrates information from clients/patients and/or relevant others									
3. Selects appropriate evaluation instruments/ procedures									
4. Administers and scores diagnostic tests correctly									
5. Administers non-standardized assessment protocols accurately and efficiently									
6. Adapts evaluation procedures to meet client/patient needs									
7. Structures the environment to meet patient and family needs									
8. Possesses knowledge of etiologies and characteristics for each communication and swallowing disorder									
9. Interprets, integrates, and synthesizes test results, history, and other behavioral observations to develop diagnoses									
10. Collects and interprets relevant behavioral observations									
11. Makes appropriate recommendations for intervention									
12. Communicates findings to patient and relevant others									
13. Completes administrative functions and documentation necessary to support evaluation									
14. Refers clients/patients for appropriate services									
TREATMENT SKILLS									
1. Develops setting-appropriate intervention plans with measurable and achievable goals. Collaborates with clients/ patients and relevant others in the planning process									
2. Structures plans for maximal efficacy									
3. Implements intervention plans (involves clients/patients and relevant others in the intervention process)									
4. Selects and uses appropriate materials/ instruments									

continues on following page

Table continued

	Speech sound production	Fluency	Voice	Lang.	Hearing	Swallowing	Cognition	Social aspects	Comm. modalities
5. Provides appropriate introduction/ explanation of tasks									
6. Provides and explains accurate and appropriate clinic rationale for all parts of session									
7. Measures and evaluates clients'/ patients' performance and progress									
8. Uses appropriate models, prompts or cues. Allows time for patient response.									
9. Modifies treatment plans, strategies, materials, or instrumentation to meet individual client/patient needs									
10. Completes administrative functions and documentation necessary to support treatment									
11. Identifies and refers patients for services as appropriate									
PROFESSIONAL PRACTICE, INTERACTION AND PERSONAL QUALITIES									
1. Demonstrates knowledge of and interdependence of communication and swallowing processes									
2. Uses clinical reasoning and demonstrates knowledge of and ability to integrate research principles into evidence-based clinical practice									
3. Adheres to federal, state, and institutional regulations and demonstrates knowledge of contemporary professional issues and advocacy (includes trends in best professional practices, privacy policies, models of delivery, and reimbursement procedures/fiduciary responsibilities)									
4. Communicates effectively, recognizing the needs, values, preferred mode of communication, and cultural/linguistic background of the patient, family, caregiver, and relevant others									
5. Establishes rapport and shows care, compassion, and appropriate empathy during interactions with clients/patients and relevant others									
6. Manages patient behaviors appropriately									
7. Uses appropriate rate, pitch, and volume when interacting with patients or others									
8. Follows standard precautions									
9. Provides counseling regarding communication and swallowing disorders to clients/patients, family, caregivers, and relevant others									
10. Collaborates with other professionals in case management									
11. Displays effective oral communication with patient, family, or other professionals									
12. Displays effective written communication for all professional correspondence									

continues on following page

Table continued

	Speech sound production	Fluency	Voice	Lang.	Hearing	Swallowing	Cognition	Social aspects	Comm. modalities
13. Adheres to the ASHA Code of Ethics and Scope of Practice documents and conducts him or herself in a professional, ethical manner									
14. Demonstrates professional- ism									
15. Is timely with all clinical requirements									
16. Recognizes own limitations, stays within the boundaries of training, and displays appropriate level of confidence									
17. Demonstrates openness and responsiveness to clinical supervision and suggestions									
18. Displays organization and preparedness for all clinical sessions									
19. Personal appearance is professional and appropriate for the clinical setting									

APPENDIX B

Expected Clinical Competencies Based on Student Level

Student's First Clinical Practicum

Student Clinician ____ Semester ________________
Supervisor _________________________________

Professional Behavior

1. _____ demonstrates all HIPAA related rules: client privacy and confidentiality
2. _____ shows respect to client regardless of client or parent's behavior
3. _____works as a team player with clients and other professionals
4. _____ dresses professionally as judged by supervisor and /or clinic director
5. _____ reads & reviews client charts
6. _____ shows up promptly
7. _____ follows client safety and infection control procedures
8. _____ demonstrates use of IPP (inter-professional practice) with professionals involved in cases
9. _____ demonstrates cultural humility

General Teaching Methods (not specific to any particular disorder)

10. _____ provides good speech models most of the time

11. _____ provides positive feedback with client
12. _____ provides positive feedback with caregiver
13. _____ demonstrates what the client is to do through examples before asking client to do it
14. _____ establishes positive social relationship with clients and client's family
15. _____ involves client's family in therapy as much as possible
16. _____ selects appropriate and sufficient materials to support the clinician's plan
17. _____ measures baseline levels for goals before teaching
18. _____ utilizes EBP (Evidence Based Practice)

Data Collection, Reporting, and Use of Data for Planning

19. _____ generates narrative descriptions of client's performance and responses (without raw data) for each therapy activity (written in progress notes)
20. _____ writes appropriate SOAP notes with moderate guidance

Changes Strategy Based on Client's Responses (clinical process analysis)

21. _____ after session with supervisor guidance

Supervisory Processes

22. _____ informs supervisor in writing of changes in client scheduling
23. _____ attempts to schedule clients and encourages them to come
24. _____ attends supervisory conferences prepared with client charts and all relevant data
25. _____ asks specific questions
26. _____ attempts to implement supervisor's suggestions
27. _____ checks with supervisor when unsure what to do
28. _____ completes a total of three self-evaluations on three sessions

Assessment/ Evaluation Activities (Please see last page for list of extensive assessment procedures)
Exhibits proficiency at administering scoring and interpreting the following types of assessment:

29. _____ assesses new clients in a timely manner when assigned by supervisor
30. _____ re-assess clients where needed for evaluation updates and/or discharge decision
31. _____ hearing screenings completed on each client
32. _____ swallowing screenings completed on each client

Student's <u>Second</u> Clinical Practicum

Student Clinician ______________________________________ Semester _______________
Supervisor ______________________________

Professional Behavior

1. _____ demonstrates all HIPAA related rules: client privacy and confidentiality
2. _____ shows respect to client regardless of client or parent's behavior
3. _____works as a team player with clients and other professionals
4. _____ dresses professionally as judged by supervisor and /or clinic director
5. _____ reads & reviews client charts
6. _____ shows up promptly
7. _____ follows client safety and infection control procedures
8. _____ demonstrates use of IPP (inter-professional practice) with professionals involved in cases
9. _____ demonstrates cultural humility
10. _____ educates client or caregiver on communication disorders

Data Collection, Reporting, and Use of Data for Planning

11._____ keeps precise data on client's responses for each therapy activity (written in progress notes)
12._____ review client goals/determine need to change goals when appropriate
13._____ uses data in support of EBP

Changes Strategy Based on Client's Responses (clinical process analysis)

14._____ after session independently
15._____ uses materials that turn out to be interesting to the client
16._____ selects activities that interest client and result in good attention and effort
17._____ establishes imitation behavior in client if needed
18._____ uses testing and teaching in appropriate proportions
19._____ keeps control of session even with clients who want to control session
20._____ shares & educates client progress with clients' parents/caregivers
21._____ write appropriate SOAP notes with minimal guidance
22._____ uses data, i.e. EBP to modify strategies when applicable

Supervisory Process

23._____ assesses own clinical performance
24._____ makes therapy suggestions
25._____ demonstrates initiative in structuring supervisory conference
26._____ uses conference time efficiently

Assessment/ Evaluation Activities (Please see last page for list of extensive assessment procedures) Exhibits proficiency at administering scoring and interpreting the following types of assessment:

27._____ assesses new clients in a timely manner when assigned by supervisor
28._____ re-assesses clients where needed for evaluation updates and/or discharge decision
29._____ hearing screenings completed on each client
30. _____ swallowing screenings completed on each client

Student's Third Clinical Practicum

Student Clinician _________________________________ Semester ______________
Supervisor __________________________

Professional Behavior

1. _____ demonstrates all HIPAA related rules: client privacy and confidentiality
2. _____ shows respect to client regardless of client or parent's behavior
3. _____works as a team player with clients and other professionals
4. _____ dresses professionally as judged by supervisor and /or clinic director
5. _____ reads & reviews client charts
6. _____ shows up promptly
7. _____ follows client safety and infection control procedures
8. _____ demonstrates use of IPP (inter-professional practice) with professionals involved in cases
9. _____ demonstrates cultural humility
10. _____ participates in mentoring/training students (underclassman)
11. _____ counsel clients and caregivers as appropriate

Data Collection, Reporting, and Use of Data for Planning

12. _____ proposes a specific plan for next session based on data and impressions from previous session (written in progress notes) (this is the heart of evidence-based practice)
13. _____ review client goals/determine need to change goals when appropriate

Changes Strategy Based on Client's Responses (clinical analysis)

14. _____ during session
15. _____ is sensitive to the power of the client's non-verbal cues
16. _____ establishes interactive and attentive behavior in client if needed
17. _____ provides rewards the client is willing to work for
18. _____ establishes effective cueing systems:
    a. uses client responses to determine which cues to use
    b. fades the cues ASAP
    c. establishes "bridging cues" (for use at higher levels)

19. _____ role playing techniques when appropriate
20. _____ programs for carryover skills/ functional activities as appropriate
21. _____ gives family carryover activities as appropriate
22. _____ exhibits flexibility with goals/ activities to adapt to client responses
23. _____ makes referrals for other assessments and interventions as appropriate
24. _____ teaches client to answer direct questions appropriately (if part of documented goals)
25. _____ teaches client to ask questions (if part of documented goals)
26. _____ teaches conversational skills (turn-taking, not interrupting, maintaining topic)

Supervisory Process

27. _____ participates effectively in staffing's/consultations with other professionals

Assessment/ Evaluations Activities (Please see last page for list of extensive assessment procedures)
Exhibits proficiency at administering scoring and interpreting the following types of assessment:

28. _____ assesses new clients in a timely manner when assigned by supervisor
29. _____ re-assess clients where needed for evaluation updates and/or discharge decision
30. _____ hearing screenings completed on each client
31. _____ swallowing screenings completed on each client

Assessment/ Evaluation Activities/ Intervention Activities/over the three semesters
Exhibits proficiency at administering scoring and interpreting the following types of assessment: (May have to set up extra adult evaluations)

I. The following should be completed at least two times over one or two semesters: (including speech, language, swallowing, hearing, and oral peripheral)
   1. Two-- full adult assessments
      a. _____, Semester _____________
      b. _____, Semester _______________
   2. Two-- full pediatric assessments
      a. _____, Semester _____________
      b._____, Semester _____________
   3. APD screening using <u>two different</u> screening protocols
      Semester ______________name of screener used _______________________
      Semester ______________name of screener used________________________
   4._____ Conduct Phonological Processing Assessment (at least over one semester)
   5._____Complete preschool speech/language/hearing screenings (may be asked to complete over 3 semesters)

II. Students must participate and facilitate one of the following (please highlight the appropriate group):
1. _____ at least one full semester of ASD Social Group
2. _____ CILLS group/aphasia group/or other focused group, i.e., ESL support, voice drop-in lab or fluency support group, Parkinson's support, hearing support group (for one semester)
3. _____ or facilitate a group from Mekeel Child Care Center, or self-generated therapy group for three to five sessions

III. Students must complete at least 10 of the items below over the three semesters:
1. ____ Complete standardized articulation assessment/ phonological assessment
2. ____ Complete Standardized Adult Language and/or Cognitive Assessment
3. ____ Complete Speech Sample Inventory/Intelligibility Rating
4. ____ Complete Informal Language Sample
5. ____ Complete Standardized Receptive/Expressive Vocabulary Tests
6. ____ Complete Functional Adult Assessment
7. ____ Complete one "Mini-Mental" State Exam (screener) on an adult participant
8. ____ Formally and /or informally assess *voice*
9. ____ Formally and /or informally assess *fluency/ rate*
10. ____ Complete oral motor examination/swallow screening/clinical swallowing examination (CSE)
11. ____ Complete a Quality-of-Life (QOL) measure on at least one client
12. ____ Explore alternative/augmentative modalities appropriate for clients
13. ____ Complete an autism screening device

APPENDIX C

Speech and Hearing Center (Summer Clinic 2020)

Student Clinician Therapy Session Evaluation Rubric (Smith, 2015)
Name of the Clinician: ____________________Client ID/Diagnosis: ________________________
Clinic Weeks: 1 2 3 4 5 6
Session Date/Time slot: ____
Sessions: 1 2
Instructions: Please use the following rubric to rate the skills after each therapy session.
Rubric: 1- Needs improvement; 2-Needs slight improvement; 3-Neutral; 4-Adequate performance; 5-Excellent performance

	SKILLS	RATING SCORE	COMMENTS
1.	Overall Performance		
2.	Gave clear directions		
3.	Gave appropriate amount of information to the client		
4.	Used language level that was appropriate for the client		
5.	Used appropriate rate and volume of speech		
6.	Appeared confident in the session		
7.	Appropriate energy level		
8.	Time was well managed		
9.	Session reflected semester goals		
10.	Used behavior management tools effectively		
11.	Used behavior management strategies effectively		
12.	Provided adequate feedback and reinforcement		
13.	Detected and utilized clinical opportunities		
14.	Used an adequate variety of therapy techniques		
15.	Used intervention techniques frequently enough		
16.	Used intervention techniques effectively		
17.	Modified plan during the session as needed		
18.	Maintained control of environment and materials		
19.	Activities and materials were appropriate for the client		
20.	Planning prior to session was adequate		
	TOTAL SESSION SCORE:		

Comments:
Clinician Signature________________________ Supervisor Signature: _________________
(**Reference:** Smith, H. C. (2015). Student Clinicians Self-Evaluating Performance by Reviewing Session Videos and Using a Rubric. *Perspectives on Administration and Supervision*, *25* (2), 78-84).

APPENDIX D

Cumulative Scores of Graduating Class 2020 (COVID-19 Pandemic Cohort)

Performance Rating Scale Grading Criteria based on a score between 1 and 4, in 0.25 increments 1. Inadequate; skill present <30% of the time; 2. Emerging; skills present 31-59% of the time; 3. Present; skill present 60-89% of the time; 4. Consistent; skill present 90-100% of the time)

	Specific Skills	Average Score Obtained
A.	Overall Evaluation Scores	3.84
B.	Overall Treatment Scores	3.86
C.	Overall Professional Practice, Interaction and Personal Qualities Scores	3.90
C.1.	Demonstrates knowledge of and interdependence of communication and swallowing processes	3.80
C.2.	Uses clinical reasoning and demonstrates knowledge of and ability to integrate research principles into evidence-based clinical practice	3.77
C.3.	Adheres to federal, state, and institutional regulations and demonstrates knowledge of contemporary professional issues and advocacy (includes trends in best professional practices, privacy policies, models of delivery, and reimbursement procedures/fiduciary responsibilities)	3.98
C.4.	Communicates effectively, recognizing the needs, values, preferred mode of communication, and cultural/linguistic background of the patient, family, caregiver, and relevant others	3.89
C.5.	Establishes rapport and shows care, compassion, and appropriate empathy during interactions with clients/patients and relevant others	4.00
C.6.	Manages patient behaviors appropriately	3.78
C.7.	Uses appropriate rate, pitch, and volume when interacting with patients or others	3.95
C.8.	Follows standard precautions	3.97
C.9.	Provides counseling regarding communication and swallowing disorders to clients/patients, family, caregivers, and relevant others	3.75
C.10.	Collaborates with other professionals in case management	3.78
C.11.	Displays effective oral communication with patient, family, or other professionals	3.94
C.12.	Displays effective written communication for all professional correspondence	3.88
C.13.	Adheres to the ASHA Code of Ethics and Scope of Practice documents and conducts him or herself in a professional, ethical manner	4.00
C.14.	Demonstrates professionalism	3.95
C.15.	Is timely with all clinical requirements	3.94
C.16.	Recognizes own limitations, stays within the boundaries of training, and displays appropriate level of confidence	3.84
C.17.	Demonstrates openness and responsiveness to clinical supervision and suggestions	3.95
C.18.	Displays organization and preparedness for all clinical sessions	3.85
C.19.	Personal appearance is professional and appropriate for the clinical setting	4.00

Chapter 4
Applying the Viable System Model to Local Government in a Post-Pandemic Context

Sheila E. Murphy
Work Transformed, USA

Tracey A. Regenold
Work Transformed, USA

ABSTRACT

The pandemic has challenged local government organizations to respond to highly dynamic environments involving health, social, educational, and economic changes. Recognizing local government organizations as systems optimizes them to respond and excel in their ability to serve the public. The purpose of this conceptual chapter is to assess the viable system model's (VSM) efficacy as a vehicle for evaluation within the context of local government organizations, and to determine what enhancements to the model might be needed for post-pandemic application. The viable system model may benefit from the integration of complementary enhancements that may humanize the VSM for use in a post-pandemic context. Ultimately, the new reality of the post-pandemic era calls for systems thinking to reveal ways to structure service delivery reliably and innovatively.

INTRODUCTION

During the pandemic, service continuity has been a major challenge for local government organizations such as counties, cities, and public school districts. The global pandemic exemplifies a new reality that calls for systems thinking to elucidate ways to structure service delivery reliably and innovatively in local government organizations. As political, social, economic, and physical environments change, the need to create new services and explore adaptive opportunities emerges.

DOI: 10.4018/978-1-6684-7285-9.ch004

Recognizing local government organizations as systems optimizes them to respond and excel in their ability to serve the public, especially in times of crisis. Such structuring and innovation of public services may begin by evaluating the local government systems that provide them. The purpose of this conceptual paper is to assess the Viable System Model's (VSM) efficacy as a vehicle for evaluation within the context of local government organizations and to determine what enhancements to the model might be needed for post-pandemic application.

As local government organizations embrace the need to evaluate their capability to deliver social services in a consistent and reliable manner, the Viable System Model (Beer, 1972) is worthy of consideration as a framework. The VSM provides a basis for flexible evaluation to ensure service delivery during and following the pandemic. Two questions guide this chapter's conceptual assessment of the Viable System Model as applied to local government organizations: 1) To what extent could the Viable System Model inform the effectiveness of the delivery of social services by local government organizations? 2) What, if any, enhancements to the Viable System Model might be needed to serve local government organizations in a post-pandemic context?

Stafford Beer, an operations research theorist and cybernetician, developed the Viable System Model. The Model is based in part on cybernetics, a form of systems thinking that emphasizes communication and control (Wiener, 1961). The Viable System Model has served as a vehicle for understanding organizations as systems since its introduction in Beer's (1972) seminal book *The Brain of the Firm.* The Viable System Model (Beer, 1972, 1979, 2007) represents a framework for evaluating a system's ability to sustain itself to perform according to its purpose. In the model, Beer (1972) identified several critical functions of an organizational system to guide the understanding of how functions can interact to establish and reinforce the strength of the organization as it continually evolves to serve its purposes.

The COVID-19 pandemic represented a significant challenge to local government organizations whose mission is to provide social services to people in need. Researchers have applied the Viable System Model to state and federal government organizations to address issues during the pandemic. These issues include social equity and mitigation of economic inequality (Deslatte et al., 2020; Kobayashi et al., 2021; Rauhaus & Johnson, 2021), public health (Greer et al., 2021; Haim et al., 2021), public education (Chernick et al., 2020; López-Santana & Rocco, 2021), economic and workforce development (Mahboubi & Moyaya, 2022), and election services (Pichhio & Santolini, 2022). While each is important in itself, services such as these are inherently interconnected, and people depend upon them for support at the local level. The nature of this support and its relationship to the health and prosperity of communities offers a reminder of the systemic nature of local social service delivery.

Such substantial, ongoing crises as the pandemic call to mind the importance of investing in the strength of social systems. How organizations ensure that their functions are defined and interconnected is part of a systemic reality (Beer, 1979). We assert that shared understanding of the nature of a system's structure and the authorized mission of a local government organization are imperative to effective service delivery, particularly during crises.

This chapter is organized as follows: The next section will briefly introduce the Viable System Model; this will be followed by a more detailed discussion of the six functions of the VSM. Next, we will discuss using the VSM to evaluate organizational systems. This will be followed by a conceptual application of the VSM to local government organizations, and then specific applications of each of the six functions will be reviewed. We will subsequently discuss our reasons for adding complementary features to the model, identify issues in government social systems during the pandemic, and then discuss our recommendations for enhancements to the VSM. Finally, we will conclude the chapter.

THE ESSENCE OF THE VIABLE SYSTEM MODEL

The Viable System Model is based on six interactive functions. These functions include productivity, coordination, operations, auditing, development, and identity (Beer, 1972, 2007; Ramírez-Gutiérrez et al., 2021; Schwaninger & Scheef, 2016). The first four of these functions (productivity, coordination, operations, and auditing) address the current system's performance, while the fifth (development) addresses strategies for improvement to meet a changing environment. The sixth function (identity) serves to balance the current and future organizational system in a manner that maintains its viability. Far from a linear construct, the Viable System Model emphasizes the interdependency of the six functions of an organization.

Beer (1972) asserted that an organizational system must be viable and maintain its separate existence and identity as it is bounded within an environment. The viability of an organization to maintain itself with independent existence depends on its possessing a sufficient range of choices with which to respond to changes in the environment (Ashby, 1957). Such viability positions the organization to perform the established functions that typify any sustainable organizational system.

The VSM provides a lens through which to examine the functions and interactions that characterize current performance while revealing the potential for organizational re-design that implicitly integrates systems thinking. By systems thinking, we mean the ability to perceive and mentally structure organizational issues and problems within an interdependent and purpose-driven framework. Specifically, each of the six primary functions identified in the Viable System Model references a dimension of performance. The production function is fundamental to the system's purpose, a central unifying feature of the system that integrates parts of the system to perform as it fulfills the system's purpose (Beer, 1972, 1979, 2007).

For instance, a community college system is responsible for ensuring high-quality education to serve the people in a given locale. From a systemic view, this is the central purpose of the college system. In support of the production function, the coordination function must be present to ensure that parts of the system are meaningfully and logically connected to sustain production or delivery of the services that represent the organization's reason for being to the public it serves. These two functions posited by Beer (1972) in the VSM integrate the primary reality of systems thinking: one central purpose and a series of interconnected parts that join to fulfill the purpose.

The VSM framework specifies six critical functions and their interaction (Beer, 1972, 2007). The VSM helps point to opportunities for change that substantiate the system's ability to exist as a viable and value-producing entity. During and after the pandemic, it is especially critical to have a framework that serves both as an evaluative model and as a learning model for effective system functionality.

SIX FUNCTIONS OF THE VIABLE SYSTEM MODEL

In this section we will clarify each of the six functions of the Viable System Model. These functions work in systemic fashion, interdependently and recursively, toward the fulfillment of an organization's mission or overarching purpose. The goal of the model is to describe for organizations the functions necessary to achieve a workable balance in the face of a fast-paced and rapidly changing environment; in other words, to attain and maintain viability.

The Productivity Function

The productivity function specifies the value of services delivered by the social system in response to the authorized mission or purpose(s) of the organization. This function "is responsible for producing and delivering the goods and services that the organization generates to the environment" (Ramírez-Gutiérrez et al., 2021, p.334). The productivity function calls attention to what the organization is actually achieving for its customers, requiring agreed-upon metrics and qualitative information. This function represents the main value that customers experience from the services provided.

The Coordination Function

The function of coordination facilitates production by linking communication channels. Ramírez-Gutiérrez et al. (2021) characterize the coordination function as "ensuring harmonious operation among the operation units that constitute system 1 [the productivity function]" (p. 336). Schwaninger (2006), in turn, emphasizes the coordination function as having responsibility for "coordination of subsystems, attenuation of oscillations between them" (p. 955). The flow of information supports the alertness of people responsible for multiple functions within the system to changes in internal operations that potentially influence the performance of the system as it fulfills its purposes.

The Operations Function

The operations function allocates resources needed to produce the services provided by the system. According to Schwaninger and Scheef (2016), the function of operations "is responsible for resource allocation and fosters cohesion, assuring a global performance optimum, which often deviates from the local optima in the primary units" (p. 3-4). In other words, the operations function discourages suboptimization, or sacrificing the whole system for the benefit of a single part of the organization. According to Hitch (1953), counteracting suboptimization requires examination of assumptions as well as criteria to guide decisions that affect all levels of operation.

The Audit Function

The audit function monitors and informs the system through the ongoing capture and communication of information about effectiveness of service delivery that strengthens the operations function. The audit function reveals both effective and problematic areas that require further attention (Keating, 2000). Overall, the audit function furnishes relevant feedback that facilitates decision making by the operations function.

The Development Function

The development function supplies the system external intelligence that keeps the services and their delivery relevant and responsive in serving its population. Schwaninger (2006) defines the development function as "management for the long term, relationships with the overall environment" (p. 955). The development function can be considered the intelligence function regarding the environment. According to Ramírez-Gutiérrez et al. (2021): "the development function takes care of the organization's future and environment" (p. 336). This requires the continual capture of emergent information about needs and potential ways to deliver services.

The Identity Function

The identify function is the ultimate decision maker in the organizational system (Ramírez-Gutiérrez et al., 2021). The identify function must also manage the organization's balance on several fronts. First, balance must be struck between the internal workings of the organization and the external environment (Ramírez-Gutiérrez et al., 2021; Schwaninger & Scheef, 2016). Next, the identity function must balance the needs and issues the organization faces now with those it may be faced with in the future (Schwaninger & Scheef, 2016). Last, but not least, the identify function is responsible for balancing its *raison d'être*, internally and externally (Schwaninger & Scheef, 2016) with the wider systems of government, community, and society.

THE VSM AS A FRAMEWORK FOR UNDERSTANDING AND EVALUATING SYSTEMS

The six functions of the VSM provide a way of examining organizations to discern their ability to serve clients effectively in the face of any perturbations from a rapidly changing environment, such as a global pandemic. It should be noted that in this way the VSM serves as "a template against which the architecture of a new operation can be designed or an existing operational structure can be diagnosed" (Keating, 2000, p.184). Millar (2009) emphasized the VSM as a "thinking framework" including a "common language" that fosters discussion and consideration of how the organization can be re-envisioned (p. 34). Espinosa and Walker (2013) found the VSM useful as an "hermeneutic enabler of the community learning process concerning their self-organization" (p. 118) in the redesign of ways to address complexity from the environment.

The VSM has served as a basis for enriching the understanding and effectiveness of organizational systems in a wide variety of settings, including nation states (Chaudry, 2022; Rodriguez-Ulloa, 2022), non-profit organizations (Reyes, et al., 2018), state-level child welfare agencies (Fitch et al., 2014), and private industry organizations encompassing a wide array of initiatives (Lamé, et al., 2017). The broad appeal of the VSM for examining organizational systems is that it provides a lens through which to examine the functions and interactions that characterize current performance while revealing the potential for change. A strong framework for understanding an organizational system offers an especially important vehicle in times of crisis such as the global pandemic.

Crises other than the pandemic and community change initiatives have supported the notion that the Viable System Model helps build system strength. Haim et al. (2021) found that building connections and relationships with village leaders during a health crisis in a conflict zone in the Phillipines improved trust in government as shown by a 20% increase in willingness to report key health information. Such an increase in trust would be especially valuable during a pandemic. Increased trust in government can help mitigate health crises in a given locale (Joyce, et al., 2020). The Viable System Model substantiates understanding that can lead to trust based on strengthening relationships of members of the system who share commitment to a purpose.

Clark (2021) conducted a study of a multi-organizational collaboration of local and central government, nonprofits, and funders formed to serve youth following the Canterbury earthquakes in New Zealand. The VSM, when combined with a Team Syntegrity model, was found to support the structuring of collaborative efforts and to help participants to understand the post-disaster problems needing to be addressed. Combining resources in the spirit of intentional collaboration fortified by an understanding of the functions of the VSM proved beneficial following a natural disaster.

It should be noted that the examples from the literature demonstrate the applicability of the Viable System Model for clarifying and enriching understanding of systems during times of crisis and large-scale change. Our emphasis in this chapter includes broad utilization of the VSM as a framework for recognizing how local government organizations presently work and might be changed to meet environmental constraints.

THE VIABLE SYSTEM MODEL'S POTENTIAL IN THE CONTEXT OF LOCAL GOVERNMENT

Based on the literature, the VSM has served as an effective framework for revealing systemic issues in government social systems (Fitch, 2014; Lamé, 2017; Parsons & Shelton, 2019). Further, the VSM has been recognized as a vehicle for guiding change by organizational systems for redesign of their structures (Millar, 2009; Schwaninger, 2006). The potential exists for applying the Viable System Model for the purposes of energizing local government organizations to shape understanding within and external to the organization, notably in the post-pandemic period. Bounding the local government system and distinguishing the environment from the internal workings begins the process of recognizing the assets within the organization and the resources available to support the organization. The six functions of the VSM provide a basis for recognition of areas of opportunity toward increasingly responsive service.

Local government faces constant challenges that distinguish it from other organizations. These challenges become especially relevant when a global crisis such as the pandemic emerges. First, local government is on the lowest rung among government organizations. In the U.S., for example, federal and state levels of government set the parameters for funding support and regulatory requirements, and local government organizations must adhere to guidelines established outside their immediate purview. In the case of the pandemic, local government had to rely on higher tiers of government for needed support, notably for small businesses and individuals facing dire health and safety issues as well as threats to economic viability. Consistent with the VSM, state and federal levels of government represent one form of perturbation from a local government's environment that must be anticipated, prepared for, and buttressed against.

Second, local government organizations consistently face complexity and even chaos from the wider external environment in which they operate. As economic, educational, social, and political pressures affect the ability of local government to perform, social isolation exacerbates the need for support services. The overwhelming variety emanating from the outer environment prompts local government to address needed calls for service.

Third, local government organizations coordinate the delivery of services directly to the public. Such coordination requires that more attention be paid to communication and collaboration, both internally and externally, necessary for the production and delivery of services in a responsive way. The stronger the connections of people and departments within the internal organization, the greater the capability of local organizations to address needs. Internal coordination, communication, and collaboration are required to respond to people and communities needing government services.

Taken together, local government is charged with ensuring mission-focused and well-connected organizational systems that position them to demonstrate not only reliable service delivery but to develop innovative services as well. Viable organizational systems position themselves as dependable service mechanisms that know their operations well enough to reinvent their structures and processes to meet dynamic outer realities, such as the global pandemic. Amid crises, local government services must respond by demonstrating new ways of supporting people in dire need of social services.

With this reality in mind, the VSM provides a clarifying framework for helping members of local government organizations recognize functions essential to viability in turbulent environments. Further, the common language provided by the VSM enables elected officials, appointed leaders, staff members, stakeholders, and clients to perceive the organization accurately and usefully. The VSM sets the stage for an energizing interplay among these functions, combined with an evaluative mechanism through which to view the ways that critical functions operate.

In addition, the VSM provides a means of crafting connections among functions within the system, while recognizing necessary supports and pressures alike coming from the outer environment. Espinoza and Walker (2013) found that using the VSM in a three-year action research project with an Irish eco-community resulted in improved roles and connection among them. The VSM guided restructuring of the community, clarifying both roles and tasks and strengthening the bonds among people to build their system toward increasing effectiveness.

The Viable System Model also supports learning throughout the organization and offers a clear basis for analysis of the organization's activities. Fitch et al. (2014) applied the VSM to substantiate the learning dimension of an organization responsible for child welfare. They were concerned with entropy and variety attenuation in the organization's culture. The authors defined entropy as: "any activity that is nonproductive in terms of time or resources," (Fitch et al., 2014, p. 132). They defined variety attenuation as: "any process that leads to the unnecessary loss or underutilization of skills or expertise" (p. 132). In their findings, the authors lauded the presence of variety amplification: "One of the most desired processes in a learning organization" (p. 131). Variety amplification denotes increasing the range of actions and directions available to the organization and its professionals.

The framework of the VSM provides a foundation for understanding the current status and future potential for strengthening local government organizations. It is important to determine which organizational activities add value, which reduce utilization of resources, and which among them result in entropy. Such clarification supports the ability of local government organizations to renew their capability to serve in new and responsive ways. Fitch et al. (2014) emphasized a key point: "Training needs to be re-conceptualized as an expertise leveraging endeavor" (p. 139). Leveraging learning through education represents a sound investment in the social systems of local government.

In addition, services such as economic and workforce development support growing local economies, most important during and after the global pandemic. Contributions by individuals seeking to build their skills and careers requires the inclusive, highly participatory buy-in of community and business leaders, nonprofit partners, and multiple government representatives working in common. Social services including library services, neighborhood services, and judicial outreach for prevention to support at-risk youth in their pursuit of education and job learning can similarly benefit from the application of the VSM for evaluating effectiveness of service delivery and design.

CONCEPTUAL APPLICATION OF THE VSM TO LOCAL GOVERNMENT

The utility of the Viable System Model in local government organizations was broadly outlined in the preceeding section. Next, we will describe in greater detail a conceptual application of each of the VSM's six functions to examples of local government agencies. This will be followed by a discussion of potential enhancements to the VSM in a post-pandemic context. As shown in Table 1, the functions of the VSM have direct application to local government.

Table 1. Functions of the viable system model and application to local government

Function	Application to Local Government
Productivity	Addresses the structural efficacy of the system; clarifies the extent the organization is delivering services according to mission; represents the main value customers receive from the social system.
Coordination	Facilitates production by effectively linking communication channels; supports attention to changes that influence system performance; guides social system delivery on the mission.
Operations	Controls day-to-day workings; allocates resources needed to produce the services provided by the system; emphasizes day-to-day performance; places primary emphasis on the total system.
Audit	Monitors operations and informs the system through the ongoing capture and communication of information about effectiveness of service delivery; facilitates decisions by the operations function; supplies fresh and ongoing performance information to members of the social system to improve performance.
Development	Supplies the system external intelligence that keeps the services and their delivery adaptive, relevant, and responsive; integrates environmental and internal intelligence for performance of the system; facilitates the system's adaptation to ensure viability.
Identity	Provides clear direction and proactive messaging directed to the public; represents the system to the wider systems in the environment; ensures policy conformance; serves as the ultimate decision maker for the system; balances the internal organization and the external environment.

The Productivity Function: Structural Efficacy in Local Government

A deliberate focus on each of the six functions of the VSM provides not only an evaluative lens for assessing structural efficacy for mission fulfillment by local government organizations, but a pathway to collaborative and creative design that enhances the ability of a viable organization to improve itself. The first of these six functions, productivity, commits a local government to posing a key question: To what extent is the organization delivering services according to mission? Organizations of all types often claim to work toward a mission or state that they are producing X results when what they are in fact producing is incongruous with their statements. It is not that they are misstating their intentions, but simply that they may not be aware of the systemic significance of aligning delivery with design.

In local government, an example of a social system that claims to produce a particular result might be found in the area of education. Community colleges may recognize a need for providing bachelor's degrees, certifications in trades, and other stackable credentials. The sheer breadth of needs, especially during periods of change and crisis, may challenge the college to live up to its desire to serve the community. The reality of the pandemic necessitates increasing rigor to support productivity during a time of crisis. The VSM guides systemic thinking by keeping the system on track with what its mission is, thereby avoiding mission creep.

The Coordination Function: Effectiveness of Communication in Local Government

The second of the six functions of the VSM, coordination, commits the organization to a key question: "How effectively do the functions of the organization communicate and coordinate their efforts?" Communication and coordination are imperative to system viability. The VSM provides recognition that parts of the organization must be connected on a constant basis to deliver on the mission. For the sake of consistency, the coordinative function of the VSM supports efficiency and effectiveness of the processes involved.

As an example, social services provided by local government, such as mental health services for justice-involved youth or adults, are often delivered by partnering organizations. It is critical that the main organization and its provider partners stay in close communication regarding resource needs and the ability to deliver services. The global pandemic has required increasing connectivity among partners in coordinating services. Connecting a local government organization with one or more service providers represents an aspect of systems thinking and is vital to the shared interests of the overall system. The VSM supports coordination among parts of the system to deliver on the mission.

The Operations Function: Control of Day-to-Day Workings in Local Government

The third of the six functions of the VSM, operations, emphasizes control. This anchoring function of running the day-to-day operations within local government cannot be overstated. Checkland (1999) defined control as: "The process by means of which a whole entity retains its identity and/or performance under changing circumstances" (p. 313). The emphasis on current ways of delivering services and on how these services might be improved upon, especially in face of restrictions required by the pandemic, reflects the importance of the operations function.

An example of the operations function in local government can be seen in the area of human services, such as providing senior day program services and home delivery of meals to homebound individuals. Control of these day-to-day operations is imperative for effective service delivery in the spirit of the organization's mission. In times of crisis, consistent delivery of services remains imperative amid restrictions on direct contact with other people, requiring creativity among system partners.

It should be noted that decisions made directly by the autonomous units of production affect the quality of day-to-day operations. The VSM stimulates a systems view of the organization, in light of its operations, clarifying the extent of connectivity among functions such as production and coordination. Seen as one integral whole, the system is positioned to fulfill its purposes.

The Audit Function: Monitoring Operations In Local Government

Included as a subset of the operations function defined by Beer (1972) is the audit function, the purpose of which is the monitoring of vertical and lateral elements of the operations function. "Sporadic" monitoring and validation, according to Schwaninger and Scheef (2016, p. 4), maintains fresh and accurate transfer of information between operating units and clarifies the status of performance. This sub-function provides communication and connection that in turn support the productive capability of the system. At no time is the audit function more necessary and more challenging to perform than during a global pandemic that potentially restricts access to locations and services by those auditing and monitoring services. Clarifying the extent to which service delivery currently works sets the stage for making necessary adjustments to effective service provision.

Closely aligned with the audit function in local government is system learning. Argyris (1991) points out that: "System learning results from the detection and correction of error. Correction may occur when staying within the current system structure (single-loop) or may shift structure based on questioning the fundamental assumptions, objectives, and norms of the system (double-loop)" (para. 4). The audit function supports system learning by supplying clear and pertinent information about resource needs and operational delivery in support of the production and operations functions.

An example of the audit function in local government is in early childhood education, with the United States' Head Start Birth to Five program. This social service represents an investment in learning readiness for children in the designated age group. The audit function provides assessments based on the national program made available locally through cities, towns, rural communities, and educational partners to determine social and emotional development, language, cognitive, early reading, science, math, and physical development.

Although the Head Start program is already audited on a regular basis, it could benefit from making changes based on double-loop learning. In times of crisis, notably a health crisis that limits direct human contact, early childhood education must be handled in new ways. The audit function supports examining patterns of performance and potential new ways of responding to the organization's purpose. The VSM supports improvement of service by integrating the audit function to supply fresh and ongoing performance information to members of the social system, providing alerts to opportunities for improved design and service.

The Development Function: Adaptive Response of Local Government Systems

The fifth of the VSM functions, development, is focused on building an adaptive response to a dynamic, larger outer environment in which the local government organization exists. The development function integrates environmental and internal intelligence for performance of the system. According to Leonard (2009), the role of the development function "is to observe the anticipated future environment and its own states of adaptiveness and act to bring them into harmony" (p. 228). Schwaninger and Scheef (2016) shared that knowledge management, strategic planning, and research and development represent associated functions that support practice optimization. As the system encounters real-time change emanating from the outer environment, the system is forced to adapt and transform itself to maintain viability. Ideally, the development function becomes the stimulus for proactive change that results in the organization's enhanced ability to serve its customers.

An example of the development function in local government occurs with the issue of homelessness, exacerbated during a time of health crisis when all people must be protected from the spread of infection. Cities and counties recognizing a need for affordable housing pursue the opportunity to reach out to businesses, community leaders, and residents to identify needs and resources. All views provide new information that guides the organization in developing an up-to-date and complete recognition of needs and resources.

At the time of the pandemic and post-pandemic, the development function has revealed especially challenging situations faced by people experiencing homelessness. Support functions including shelter, food, health care, child care, and transportation, were limited by health-based restrictions. Starting with listening, observing, and convening multiple service provision partners, social systems sought to translate needs into real services through an array of resources. Among the solutions for providing shelter was discovering and including which nonprofit organizations could play a role in supporting individuals experiencing homelessness. The VSM supports the gaining of intelligence from the outer environment and informing the system of needs and challenges.

The Identity Function: Clear Direction and Proactive Messaging by Local Government

The sixth function of the VSM is identity. This function provides clear direction and proactive messaging directed to the public. The identity function involves the representation of the system to wider systems and ensures policy conformance. According to Ramírez-Gutiérrez et al. (2021), the system's identity "constitutes the highest authority in the organization. It balances the organization's present and future, taking into account the internal and external factors that affect it" (p. 336). Schwaninger and Scheef (2016) state that identity includes "the values, norms and principles, which govern the system, i.e., the ethos of the organization" (p.4).

Local government is represented by its elected and appointed officials who collectively handle the identity function of the organization to the outer environment. The identity function serves as a reminder that building a strong bridge between the system and its outer environment increases the ability of the organizational system to become effective for the people served. Communicating the system's identity within and outside of the system is vital to energizing the people within the system and those who depend on its services.

As an example, a juvenile detention center serving a county may place strong emphasis on high quality education for young offenders. The identity function supports the organization's identity and policy by ensuring strong messaging about prevention and services to help youth become contributing members of the community. In face of some public skepticism about investing education dollars in youth and adult offenders, the county government might take the initiative to proclaim the value of offering youth a way to improve their lives and the lives of others. Recognition of the economic and social benefits to the larger community becomes part of the messaging.

It should be noted that during a time of crisis, communication within and outside the system become especially vital for purposes of clarity and commitment to the system's purposes. Communication has been described as one of the most important principles of systems thinking (Checkland, 1999). People need reassurance to guide both performance and participation in government services in challenging times. The VSM supports clarification of the system's identity to enable the social system to deliver services according to its authorized purposes.

OPPORTUNITY TO ENHANCE THE VSM FOR A POST-PANDEMIC CONTEXT

In the preceding sections, we addressed the first guiding question: To what extent could the Viable System Model inform the effectiveness of the delivery of social services by local government organizations? In this section, we address the second guiding question: What, if any, enhancements to the Viable System Model might be needed to serve local government organizations in a post-pandemic context? As indicated throughout this chapter so far, the VSM provides a strong framework for assessing the efficacy of the organizational structure.

Providing a lens for stimulating a shared understanding of the system and for transforming the organizational system, the VSM framework is essentially analytical and positions social systems to develop and improve continually over time. In this section we will review issues in government social systems during the pandemic, then discuss potential enhancements to the VSM for a post-pandemic context. Finally, we draw conclusions regarding application of the VSM for enhancing government social systems in times of crisis.

Issues in Government Social Systems During the Pandemic

The pandemic highlighted multiple issues concerning the provision of government services during the pandemic. These issues included inequality, public health and education, economic and workforce development, and election services (Chernick et al., 2020; Deslatte et al., 2020; Greer et al., 2021; Haim et al., 2021; Kobayashi et al., 2021; López-Santana & Rocco, 2021; Mahboubi & Moyaya, 2022; Pichhio & Santolini, 2022; Rauhaus & Johnson, 2021). One clear example involved the integration of health policies and social policies (Greer et al., 2021; Manthorpe et al., 2022).

Greer et al. (2021) compared early responses to the COVID-19 pandemic by Brazil, Germany, India, and the United States to examine how effective each nation was in containing the virus. The most successful of the four nations was Germany, as it effectively blended social and healthcare policies. Brazil and the United States emphasized social policies separate from public health restrictions, resulting in a failure to stop the spread of the virus. Greer et al. (2021) concluded that "Separating social policy and public health, in theory or in practice, undermines both, and heightens the risk that both will fail" (p. 1219).

In the United States this separation of policies during the pandemic demonstrated the limits of federalism (López-Santana & Rocco, 2021; Patton et al., 2022; Weissert et al., 2021) and revealed the strain placed on state and local government organizations. Although federal support during the pandemic exceeded that of the great recession, it was beset by obstacles in a conditional, "fend-for-yourself federalism" (López-Santana & Rocco, 2021, p. 365). López-Santana and Rocco (2021) asserted that the pandemic broadcasted the fissures within U. S. systems of governance and politics as well as those within U. S. social systems. India, in turn, pursued strong public health measures without social policy support, resulting in failure to contain the spread of the disease.

Perspectives on patterns of fiscal policy and practice (Chernick et al., 2020; Joyce & Prabowo, 2020) offer further issue salience regarding the pandemic. Chernick et al. (2020) emphasized that costs of the pandemic are wide-ranging and transcend healthcare costs. The higher costs of delivering education and other services, with attention to social distancing and alternative delivery mechanisms, added to the weight of fiscal responsibility for local government. The pandemic brought to light a strong connection between education and healthcare. Further, the pandemic revealed the importance of establishing and maintaining effective services that inspire public trust. Amid concerns about fiscal viability of local government in face of a crisis such as the pandemic (Chernick et al., 2020; Deslatte et al., 2020; López-Santana & Rocco, 2021; Joyce & Prabowo, 2020), the VSM has an important role to play, specifically in supporting decision making that enriches mission fidelity for local government social systems.

Public trust in social systems (Kye & Hwang, 2020; Haim et al., 2021; Picchio & Santolini, 2022) represents a critical feature of the efficacy of government. In a study of South Korea during the pandemic, Kye & Hwang (2020) identified "a causal effect of crisis management on social trust" (p. 1). Using a nationally representative sample survey, the authors analyzed 10 social trust measures. The authors found that during the pandemic, trust in local government organizations increased, while trust in other institutions such as religion and the media decreased (Kye & Hwang, 2020). Kye and Hwang (2020) posited that the increases or decreases in trust were a result of taking action or not.

A similar position was taken by Joyce et al. (2020), who emphasized government as an investment in social systems. The authors examined patterns and protocols by various nations during the COVID-19 pandemic (Joyce et al., 2020). They found that countries that took bold steps at the outset of the pandemic produced better containment results (Joyce et al., 2020). During times of crisis such as the global pandemic there is potential value in using the six interactive functions of the VSM for evaluation of the system and identifying structural advances needed for active, effective local government.

Proposed Enchancements to the Viable System Model

While the Viable System Model has been used in government organizations for many years, the disruption of services and requisite creativity of responses during the pandemic suggest the need for improvements to the model. Use of the VSM for local government systems, especially in a post-pandmenic context, may benefit from complementary enhancements that increase the probability of its success. The authors recognize the potential benefits of humanizing the VSM for use in local government. By humanizing, we mean transitioning the VSM from the conceptual, analytical construct applicable to any organization, to a more specific, concrete model represented by social systems, notably those found in local government. Proposed enhancements to the VSM for supporting local government services in a post-pandemic period fall into two main categories: 1) Embedding the VSM into management practices, and 2) Ensuring the VSM is inclusive of cultural elements.

Embedding the VSM into Management Practices

The first of these enhancements involves management practices. There are two proposed management practices: participatory involvement of stakeholders and change management. Participatory involvement of stakeholders enhances both representativeness of experience and quality of thinking when designing processes and the system itself. Active engagement of stakeholders in the local government system, such as funders, decision makers, service providers, and customers, facilitates the system's making full use of the quality of thinking, skill capability, and resource contribution to enrich its service potential.

Participatory involvement of these stakeholders using the VSM provides a rich and clarifying picture of what is true now and what is possible for a future-enhancing design of the local government organization. Participatory involvement ultimately builds shared investment in what people working together design. Designed meetings and conversations that include diverse perspectives raise the probability of achieving results.

For example, when providing neighborhood services during the pandemic, it became necessary to ensure accessible means by which people could reach professionals in city departments who could assist in dispensing resources. This required funding authorization, the design of methods for ensuring public access to professionals, and building a supply of personal protective equipment. Such efforts extended beyond any one service department and reflected many of the VSM's six functions within the social service organization.

The second management practice, change management, represents a structured approach to discovering and integrating new ways for a local government organization to perform its functions. Change management means guiding and shaping projects within local government using thinking that results in increasingly improved service. During the pandemic, delivering social services such as family support in cases of child abuse or neglect necessitated different modes of service delivery, based on health and safety requirements. Case management for parents, providing safe settings for children at risk, and integrating mental and physical health at a time of restricted contact, posed challenging environmental variations.

Change management is used to design and deliver services in a reliable and responsive way, supporting staff readiness for change and multi-organizational collaboration. Ensuring individual and organizational readiness for change (Wang, et al., 2020) contributes to effective change management. Further, engaging with social networks (Stouten, et al., 2018) represents an effective direction for change management. Embedding the VSM into the practice of change management can enhance the effectiveness of the VSM in a post-pandemic context.

It is instructive to note that change management provides practical focus fortified by continuous learning and an understandable direction for participating stakeholders that also enriches the VSM. In turn, the functions of the VSM are applied as interactive criteria that support a change initiative. Applying the VSM as a framework in situations that can benefit from change management potentially enhances the ability to recognize a need for innovation in local service delivery.

Ensuring the VSM is Inclusive of Cultural Elements

The second of these proposed enhancements, ensuring that the VSM is inclusive of cultural elements, means improving the quality of services by representing and understanding the clients receiving services. Diversity, equity, inclusion, and accessibility as primary cultural elements strengthen the local government organization's ability to identify and build upon different courses of action that members of the system can pursue. Ashby (1957) emphasizes that a system must possess at a minimum the number of possible courses of action (variety) that are present in the environment.

When the complexity of the environment is greater than what the system itself possesses, the system will not be strong enough to withstand the pressures that emerge, and as a result, the environment alone will dictate the organization's outcomes. If the environment were left to dictate outcomes at the level of local government, counties, cities, towns, school boards, etc., would lose their autonomy. Fully addressing challenges posed by the environment helps the system develop and maintain a range of choices at least equal to the challenges posed.

Far from a solely conceptual understanding of social systems, ensuring the inclusion of cultural elements demonstrates the full scope of potential that the VSM brings to organizations dedicated to human service. A post-pandemic context requires fidelity to service provision and enhancing the system's ability to innovate to meet the needs of human beings. Ensuring diversity, equity, inclusion, and accessibility system-wide enriches the quality of thinking and understanding about whom we serve and how we do so.

For local government in particular, placing emphasis on culture, fortified by learning, positions the system to recraft itself in meaningful ways. An holistic emphasis allows employees and stakeholders of the organizational system to discover new ways to support clients, especially in times of emergency. Smuts (1936) explored the concept of holism in a revolutionary manner, describing it as "a process of creative synthesis" in which "the resulting wholes are not static but dynamic, evolutionary, creative" (p.

86). Holism suggests that the essence of the system supports integrating continuous shared learning by diverse stakeholders to build and stabilize the organization as an instrument of innovation and ingenuity.

CONCLUSION

The pandemic has challenged local government organizations to respond to highly dynamic environments involving health, social, educational, and economic changes. The Viable System Model developed by Beer (1972) provides a framework that has guided numerous organizations worldwide in evaluating and structuring themselves to meet dynamic needs. Local government organizations can benefit from examining their systems by applying the Viable System Model toward optimizing consistency and delivery of services.

This chapter was guided by two questions: 1) To what extent could the Viable System Model inform the effectiveness of delivery of social services by local government organizations? 2) What, if any, enhancements to the Viable System Model might be needed to serve local government organizations in a post-pandemic context? Addressing the first question, this chapter argued for the Viable System Model as an effective prescriptive framework and evaluative lens through which local government organizations can position themselves to deliver consistent, effective services over the long term. Addressing the second question, this chapter specified complementary enhancements to the VSM to strengthen the framework, including embedding the VSM into management practices and ensuring the VSM is inclusive of cultural elements, to ensure relevance in a contemporary local government setting.

Applying the Viable System Model to local government shows strong potential for integrating sufficient variety in structural and human terms. In the face of dynamic, rapidly-changing environments, local government organizations can benefit from being viewed through the VSM to explore their efficacy and potential for development. The evaluation of local government service delivery through the lens of the Viable System Model would support organizational productive capacity, staffing, coordination, management, collaboration, communication, policies, practices, and funding.

Initially established in a command-and-control context, the Viable System Model may benefit from the integration of complementary enhancements pertinent to the delivery of social services in the public sector. The enhancements of management practices and cultural elements may humanize the VSM for use in a post-pandemic context. Ultimately, the new reality of the post-pandemic era calls for systems thinking to reveal ways to structure service delivery reliably and innovatively in local government.

REFERENCES

Argyris, C. (1991). Teaching smart people how to learn. *Harvard Business Review*, *4*(2). https://hbr.org/1991/05/teaching-smart-people-how-to-learn

Ashby, W. R. (1957). *An Introduction to Cybernetics*. Chapman & Hall.

Beer, S. (1972). *The brain of the firm*. John Wiley & Sons.

Beer, S. (1979). *The heart of enterprise: The managerial cybernetics of organization*. John Wiley & Sons.

Beer, S. (2007). *Diagnosing the system for organizations*. John Wiley & Sons.

Chaudhry, I. S. (2022). Viable System Model: A tool for managing sustainable development holistically. *Management & Sustainability: An Arab Review, 1*(1), 50–65. doi:10.1108/MSAR-01-2022-0008

Checkland, P. (1999). *Systems thinking, systems practice: Includes a 30-year retrospective*. John Wiley & Sons.

Chernick, H., Copeland, D., & Reschovsky, A. (2020). The fiscal effects of the COVID-19 pandemic on cities: An initial assessment. *National Tax Journal, 73*(3), 669–732. doi:10.17310/ntj.2020.3.04

Clark, A. J. (2021). *Dealing with pluralism and marginalization processes in multi-organizational collaboration: Augmenting the Viable System Model with team syntegrity*. [Doctoral dissertation, Victoria University of Wellington]. doi:10.26686/wgtn.16960747.v1

Deslatte, A., Hatch, M. E., & Stokan, E. (2020). How can local governments address pandemic inequities? *Public Administration Review, 80*(5), 827–831. doi:10.1111/puar.13257 PMID:32836452

Espinosa, A., & Walker, J. (2013). Complexity management in practice: A Viable System Model intervention in an Irish eco-community. *European Journal of Operational Research, 225*(1), 118–129. doi:10.1016/j.ejor.2012.09.015

Fitch, D., Parker-Barua, L., & Watt, J. W. (2014). Envisioning public child welfare agencies as learning organizations: Applying Beer's Viable System Model to Title IV-E program evaluation. *Journal of Public Child Welfare, 8*(2), 119–142. doi:10.1080/15548732.2013.879089

Greer, S. L., Jarman, H., Falkenbach, M., Massard de Fonseca, E., Raj, M., & King, E. J. (2021). Social policy as an integral component of pandemic response: Learning from COVID-19 in Brazil, Germany, India, and the United States. *Global Public Health: An International Journal for Research, Policy and Practice, 16*(8-9), 1209–1222. doi:10.1080/17441692.2021.1916831 PMID:33876715

Haim, D., Ravanilla, N., & Sexton, R. (2021). Sustained government engagement improves subsequent pandemic risk reporting in conflict zones. *The American Political Science Review, 115*(2), 717–724. doi:10.1017/S0003055420001148

Hitch, C. (1953). Sub-optimization in operations problems. *Journal of the Operations Research Society of America, 1*(3), 87–99. doi:10.1287/opre.1.3.87

Joyce, P., Maron, F., & Reddy, P. S. (2020). Managing global pandemics: Public governance matters. In P. Joyce, F. Maron, & P. S. Reddy (Eds.), *Good public governance in a global pandemic* (Vol. 1, pp. 567–591). The International Institute of Administrative Sciences. https://www.iias-iisa.org/page/ebook doi:10.46996/pgs.v1e1

Joyce, P. G., & Prabowo, A. S. (2020). Government responses to the coronavirus in the United States: Immediate remedial actions, rising debt levels and budgetary hangovers. *Journal of Public Budgeting, Accounting & Financial Management, 32*(5), 745–758. doi:10.1108/JPBAFM-07-2020-0111

Keating, C. B. (2000). A systems-based methodology for structural analysis of health care operations. *Journal of Management in Medicine, 14*(3/4), 179–198. doi:10.1108/02689230010359183 PMID:11142060

Kobayashi, Y., Heinrich, T., & Bryant, K. A. (2021). Public support for development aid during the COVID-19 pandemic. *World Development, 138*, 105248. doi:10.1016/j.worlddev.2020.105248 PMID:33106725

Kye, B., & Hwang, S.-J. (2020). Social trust in the midst of pandemic crisis: Implications from COVID-19 of South Korea. *Research in Social Stratification and Mobility*, *68*, 100523. doi:10.1016/j.rssm.2020.100523 PMID:32572302

Lamé, G., Jouini, O., & Stal-Le Cardinal, J. (2017). *Combining the Viable System Model and Kotter's 8 Steps for multidepartment integration in hospitals.* 12ème Conférence Internationale de Génie Industriel, Compiègne, France. https://hal.archives-ouvertes.fr/hal-01519593

Leonard, A. (2009). The Viable System Model and its application to complex organizations. *Systemic Practice and Action Research*, *22*(4), 223–233. doi:10.100711213-009-9126-z

López-Santana, M., & Rocco, P. (2021). Fiscal federalism and economic crises in the United States: Lessons from the COVID-19 pandemic and great recession. *Publius*, *51*(3), 365–395. doi:10.1093/publius/pjab015 PMID:34394701

Mahboubi, P., & Mokaya, M. (2022, April 29). The skills imperative: Workforce development strategies post-COVID. *C.D. SSRN*, *609*. doi:10.2139srn.4096106

Manthorpe, J., Woolham, J., Steils, N., Stevens, M., Marntineau, S., Owen, J., & Tinelli, M. (2022). Experiences of adult social work addressing self-neglect during the COVID-19 pandemic. *Journal of Social Work*, *22*(5), 1227–1240. doi:10.1177/14680173221083446

Millar, G. (2009). *The viable governance model: A theoretical model of IT governance with a corporate setting* [Doctoral dissertation, The University of New South Wales]. doi:10.26190/unsworks/22565

Parsons, P., & Shelton, K. (2019). Organizational sustainability in online higher education: Reframing through the Viable System Model. *Online Journal of Distance Learning Administration*, *22*(3), 1–14. https://ojdla.com/archive/fall223/parsons_shelton223.pdf

Patton, D., Durand, A., Whipple, K., & Albright, D. L. (2022). Local autonomy and response to the COVID-19 pandemic. *State & Local Government Review*, *54*(2), 165–173. doi:10.1177/0160323X221089661

Picchio, M., & Santolini, R. (2022). The COVID-19 pandemic's effects on voter turnout. *European Journal of Political Economy*, *73*, 102161. doi:10.1016/j.ejpoleco.2021.102161 PMID:34975184

Ramírez-Gutiérrez, A. G., Cardoso-Castro, P. P., & Tejeida-Padilla, R. (2021). A methodological proposal for the complementarity of the SSM and the VSM for the analysis of viability in organizations. *Systemic Practice and Action Research*, *34*(3), 331–357. doi:10.100711213-020-09536-7

Rauhaus, B. M., & Johnson, A. F. (2021). Social inequities highlighted by the prolonged pandemic: Expanding sick leave. *Journal of Public and Nonprofit Affairs*, *7*(1), 154–163. doi:10.20899/jpna.7.1.154-163

Reyes, R. R., García, S. G., & García, M. G. (2018). Applying the Viable System Model to an organization with CSR goals: The case of a charity organization. *European Journal of Economics and Business Studies*, *4*(3), 7–21. https://econpapers.repec.org/article/eurejesjr/267.htm. doi:10.2478/ejes-2018-0055

Rodriguez-Ulloa, R. (2022). Cybernetic governance of the Peruvian State: A proposal. *AI & Society*, *37*(3), 1207–1229. doi:10.100700146-021-01329-3 PMID:35309097

Schwaninger, M. (2006). Design for viable organizations: The diagnostic power of the Viable System Model. *Kybernetes*, *35*(7/8), 955–966. doi:10.1108/03684920610675012

Schwaninger, M., & Scheef, C. (2016). A test of the Viable System Model: Theoretical claim vs. empirical evidence. *Cybernetics and Systems*, *47*(7), 1–26. doi:10.1080/01969722.2016.1209375

Smuts, J. C. (1936). *Holism and Evolution* (3rd ed.). MacMillan & Co.

Stouten, J., Rousseau, D. M., & De Cremer, D. (2018). Successful organizational change: Integrating the management practice and scholarly literatures. *The Academy of Management Annals*, *12*(2), 752–788. doi:10.5465/annals.2016.0095

Wang, T., Olivier, D. F., & Chen, P. (2020). Creating individual and organizational readiness for change: Conceptualization of system readiness for change in school education. *International Journal of Leadership in Education*, 1–25. doi:10.1080/13603124.2020.1818131

Weissert, C. S., Uttermark, M. J., Mackie, K. R., & Artiles, A. (2021). Governors in control: Executive orders, state-local preemption, and the COVID-19 pandemic. *Publius*, *51*(3), 396–428. doi:10.1093/publius/pjab013 PMID:34394702

Wiener, N. (1961). *Cybernetics: or control in the animal and the machine*. The MIT Press. doi:10.1037/13140-000

Chapter 5
Left Behind in COVID-19 Times:
The Effect of Income Support Measures on Vulnerable Groups in Serbia

Marko Vladisavljevic
https://orcid.org/0000-0001-6020-1355
Institute of Economic Sciences, Belgrade, Serbia

ABSTRACT

The chapter analyses the effects of income support measures implemented during the first year of the COVID-19 pandemic on poverty and income inequality in Serbia and proposes more efficient alternatives. Results indicate a decrease in poverty and inequality in 2020, compared to 2019, with much of the observed reduction due to the universal cash transfer to the adult population (UCT). However, UCT had high expenditures and contributed to the between-group inequalities increase. The second main measure: transfer to pensioners and social assistance recipients was poorly targeted, as pensioners had below-average poverty rates before the pandemic. Children and the unemployed, who had the highest pre-pandemic poverty rates, were not targeted by specific support measures. In the case of future extreme events, when formulating income support measures, the government should consider pre-crisis differences in income distribution and assess the impact of the crisis on different groups.

LEFT BEHIND IN COVID TIMES: SYSTEMS THINKING LESSONS FOR INCOME SUPPORT MEASURES IN PANDEMICS

The COVID-19 pandemic highlighted the importance of systems thinking in addressing complex problems such as economic inequality. Systems thinking examines a phenomenon that involves examining its parts and the relationships between them rather than breaking the system down into parts and focusing on one part (Ramage & Shipp, 2009). As we move toward a post-pandemic world, systems thinking can help us build a more resilient and equitable society, for example, in the form of a holistic approach to

DOI: 10.4018/978-1-6684-7285-9.ch005

health care (Butler, 2020), fair economic systems (Perugini & Vladisavljević, 2021), sustainable supply chains and adaptive infrastructures (Tsolakis et al., 2021), collaborative governance between and within countries, and more (Lin et al., 2021).

Systems thinking can help us design economic systems that prioritize sustainability and equity and are resilient in the face of shocks and disruptions. From this perspective, economic inequality is not just the result of individual actions but also a product of broader social, political, and economic systems. Lockdowns implemented to prevent contagion and the collapse of the healthcare systems have had severe consequences that disproportionately affect low-income people (Perugini & Vladisavljević, 2021). Systems thinking also suggests that the pandemic provided an opportunity to restructure our economic systems to make them more equitable and sustainable.

Systems thinking can provide valuable insights into income support measures that can address the economic inequality caused by the COVID-19 pandemic (Boes et al., 2021). These transfers can immediately relieve those in financial distress while boosting the economy by increasing consumer spending. Vulnerable groups that can benefit most from the systems approach include women and children ADDIN (Omukuti et al., 2021), youth (Cook et al., 2021), people with low levels of education, and migrants (Che et al., 2020). However, it is essential to ensure that these transfers benefit those who need them most, such as low-income households and people who have lost their jobs due to the pandemic, and that they do not place an excessive fiscal burden on the government.

Support Measures Implemented in Serbia During COVID Pandemic

Although the COVID-19 pandemic caused only a mild and temporary decrease in GDP in Serbia, the impact could differ in different groups. The COVID-19 pandemic could deepen pre-existing economic inequalities in Serbia and worsen the situation of the most vulnerable. Given the significant disparities that already existed in Serbia (Žarković-Rakić & Vladisavljević, 2021) and the fact that the most vulnerable groups were the most affected by economic losses during the pandemic, this crisis could have seriously challenged social stability in Serbia. Before the pandemic, children and the unemployed had the highest poverty rates (SORS 2021).

Perugini & Vladisavljević (2021) suggested that in the absence of proper government interventions, the situation of vulnerable groups will worsen during an economic downturn resulting from the COVID pandemic. Although the Serbian government provided generous support measures to mitigate the financial consequences of the COVID-19 pandemic (ILO 2020), those most at risk of poverty, namely children and the unemployed, were left out or not adequately addressed (Vladisavljević & Lebedinski, 2023). The main income support measure in the first year of the pandemic was the universal cash transfer (UCT) for the adult population, while those with the highest poverty rates–children-were left out. This continues the trend that policies to address child poverty in Serbia' are poorly designed (Clavet et al., 2019)

On the other hand, almost all businesses were supported to maintain permanent jobs. This means that individuals in precarious employment, such as temporary and informal workers, are left without protection. Since these jobs are usually held by vulnerable groups such as the low-skilled, youth, people from underdeveloped regions, etc., the pandemic may exacerbate existing labor market inequalities. Labor market inequalities in Serbia have already been created, at least in part, by the government's tax policies, which are not aimed at equality and higher employment, but rather at fiscal outcomes (Žarković Rakić et al., 2016; Randjelović & Vladisavljević, 2020; Žarković-Rakić & Vladisavljević, 2021)

One group targeted by one-time income transfers was pensioners. A transfer for pensioners and social assistance recipients (TPS) was the main income support measure targeting a specific group. It was primarily targeted at retirees, who were not among the most vulnerable groups because their poverty rate was lower than average before the pandemic. Thus, the unemployed, who faced unprecedented barriers to finding work (Balgova, 2021; Hensvik et al., 2021) and fewer job vacancies (Holgersen et al., 2020; Hensvik et al., 2021), as well as high poverty rates before the pandemic, were not protected by these measures.

Thus, the design of the main employment and income support measures during the COVID period in Serbia did not consider all aspects of the problem caused by the pandemic, as systems theory would suggest. While the government focused on the pandemic's health aspects and the interventions' effectiveness, less attention was paid to equity, as the two groups that were most affected were effectively excluded from the interventions or not sufficiently included.

Purpose of the Present Study

Against this background, this paper aims to understand how income inequality changed after the first year of the COVID-19 pandemic, examine the role of two income support measures (UCT and TPS) in the overall changes in poverty and income inequality, and propose alternatives to the measures implemented. In addition to children and the unemployed, we also analyze intergroup inequality changes among other vulnerable groups during the COVID-19 pandemic, such as youth, the low-educated, and women. Youth and low-educated are at risk because they are more likely to be unemployed or work in vulnerable jobs. At the same time, COVID-19 domestic work such as childcare and home-schooling has increased, typically disproportionately burdening women (Alon et al., 2020; Del Boca et al., 2020; Farré et al., 2020).

The paper contributes to the literature in several ways. First, it represents a methodological innovation by applying nowcasting projections for Serbia consistent with Eurostat projections (2021) for EU countries. Second, it analyzes in detail the impact of income support measures implemented by the government: a one-time universal cash transfer for the adult population (UCT) and a one-time transfer for pensioners and social assistance recipients (TPS). The performance of these two measures is analyzed from a systems theory perspective: by analyzing the targeting of the measure, its ability to reduce poverty and income inequality, and its cost-effectiveness at the same time. Third, the paper simulates the impact of two alternative measures that address the needs of those left behind during the pandemic: children and the unemployed.

The research relies on data from the 2020 Survey on Income and Living Conditions (SILC) and the methodology known as flesh estimates (Eurostat, 2021) to nowcast indicators of income inequality during the pandemic. While flesh estimates are available for EU countries, they are not available for Serbia. In addition, we use the microsimulation model INEQ-RS-COVID -19 to simulate the changes in inequality due to the implemented transfers and the potential impact of the proposed alternative transfers. An accurate estimate of the effect of the first year of the pandemic on poverty and inequality, as well as the impact of income support measures, will not be available until the end of 2022 when microdata from the 2021 Survey of Income and Living Conditions (SILC) will be available. This is due to the SILC methodology, as the 2021 income data (and thus the poverty and inequality indicators) refer to 2020. However, since much information is collected before this period, such as Labor Force Survey (LFS) data, administrative data, and information on social transfers, there is an opportunity to provide estimates of the inequality indicators almost two years before they are available (Eurostat, 2021).

The results suggest that the UCT has undoubtedly contributed to reducing poverty and narrowing regional disparities. However, children–the age group with the highest poverty rate before the pandemic-were excluded from the measure, worsening their relative position. In addition, there are questions about whether it was necessary to apply the transfer to all adult citizens rather than focusing on vulnerable groups, given the very high cost of the measure. On the other hand, the large-scale income support measure focused on retirees whose incomes were not at risk from the pandemic rather than focusing on the unemployed, who not only had the highest pre-pandemic poverty rates but also faced unprecedented barriers to finding work. From systems thinking perspective, to improve the program effectiveness of income support programs in the event of future similar shocks, the government should 1) take into account pre-crisis differences in income distribution and 2) assess the impact of the crisis on different groups.

This chapter is structured as follows. After this introduction, in the next chapter, we explain the methodology and data used for the research. In the third chapter, we present the results of the estimation of the poverty and inequality effects of the COVID-19 pandemic in Serbia. Chapter four provides evidence of the role that two income support measures (UCT and TPS) had in poverty and inequality changes and proposes alternatives. Chapter five concludes.

METHODOLOGY AND DATA

As mentioned in the introduction, the methodology for estimating the impact of the COVID -19 pandemic on income inequality and poverty consists of two pillars. Under the first pillar, we use LFS data from 2019 and 2020 to assess labor market changes and the corresponding impact on respondents' labor market income. Under the second pillar, we use administrative data sources and the microsimulation tool INEQ-RS-COVID -19 to assess changes in social transfers and other non-labor income sources in Serbia. In this chapter, we explain each of these two steps in detail.

Estimation of Changes in Labor Income

The methodology is based on updating SILC labor income data in three steps: 1) estimating net stock changes in employment by demographic clusters; 2) estimating the probability of losing a job and finding a job to determine which person from the cluster has the highest probability of losing/gaining a job; and 3) updating labor income variables based on the calculated probability of losing a job and changes in wages between 2019 and 2020. Considering that COVID -19 has caused structural changes in the Serbian labor market (Vladisavljević & Lebedinski, 2022), the estimation explicitly simulates transitions between labor market statuses in the first year of the pandemic rather than just reweighting the data. A similar methodology has been used in a number of EU countries (Avram et al., 2011; Figari et al., 2011; Fernandez Salgado et al., 2013; Rastrigina, O., et al., 2016).

In the first step, detailed quarterly LFS data are used to calculate the quarterly changes in net employment that occurred between 2019 and 2020. Net changes are analyzed in clusters based on the five characteristics. These characteristics are: age (groups 15-24, 25-54 and 55-64), gender, education group (primary, secondary and tertiary), settlement type (urban or rural) and NUTS2 regions (4 regions). If the cluster was too small to estimate changes, individuals were grouped into one cluster until the minimum number of units per cluster was 20 by grouping different age groups and keeping other characteristics constant. For example, a cluster consists of men aged 25-54 with secondary education from the Belgrade urban area. The net change in employment represents the difference in the number of people employed in this cluster in 2019 and 2020. The information from the LFS data is then applied to the SILC data to account for changes in the labor market that occurred between years.

In the second step of the analysis, the LFS panel data is used to estimate the job loss and job finding models. The estimated coefficients from the job loss and job finding models from the LFS are applied to the SILC data to find those within the cluster most likely to lose a job when the net change in employment in the cluster was negative and those most likely to find a job when the net change in employment in the cluster was positive.

The job loss and job finding model can be represented as

$$P(\Delta E_i) = \alpha + \gamma_1 P_i + \gamma_2 J_i + \varepsilon_i \tag{1}$$

where ΔE_i is the change in labor market status between the years which is regressed on the set of personal (P) (sex, age, education, marital status presence of children, household size, region and settlement) and job (J) characteristics (employment status, part-time – full time work, and sector of activity) in the year before the crisis. For a job finding model, which represents transitions from non-employment to employment, job characteristics in the year before the crisis cannot be observed, so the model is limited to determinants from the set of personal characteristics. In the job loss and job finding model e variables which are available both in LFS and SILC data are used, so that the coefficients γ_1 and γ_2 from these models, estimated in LFS, can be applied to SILC data.

As we analyze the quarterly changes in net employment for each cluster, the net changes in employment in each quarter are used, selecting those with the highest probability of job loss/gain to account for the net change in a cluster for a given quarter.

The third step is to calculate the proportion of income from the previous year in SILC that the individual lost/gained. Since there are four quarters, an individual may lose 25%, 50%, 75%, or 100% of their income depending on how many quarters the net change in their cluster was negative. Similarly, in clusters where the number of jobs has increased, 25%, 50%, 75%, or 100% of the reference year income is imputed depending on how many quarters the net stock change in that cluster was positive. The reference year income is set as 80% of the median income of that cluster in the previous year for those individuals who had nonzero income in that cluster. The 20% reduction in median income in this category takes into account the fact that the individual has a lower tenure. A similar approach is used in Eurostat (2020). Wages are also adjusted upward to account for growth in wages between 2019 and 2020. To calculate the 2020 increase in reference income, wages are increased for each employed person in the data. For each person, the increase is equal to the wage increase in the industry in which they work. The average nominal annual growth of wages in Serbia in 2020 was 9.6 percent

Estimation of the Changes in Other Income Sources

The previous section described the methodology used to simulate labor income losses and gains in the SILC data. To assess the impact of the COVID -19 pandemic on poverty and income inequality, we also need to update other income sources in SILC. Income sources such as non-labor income, pensions, and non-means-tested social transfers were simply updated using administrative data. The value of pensions and non-means-tested social transfers was updated based on the average increase in social transfers between 2019 and 2020. On the other hand, we assume that non-labor income such as rents and capital income increased at the same rate as wages (by 9.6%).

On the other hand, microsimulation techniques are used to account for changes in means-tested transfers and policy measures applied by the Serbian government to mitigate the impact of COVID -19, as well as to estimate the impact of alternative policy measures. Microsimulation techniques have been used for decades in analyzing the impact of economic and social policy reforms on inequality and poverty and other indicators of household well-being (Bourguignon and Spadaro, 2006). The micro-simulation tool developed for the purposes of this research was inspired by microsimulation systems such as EUROMOD - the European microsimulation model (Sutherland and Figari, 2013) and the CEQ package (Lustig, 2018). [1]

Among means-tested social transfers, we simulate the effects of monetary social assistance and child benefits, since these two transfers are means-tested and act as automatic stabilizers in the event of an economic downturn. On the other hand, the relief measures from COVID are one-time support measures that are not present in the SILC data. The main income support measure during the first period was the universal cash transfer to the adult population of €100. In addition, pensioners and temporary benefit recipients received one-time transfers of RSD 4,000 (about €34).[2] We also use microsimulation techniques to estimate the effects of these policies.

THE EFFECTS OF COVID-19 ON POVERTY AND INEQUALITY

Changes in the Labour Market Income

Table A 1 in the Appendix shows the overall changes in the number of workers by five demographic characteristics used to form the clusters. While the average decline in the number of workers was small, the decline in the number of jobs lost was highest among workers with primary education, workers from urban areas, youth, men, and in Southeastern Serbia. Broken down by quarters and in line with the results of the study on the impact of COVID-19 pandemic on the labor market (IES, 2022), there was an increase in employment of almost 80,000 workers in Q1 compared to the same quarter of the previous year (i.e. an average increase of 570 workers per cluster), which represents a continuation of the previous positive trends in the labor market in Serbia (Table A 2).

The results of the estimation of the models are presented in Table A 3 in the Appendix and largely in line with the expectations. The likelihood of job loss, ceteris paribus, was higher among younger workers, men, workers with primary or secondary education, and those with young children. The likelihood of losing a job was lower among married people than among all other groups and in larger households. By status, the likelihood of losing a job was higher if one was a temporary or informal worker (either self-employed or employed) than if one was a permanent or self-employed worker, and also higher if one worked part-time. When broken down by sector, the likelihood of losing a job was higher in Construction, Transport, Food and accommodation services, and Arts, entertainment, and recreation, and lower in the government sector. On the other hand, the job finding model suggests that the likelihood of finding a job during the pandemic was lowest among 15- to 19-year-olds and the oldest members of the working-age population. Women, those with primary education, and single people were less likely to find a job during the first year of the pandemic, while the same was true for those from smaller households.

The results in Table A 4 and Table A 5 represent the results of the simulation of the labour income lusts and gains. For about 88.5 percent of the employed, there is no loss of income. According to the simulations, about 6.6% of the employed lost 25% of their labor income, 3% lost 50% of their income, 1.4% lost 75%, and 0.6% of the employed lost all of their income in 2019. On the other hand, about 75.4% of the unemployed would have no labor income, while 24.6% of them would have some labor income in 2020. Of these, 13.5% would earn reference income for one quarter (i.e., 25% of annual reference income), 6.3% for two quarters, 3.3% for three quarters, and 1.5% would earn reference income for the entire year. As explained in the methodology section, the reference income is calculated based on the median income of the employed members of the cluster.

Table 1 shows the simulated change in labor income (including income from employment, self-employment, and agriculture) for each group. Although all groups show an increase in total net earnings, their trends differ substantially. By age group, the increase in earnings is smallest among younger people - by 5.6%, which is consistent with the observed negative change in employment. On the other hand, the increase is highest in the prime-age group (between 25 and 54 years old), – by 10.4%, as well as for older workers. As mentioned earlier, there was an increase in employment among older workers, while there was stagnation among prime-age workers. Thus, the sharpest increase among prime-age workers is not the result of their increasing employment, but of the fact that workers in this group, who were most likely to lose their jobs, were more likely to have lower wages than older workers. Similarly, men experienced a slight decrease in employment and women a slight increase, which results in slight difference in their income changes.

On the other hand, differences in employment trends have largely determined differences in changes in average earnings by education groups. Namely, since individuals with primary education experienced the largest decline in employment, their total wages also increased the least – by only 3.5%. On the other hand, the increase is highest for individuals with secondary education, as they experienced a 10.3% increase, which is the result of both annual wage increases and the increase in the number of employees with secondary education. Similarly, urban workers experienced a significant decrease in employment and lower income increase. Finally, regional differences in the increase in total income almost entirely reflect differences in net changes in employment, except in Vojvodina, where the wages increase is the lowest, while the region experienced only a slight decline in net employment.

Changes in Non-Labour Income, Social Transfers, and Pensions

Table A 6 in the appendix presents the coefficients used to uprate the values of pensions and social transfers that are non-means tested, which are based on the administrative data. Due to lack of information for the uprate of non-labour income sources such as income from capital and rent we used the growth rate of wages in the formal employment of 9.6%.[3] For the means tested benefits: child allowance and monetary social assistance we used microsumulation techniques to calculate the changes between the years.

Table 2 presents the results of changes in the various sources of income used in the calculation of disposable income. The largest increases in income were in labor and capital income, which increased by about 9%. As we saw in the previous chapter, the income growth was due to the growth in wages in 2019/20, as there were no significant changes in employment during that period. Of the other sources of income, pensions grew by an average of 5.3%, while social transfers (benefits) grew by an average of about 4.3%. Since income from labor and capital accounts for the largest share of household income, the growth rate of total net disposable income is much closer to the growth rate of income than that of the other income sources. On average, net disposable income grew by about 7.7%.

Table 1. Average change in labor earnings (pop. 15-64, without unpaid family members)

	2019	2020	change
Age group			
16-24	105,010	110,850	5.6%
25-54	436,741	481,996	10.4%
55-64	400,073	430,487	7.6%
Gender			
Male	457,740	498,392	8.9%
Female	295,548	322,924	9.3%
Education group			
Primary	175,517	181,674	3.5%
Secondary	349,222	385,110	10.3%
Tertiary	627,781	679,985	8.3%
Settlement type			
Urban	427,016	459,362	7.6%
Rural	305,797	342,681	12.1%
NUTS2 regions			
Belgrade	523,567	568,417	8.6%
Vojvodina	304,360	328,549	7.9%
West Serbia	378,674	418,339	10.5%
South-East Serbia	306,994	333,946	8.8%
Total	**378,465**	**412,628**	**9.0%**

Source: Simulation based on the SILC data.

Table 2. Changes in net disposable income and its components

Income source	2019 data		2020 simulated data		2020/2019 change
	Amount (in RSD)	percent	Amount (in RSD)	percent	
Labour / capital income	660,148	70.7%	718,414	71.5%	9.0%
Pensions	217,356	23.3%	228,769	22.8%	5.3%
Social transfers (benefits)	50,153	5.4%	52,316	5.2%	4.3%
Net transfers between households	5,565	0.6%	5,565	0.6%	0.0%
Net disposable income	933,221		1,005,064		7.7%

Notes: Own calculation based on the 2019 SILC data and simulated 2020 data. N(households) = 2,388,975.

On the other hand, the COVID relief measures are one-off assistance measures which have not existed in SILC data and we had to simulate their introduction within the microsimulation package. Main income support measure during the first period was the universal cash transfer to adult population (UCT), in the amount of 100€. According to the Minister of Finance about 6.15 million people have received the transfer, with estimated of about 615 million euros (or about 72 billion RSD). Additionally, pensioners and monetary social assistance recipients received one-off transfer (TPS) in the amount of 4,000 RSD (about 34 euros), with the estimated cost of this measure was about 58 million euros.

Table 3 shows the results of simulating these one-time transfers and their impact on household disposable income. According to our estimates, households received on average about 28,400 RSD (about 250 EUR) from the UCT. This means that, on average, about 2.5 adult household members received the UCT. The total cost estimated in our data is RSD 68 million (or about EUR 575 million), which is consistent with the cost estimated by the Ministry.

Table 3. Income support measures implemented during COVID-19 pandemic

	Average amount (in RSD)	Total cost in mln RSD	Total cost in mln EUR
Net disposable income before COVID 19 transfers	1,005,064		
Universal cash transfer to adult population (UCT)	28,395	67,835	574,9
One-off transfer to pensioners and monetary social assistance recipients (TPS)	2,577	6,156	52,19
Net disposable income after COVID 19 transfers	1,036,036		

Notes: Own calculation based on the 2019 SILC data and simulated 2020 data. N(households) = 2,388,975.

On the other hand, the one-time TPS resulted in an average increase in disposable income of RSD 2,577, and the estimated cost of this program is RSD 52.2 million, according to our data. Overall, the average annual net disposable income increased by about RSD 31 thousand or by 3.3% due to the implementation of these two programs. Overall, the estimated net disposable income in 2020 (through simulation) is about 11% higher than that of 2019.

Impact of COVID 19 Pandemic on Poverty and Inequality

According to SORS (2021), poverty rate[4] in Serbia in 2020 stood at 21.7 percent. However, we should note that SILC 2020 use 2019 as the reference year and that thus this poverty rate refers to 2019[5] and in our analysis we will refer to these indicators as indicators for 2019. Table 4 compares 2019 poverty indicators with poverty indicators based on our 2020 simulations. The data show that the poverty rate decreased from 21.7% in 2019 to 20.2% in 2020, in line with the observed increase in net disposable income, including the impact of two one-time transfers applied in the period COVID -19.

Table 4 also indicates how different groups have been hit by the pandemic. The largest decline in the poverty rate by age group occurred in the 55-64 group, where the poverty rate declined by about 3.2 percentage points (p.p.). This decline is likely due to a substantial increase in the number of employed persons in this age group, which has already been described in more detail in section 3.1, as well as an increase in pensions and social transfers, since this group had one of the highest poverty rates before the pandemic. The smallest decrease in the poverty rate occurred among minors (ages 0-17), by only 0.6 percentage points, almost one percentage point less than for the population as a whole. This is probably because they were not eligible for two major social transfers during the COVID -19 pandemic, as

UCT was reserved for the adult population and TPS for pensioners and recipients of monetary social assistance. Considering that this group had the highest poverty rate before the pandemic, this indicates that the relative position of this group worsened, as the difference between the average poverty rate and the poverty rate of the group increased from 2.5% (24.2% versus 21.7%) to 3.4% (23.6% versus 20.2%). Thus, although the poverty rate in this group has decreased, its relative position compared to the population average has worsened. The same is true for youth (18-24), but to a much lesser extent, as the decline in the poverty rate for this group was 1.2 percentage points. Further evidence for this argument can be found in Table 5, which shows that inequality between age groups, as measured by the generalized entropy index, increased from 2019 to 2020.

Table 4. Changes in the at risk of poverty (AROP) rate (population: all)

	2019	2020 (simulated data)	Change in AROP rate (percentage points)
Total	21.7%	20.2%	1.5
Age group			
0-17	24.2%	23.6%	0.6
18-24	23.6%	22.4%	1.2
25-54	19.6%	18.0%	1.6
55-64	23.0%	19.8%	3.2
65+	22.0%	20.9%	1.1
Gender			
Men	21.3%	19.7%	1.6
Women	22.1%	20.7%	1.4
Region			
Belgrade	11.3%	10.3%	1.0
Vojvodina	19.7%	18.2%	1.5
West Serbia	27.1%	25.8%	1.3
South-East Serbia	29.3%	27.0%	2.3
Settlement			
Urban	16.1%	15.7%	0.4
Rural	30.7%	27.5%	3.3

Notes: Own calculation based on the 2019 SILC data and simulated 2020 data. Poverty rate calculated a share of individuals below the poverty threshold (60% of the median equvalized net disposable income for each year).

While there were no gender differences in the reduction in the poverty rate, the data suggest that residents of Southeastern Serbia experienced the largest decline in the poverty rate. Since the increase in labor income was at the average national level, the increase in social transfers and the introduction of COVID -19 are likely responsible for the largest decrease in poverty rates. Due to the higher increase in employment and labor income in rural areas, the decrease in poverty was much more significant in rural areas than in urban areas. Since residents of Southeastern Serbia and rural populations had the highest poverty rates, this suggests that after the first year of the COVID -19 pandemic, income inequalities between regions and settlements decreased. It should also be noted that much of the observed decrease in poverty rate of the final simulated income from 2020 is due to UCT, as presented in Table A 7 in the appendix, which splits the decreases in poverty to pre-transfer effects and effects of each transfer.

Table 5 presents the overall indicators of general income inequality and the indicators of inequality between groups. All indicators show that the simulated income distribution for 2020 is more equal than that before the pandemic. In the next chapter, we will take a closer look at the reasons for this decline in inequality.

Table 5. Changes in the overall inequality indicators and between group inequality

	p90/p10	**p90/p50**	**p10/p50**	**p75/p25**	**Gini**
Inequality indicators					
2019	4.631	1.931	0.417	2.116	0.333
2020 (simulated data)	4.237	1.881	0.444	2.060	0.320
	Age groups	Gender	Region	Settlement	
Changes in the between group inequality (general entropy GE(2) index)					
2019	0.0015	0.0001	0.0197	0.0095	
2020 (simulated data)	0.0016	0.0001	0.0187	0.0080	

Notes: Own calculation based on the 2019 SILC data and simulated 2020 data. N(households) = 2,388,975.

Similar to the conclusions from the poverty analysis, the results suggest that income inequality is decreasing across groups in different regions and between urban and urban-rural areas. While gender differences are not affected by COVID -19, differences between age groups have increased, albeit only slightly.

THE ANALYSIS OF IMPLEMENTED AND ALTERNATIVE INCOME SUPPORT MEASURES

As mentioned earlier, the two main income support measures were a one-time universal cash transfer for the adult population (UCT) of €100 and a one-time transfer for pensioners and welfare recipients (TPS) of RSD 4,000 (about €34). Although we analyze individual welfare, impacts are calculated based on household disposable income (rather than individual income). Therefore, policies that do not target a specific population group may also have "spill-over effects" on those groups through other household members.

Two hypothetical policies that could have been implemented in the first year of the COVID-19 pandemic are analyzed and their fiscal and poverty effects are examined. First, since it was already known that children and youth were at the highest risk of poverty, the decision to exclude the most vulnerable group from UCT was questionable. Therefore, as a first measure, we examine the impact of UCT if it were applied to all citizens and not only to the adults. Second, UCT has often been criticized by the public as fiscally irresponsible, and it would work much better if it were targeted to a particularly vulnerable population such as the unemployed and welfare recipients. On the other hand, TPS targeted retirees who had below-average poverty rates before the pandemic (SORS, 2021), while the unemployed, who had the highest poverty rates before the pandemic and had difficulty finding a job during the pandemic were not targeted by additional assistance. Thus, in this chapter, we examine the effects of two alternative policies that address these two issues:

1. Universal cash transfer for all citizens (including those under 18) – UCT-ALL
2. Transfer to unemployed and social welfare recipients - – TUS

We examine the fiscal cost, targeting, impact on poverty and inequality and compare the fiscal cost and the impact on poverty to arrive at a cost-benefit measure with which we can compare them.

Table 6 shows the fiscal costs of two implemented and two proposed programs in this study. According to expectations, UCT-all would have higher costs than UCT-more costs–by about 113 million euros. On the other hand, TUS costs are significantly lower, with total estimated expenditures of 112.3 million euros.

Table 6. Fiscal costs of implemented and proposed measures

	Average amount (in RSD)	Total cost in mln RSD	Total cost in mln EUR
Universal cash transfer to adult population (UCT)	28,395	67,835	574,9
One-off transfer to pensioners and social assistance recipients (TPS)	2,577	6,156	52,1
Universal cash transfer to all citizens (UCT-All)	33,934	81,068	687,0
Universal cash transfer to unemployed and social assistance recipients (TUS)	5,549	13,256	112,3

Notes: Own calculation based on the 2019 SILC and simulated 2020 data. N(households) = 2,388,975.

Figure 1 represents the targeting of two implemented and two alternative measures, i.e., the share of the transfer that was directed towards each of the pre-transfer income deciles. The analysis shows that UCT and UCT-All were distributed equally across all deciles and were therefore distributionally neutral. On the other hand, TPS was poorly targeted, with only 7% of the total transfer going to people in the lowest decile. However, half of the transfer remains in the lower parts of the income distribution, with the highest gains in the third and fourth quintiles. On the other hand, TUS performs very well from a targeting perspective, with 22.5% of the transfer going to the bottom decile and a total of 37.5% going to those in the poorest 20% of the population (which roughly corresponds to those at risk of poverty). Moreover, the bottom half of the income distribution would receive about 70% of the transfer. Therefore, unlike other measures, TUS can be considered a progressive transfer.

To analyze the impact on poverty of the two policies implemented by the Serbian government and two alternative policies, we take a different approach than in the previous section. Table 7 presents the poverty analysis based on the absolute rather than the relative poverty line. More specifically, we first analyzed disposable income before COVID 19 transfers (column 1) and set the poverty line at 60 percent of the median equivalized income of that income. We then analyzed income changes by adding income from UCT, TPS, UCT_All, and TUS separately to disposable income before COVID 19 transfers, while leaving the poverty line at its pre-transfer level. We are interested in the decline in poverty rates caused by UCT and TPS separately, apart from other factors that may have affected income and poverty.

Figure 1. Distribution of UCT, TPS, UCT-all and TUS, by pre-transfer income deciles
Notes: Own calculation based on the 2019 SILC data and simulated 2020 data.

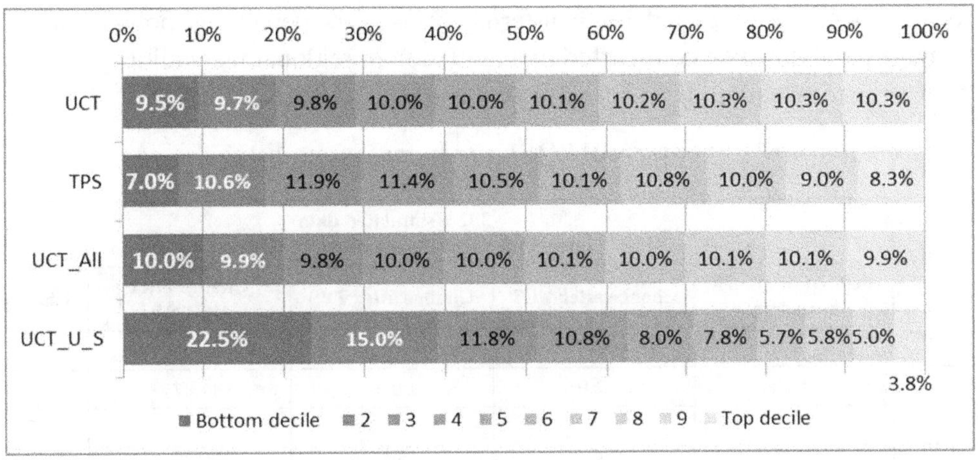

Column 1 of Table 7 (compare with Table 1, column 1) suggests that the poverty rate would decrease by about 0.4 percentage points (p.p.) if we consider net disposable income before the income support measures. These results also include the effects of the private sector job retention measures, as these effects are included in the stagnation of the employment rate. In the absence of UCT and TPC, those aged 65 and older would have faced increasing poverty, likely due to the fact that the increase in pensions was smaller than the observed increase in earnings. On the other hand, the poverty rate among youth remained unchanged.

The overall effect of the UCT on the poverty rate in absolute terms is a decrease of 2.9 percentage points. Because the UCT was applied to the entire adult population, it affected all categories, with an almost equal impact on poverty reduction in all age groups. The reduction in poverty was prominent even among children (ages 0-17), since they tend to live in households with adults. In other words, UCT reduced poverty rates for all age groups, regardless of how high poverty was before the transfer, so it did not improve the situation of the most vulnerable. On the other hand, the largest reduction in poverty after UCT is observed in less developed regions, such as Western and Southeastern Serbia, and in rural areas, where UCT has helped to reduce disparities in poverty rates between regions and settlements.

The overall effect of the UCT_all on the poverty rate is a decrease of 3.7 percentage points, a significant additional decrease compared to 2.9% after applying the UCT. The UCT_all also performs better than the UCT across all age groups, with the largest decline in the poverty rate – by 5.3 percentage points - among those who had the highest poverty rate before the pandemic - children ages 0 to 17. This is 3 percentage points higher than the UCT. Similar to the UCT, the UCT_all also shows the largest decrease in poverty in less developed regions such as western and southeastern Serbia and in rural areas. Compared to the UCT, the differences in poverty rate reduction by region and settlement are similar.

TPS had a much more moderate effect, as the reduction in poverty was 0.3 percentage points, but at the same time it was much less costly. As expected, because TPS was applied to retirees and welfare recipients, the policy had a larger impact on reducing poverty among those aged 65 and older and those aged 55 to 64 (by 0.7 and 0.4 percentage points, respectively), while the impact on other age groups was smaller (by 0.1 percentage point). Therefore, TPS helped reduce the poverty gap in the elderly population, but without the effect of closing the gap among the youth, albeit it was not its goal. On the other hand, similar to UCT, the reduction in poverty resulting from TPS was larger among people from less developed regions such as Western and Southeastern Serbia (by 0.4 percentage points) and in rural areas (by 0.6 percentage points), which contributed to a reduction in regional inequalities.

Table 7. Changes in the at risk of poverty (AROP) rate (population: all)

	2020 (simulated data) Change in the AROP rate (in percentage points)				
	before COVID 19 transfers	Change after UCT	Change after TPS	Change after UCT-All	Change after TUS
	1	2	3	4	5
Total	**21.3%**	**-2.9**	**-0.3**	**-3.7**	**-1.1**
Age group					
0-17	24.2%	-3.4	-0.2	-5.3	-1.2
18-24	22.7%	-2.8	-0.2	-3.3	-1.2
25-54	19.2%	-2.9	-0.1	-3.7	-1.1
55-64	21.7%	-3.3	-0.4	-3.7	-1.4
65+	22.4%	-2.5	-0.7	-2.7	-0.6
Gender					
Men	20.9%	-3.0	-0.3	-3.8	-1.1
Women	21.8%	-2.8	-0.3	-3.7	-1.0
Region					
Belgrade	10.8%	-2.6	-0.2	-3.4	-0.5
Vojvodina	19.3%	-1.8	-0.2	-2.6	-0.6
West Serbia	27.3%	-3.6	-0.4	-4.3	-1.6
South-East Serbia	28.6%	-3.9	-0.4	-4.8	-1.8
Settlement					
Urban	16.4%	-2.2	-0.2	-3.0	-0.7
Rural	29.3%	-4.1	-0.6	-5.0	-1.7

Notes: Own calculation based on the 2019 SILC data and simulated 2020 data. Poverty rate calculated a share of individuals below the poverty threshold. Poverty threshold is set at 60% of the median equvalized net disposable income of the pre-transfer net disposable income.

Table 8. Changes in the overall inequality indicators

	p90/p10	p90/p50	p10/p50	p75/p25	Gini
2019	4.631	1.931	0.417	2.116	0.333
Before transfers	4.490	1.912	0.426	2.117	0.329
After UCT	4.263	1.888	0.443	2.067	0.321
After TPS	4.464	1.909	0.428	2.112	0.328
After UCT_All	4.196	1.870	0.446	2.059	0.318
After TUS	4.375	1.904	0.435	2.099	0.326

Notes: Own calculation based on the 2019 SILC data and simulated 2020 data.

On the other hand, TUS would reduce poverty by 1.1 percentage points, but at the same time was significantly less costly than UCT or UCT-All. As expected, because TUS was applied to the unemployed and welfare recipients, it had a larger impact on the working-age population than on those 65 and older. Therefore, TUS did not reduce the poverty gap for the elderly population, however this was not the goal of the policy. On the other hand, similar to UCT, the reduction in poverty due to TUS was greater (by 1.7 percentage points) for people in less developed regions, such as Western and Southeastern Serbia, and in rural areas, which contributed to the reduction in regional inequalities.

From the inequality perspective, UCT improved all the inequality indicators analyzed, especially the p90/p10 ratio, which decreased by 0.227, with the Gini coefficient decreasing by about 0.8 percentage points (Table 8). On the other hand, TPS had no significant impact on inequality indicators. From the inequality perspective, UCT_all means a further improvement of all analyzed inequality indicators compared to UCT, especially the p90/p10 ratio, which decreased by another 0.7 points, and the Gini coefficient, which decreased by another 0.03%. On the other hand, although the TUS targets the unemployed and welfare recipients, it performs worse than the UCT on inequality indicators.

Cost-Benefit Analysis of the Implemented Measures

The previous chapters analyzed the cost, targeting and ability to reduce poverty and inequality of the two measures implemented by the Serbian government during the pandemic. These two measures-universal cash transfers to the adult population (UCT) and transfers to pensioners and social assistance recipients (TPS) –are analyzed in comparison with two alternative measures that could have been implemented: Universal cash transfer to all citizens, including those under 18 (UCT-ALL) and transfer to the unemployed and social assistance recipients (TUS). In this chapter we summarize the results from the previous chapters and introduce new cost-benefit indicators which refer to efficiency of the policies in lowering poverty and inequality: Poverty and inequality reduction per billion euros spent[6].

Table 9 summarizes key indicators on costs, targeting, poverty and inequality reduction, and effectiveness. We first compare two programs targeting the total population, UCT and UCT-All, and then compare programs targeting vulnerable groups, TPS and TUS. Finally, at the end of this section, we compare the effectiveness of these two groups of programs.

According to the Serbian government, UCT was the financial support for all adult citizens of the Serbian population in the face of economic uncertainty, with the additional goal of maintaining aggregate demand. Given that children (0/17) were at the highest risk of poverty before the pandemic, the decision to exclude them from assistance was surprising, especially since many of them would have to start online schooling in the face of the COVID -19 crisis, which would require additional resources, while at the same time their parents would have to reduce their labor supply to take care of their children. In addition to this equity argument, we also compared the performance of the universal cash transfer if it were paid only to adults (UCT) and if it were paid to all citizens, including minors (UCT-All). Our results clearly show that UCT-All outperforms UCT in terms of poverty and inequality reduction and program efficiency. According to our calculations, UCT-All would reduce overall poverty by 3.7 percentage points and child poverty by 5.3 percentage points, compared with the UCT's reductions of 2.9 percentage points and 3.4 percentage points (Table 9, row Poverty reduction). Better performance of UCT-All is not due the higher costs of this program (687 million vs. 574 million euros), as even when costs are accounted for UCT-All's unit decrease of poverty remains higher than the one for UCT (Table 9, row Poverty reduction per billion euro). Similarly, UCT-All would reduce inequality more than UCT and higher efficiency of UCT-All again remains when costs are accounted for (Table 9, row Inequality reduction per billion euros). Since both programs are universal programs, they do not differ significantly in targeting since transfers are distributed to all deciles of the income distribution.

Table 9. Cost benefit analysis of the implemented and proposed measures

	UCT	TPS	UCT-All	TUS
Total cost	574.9	52.1	687	112.3
Targeting				
to bottom decile	9.5%	7.0%	10.0%	22.5%
to bottom half	49.0%	51.4%	49.8%	68.0%
Poverty reduction				
All	-2.9	-0.3	-3.7	-1.1
Children	-3.4	-0.2	-5.3	-1.2
Inequality reduction				
P90/p10	-0.227	-0.026	-0.294	-0.115
Gini	-0.008	-0.001	-0.011	-0.003
Poverty reduction per billion euro				
All	-5.0	-5.8	-5.4	-9.8
Children	-5.9	-3.8	-7.7	-10.7
Inequality reduction per billion euro				
P90/p10	-0.395	-0.499	-0.428	-1.024
Gini	-0.014	-0.019	-0.016	-0.027

Notes: Own calculation based on the 2019 SILC data and simulated 2020 data.

Second, we compare two programs that focus on helping vulnerable groups: transfer to pensioners and social assistance recipients (TPS), introduced by the government, and transfer for the unemployed and social assistance recipients (TUS), which we propose here. Our results suggest that TUS is a more efficient program than TPS in terms of targeting, poverty and inequality reduction, and cost-effectiveness, although these two programs are not mutually exclusive and could be used simultaneously because they target different vulnerable groups. TUS targets the unemployed, while TPS targets retirees, while both groups also target welfare recipients. TUS has better targeting, as 22.5% of the total amount of assistance goes to the bottom 10% (compared to 7% for TPS) and the bottom half of the distribution would receive 68% of the assistance (compared to 51.4% for TPS). TUS also outperforms TPS in reducing poverty inequality, both in absolute numbers and in relative numbers when compared to its cost.

Finally, because UCT has often been criticized by the public as fiscally irresponsible, universal programs are compared to programs that focus more on vulnerable groups from a targeting and cost-effectiveness perspective. The analysis shows that programs that focus on vulnerable groups, such as TUS, are more cost-effective and have better targeting than universal programs such as UCT or UCT-All. The efficiency indicators of poverty and inequality reduction per billion euros are twice as high for TUS than for universal programs like UCT for almost all poverty and inequality indicators. In other words, the unit change in reducing poverty and inequality through a policy targeted at vulnerable groups such as TUS can be achieved at about half the cost of a universal program (UCT). On the other hand, UCT or UCT-All, because they target the entire population, are better in terms of reducing poverty and inequality overall, although they have much higher costs.

CONCLUSIONS AND POLICY IMPLICATIONS

The COVID-19 pandemic had a significant impact on the global economy and highlighted existing economic inequalities that were exacerbated by the crisis. Systems theory provides a helpful framework for understanding the complex interrelationships among various economic, social, and political factors that contribute to these inequalities. The objective of this study is to 1) assess changes in poverty and income inequality during the first year of the pandemic COVID -19, 2) assess targeting and equity and conduct a cost-benefit analysis of the financial assistance interventions implemented, and 3) propose alternative financial assistance interventions that could better meet the needs of the population during the first year of COVID-19.

The results show that income is more evenly distributed in 2020 than in 2019, with the at-risk-of-poverty rate decreasing from 21.7% in 2019 to 20.2%, and inequality indicators also decreasing (Gini by 1.3 percentage points and p90/p10 index by 8.5%). Moreover, regional differences in income inequality were reduced after the first year of the pandemic. However, inequality between groups increased for age groups and education levels. The smallest decline in poverty was among children (up to age 17), who had the highest poverty rate before the pandemic. Their poverty decline was one percentage point lower than in other groups since they were the only group not receiving additional income support during this period. Their poverty rate was reduced, but their relative position compared with the population average worsened.

The two main income support measures implemented in the first year of the pandemic were the Universal cash transfer for the adult population (UCT) and the Transfer to pensioners and social assistance recipients (TPS). While the UCT undoubtedly helped reduce poverty and narrow regional disparities in poverty rates and disparities among educational groups, it had two negative aspects. First, the decision to exclude children from UCT (or any other assistance) is highly questionable, as they had the highest poverty rates before the crisis, and home-schooling likely put additional strain on household budgets. Second, the UCT generated high expenditures, and questions arose about applying the transfer to all adult citizens rather than focusing on vulnerable groups. On the other hand, TPS was much less costly (about tenfold), reduced old-age poverty, and reduced regional disparities in poverty rates. However, TPS could have been better targeted, with only 7% of the transfer going to the bottom 10% of income earners.

From a systems theory perspective, in this paper, we show how political factors (the government), by failing to consider various social and economic factors (pre-existing inequalities and risks faced by different groups during crises), created income support policies that, contrary to their purpose, increase rather than reduce disparities between groups. To improve the program effectiveness of income support programs in the event of future similar shocks, the government should 1) consider pre-crisis differences in income distribution and 2) assess the impact of the crisis on different groups.

ACKNOWLEDGMENT

This paper is a result of the research project "Social Stability in Serbia Challenged? Pandemics, Economic Losses, Inequality, and Policy Responses - INEQ RS COVID-19", no. 7552225, financed by the Science Fund of the Republic of Serbia, within the framework of Special Research Program on COVID-19. Work of Marko Vladisavljević is further supported by the Ministry of Science, Technological development and Innovation of the Republic of Serbia.

Correspondence concerning this chapter should be addressed to Marko Vladisavljević, Institute of Economic Sciences, Zmaj Jovina 12, Belgrade, Serbia. Email: marko.vladisavljevic@ien.bg.ac.rs;

REFERENCES

Alon, T., Doepke, M., Olmstead-Rumsey, J., & Tertilt, M. (2020). *The impact of COVID-19 on gender equality* (Issue w26947). National Bureau of economic research.

Balgova, M. (2021). Job Search during a Pandemic Recession: Survey Evidence from the Netherlands". *IZA Discussion Papers, 14180.*

Boes, S., Sabariego, C., Bickenbach, J., & Stucki, G. (2021). How to capture the individual and societal impacts of syndemics: The lived experience of COVID-19. *BMJ Global Health, 6*(10), 006735. doi:10.1136/bmjgh-2021-006735 PMID:34706878

Bourguignon, F., & Spadaro, A. (2006). Microsimulation as a tool for evaluating redistribution policies. *The Journal of Economic Inequality, 4*(1), 77–106. doi:10.100710888-005-9012-6

Butler, S. M. (2020). After COVID-19—Thinking differently about running the health care system. *JAMA Health Forum, 1*(4), 200478–200478. doi:10.1001/jamahealthforum.2020.0478 PMID:36218622

Che, L., Du, H., & Chan, K. W. (2020). Unequal pain: A sketch of the impact of the Covid-19 pandemic on migrants' employment in China. *Eurasian Geography and Economics, 61*(4–5), 448–463. doi:10.1080/15387216.2020.1791726

Clavet, N. J., Tiberti, L., Vladisavljević, M., Rakić, J. Ž., Anić, A., Krstić, G., & Ranđelović, S. (2019). Reduction of child poverty in Serbia: Benefit or employment strategy? *Economics of Transition and Institutional Change, 27*(3), 615–645. doi:10.1111/ecot.12197

Cook, J., Threadgold, S., Farrugia, D., & Coffey, J. (2021). Youth, precarious work and the pandemic. *Young, 29*(4), 331–348. doi:10.1177/11033088211018964

Del Boca, D., Oggero, N., Profeta, P., & Rossi, M. (2020). Women's and men's work, housework and childcare, before and during COVID-19. *Review of Economics of the Household, 18*(4), 1001–1017. doi:10.100711150-020-09502-1 PMID:32922242

Eurostat. (2020). *Flash estimates of income inequalities and poverty indicators for 2019.* https://ec.europa.eu/eurostat/documents/7894008/8256843/Flash-estimate-of-income-inequalities-and-poverty-indicators-experimental-results-2019.pdf

Eurostat. (2021). *Nowcasting income indicators during the pandemic (FE 2020) Methodological developments*. https://ec.europa.eu/eurostat/documents/7894008/8256843/Methodological-note-2020.pdf/9a70fb55-ceb7-d25a-1b31-ab0c030095d2?t=1625228771763

Farré, L., Fawaz, Y., González, L., & Graves, J. (2020). *How the COVID-19 lockdown affected gender inequality in paid and unpaid work in Spain*.

Figari, F., Salvatori, A., & Sutherland, H. (2011). Economic downturn and stress testing European welfare systems. *Research in Labor Economics, 32*, 257–286. doi:10.1108/S0147-9121(2011)0000032012

Forsythe, E., Kahn, L. B., Lange, F., & Wiczer, D. (2020). Labor demand in the time of COVID-19: Evidence from vacancy postings and UI claims. *Journal of Public Economics, 189*, 104238. doi:10.1016/j.jpubeco.2020.104238 PMID:32834178

H., S., & Figari, F. (2013). EUROMOD: the European Union tax-benefit microsimulation model. *International Journal of Microsimulation, 6*(1), 4–26.

Hensvik, L., Le Barbanchon, T., & Rathelot, R. (2021). Job search during the COVID-19 crisis. *Journal of Public Economics, 194*, 104349. doi:10.1016/j.jpubeco.2020.104349

Holgersen, H., Jia, Z., & Svenkerud, S. (2020). *Labor demand during the covid-19 crisis in norway: Evidence from vacancy posting data*.

I.L.O. (2020). *COVID-19 and the World of Work Rapid Assessment of the Employment Impacts and Policy Responses*. International Labour Organization.

Lin, X., Yang, H., Wu, Y., Zheng, X., Xie, L., Shen, Z., & Hu, S. (2021). Research on international cooperative governance of the COVID-19. *Frontiers in Public Health, 9*, 566499. doi:10.3389/fpubh.2021.566499 PMID:33996704

Lustig, N. (Ed.). (2018). *Commitment to equity handbook: Estimating the impact of fiscal policy on inequality and poverty*. Brookings Institution Press.

O.E.C.D. (2013). *Framework for integrated analysis", in OECD Framework for Statistics on the Distribution of Household Income, Consumption and Wealth*. OECD Publishing. https://doi.org/doi:10.1787/9789264194830-11-en

Omukuti, J., Barlow, M., Giraudo, M. E., Lines, T., & Grugel, J. (2021). Systems thinking in COVID-19 recovery is urgently needed to deliver sustainable development for women and girls. *The Lancet. Planetary Health, 5*(12), 921–928. doi:10.1016/S2542-5196(21)00232-1 PMID:34895499

Perugini, C., & Vladisavljević, M. (2021). Social stability challenged by Covid-19: Pandemics, inequality and policy responses. *Journal of Policy Modeling, 43*(1), 146–160. doi:10.1016/j.jpolmod.2020.10.004 PMID:33311816

Ramage, M., & Shipp, K. (2009). *Systems thinkers*. Springer. doi:10.1007/978-1-84882-525-3

Randjelović, S., & Vladisavljević, M. (2020). Social Welfare Effects of Progressive Income Taxation under Increasing Inequality. *Prague Economic Papers, 2020*(5), 575–599. doi:10.18267/j.pep.750

Rastrigina, O., Leventi, C., Vujackov, S., & Sutherland, H. (2016). *Nowcasting: Estimating developments in median household income and risk of poverty in 2014 and 2015* (Issue EM8/16)). Euromod Working Paper.

Rawal, V., Kumar, M., Verma, A., & Pais, J. (2020). COVID-19 lockdown: Impact on agriculture and rural economy. *Society for Social and Economic Research, 13*(2), 34.

S.O.R.S. (2021). *STATISTICAL RELEASE PD10 Survey on income and living conditions* (Issue 282 year LXXI). https://publikacije.stat.gov.rs/G2021/PdfE/G20211282.pdf

Tsolakis, N., Zissis, D., & Tjahjono, B. (2021). Scrutinising the interplay between governance and resilience in supply chain management: A systems thinking framework. *European Management Journal.*

Vladisavljević, M., & Lebedinski, L. (2023). *Effects of the COVID-19 pandemic on the labour market and the position of vulnerable groups in Serbia.* Institute of Econonomic Sciences. Belgrade https://www.library.ien.bg.ac.rs/index.php/monog/article/view/1586/1299

Žarković Rakić, J., Randjelović, S., & Vladisavljević, M. (2016). Labour market effects of social security contributions reform in Serbia. *Economic Annals, 61*(208), 73–91. doi:10.2298/EKA1608073Z

Žarković-Rakić, J., & Vladisavljević, M. (2021). Tax reforms and income inequality in former Yugoslav countries: Escaping the avant-garde neoliberalism in the income tax policy. *Panoeconomicus, 68*(2), 231–252. doi:10.2298/PAN2102231Z

ENDNOTES

1. This tool represents a system of syntaxes (so-called files), programmed in the STATA software package that allows analysis of the effects of reforming existing policies or introducing new policies in the field of taxation and social benefits. The tool is available online at https://www.ien.bg.ac.rs/en/project-ineq-rs-covid-19/ (Microsimulation package INEQ-RS-COVID-19 - Version 1)
2. https://www.srbija.gov.rs/vest/en/152484/government-recommends-payment-of-full-amount-of-pensions.php
3. Income capital data are likely underestimated in SILC. According to 2019 data they represent less than 1% of the total labour and non-labour income. Therefore, we do not show this category separately from overall income in Table 6.
4. In order to simplify the presentation of the results, we will only use the terms poverty rate and at risk of poverty rate interchangeably.
5. Data for 2021 with 2020 income data will be released in late 2022.
6. We opt for billion euros reference, as the lower costs (such as per one million would be too small to be comparable.

APPENDIX

Table 10. Changes in the number of the employed between 2019 and 2020 by demographic groups (population 15-64, without unpaid family members)

	2019	2020	change
Age group			
15-24	141,713	133,960	-7,752.70
25-54	2,010,464	2,008,091	-2,373.50
55-64	471,137	477,197	6,060.30
Gender			
Male	1,479,850	1,473,290	-6,560.70
Female	1,143,464	1,145,959	2,494.80
Education group			
Primary	337,846	312,239	-25,607.20
Secondary	1,547,364	1,570,717	23,352.70
Tertiary	738,104	736,292	-1,811.30
Settlement type			
Urban	1,623,067	1,587,234	-35,833.00
Rural	1,000,247	1,032,014	31,767.20
NUTS2 regions			
Belgrade	712,434	710,164	-2,269.30
Vojvodina	723,618	721,785	-1,833.30
West Serbia	676,645	682,766	6,121.20
South-East Serbia	510,617	504,533	-6,084.50
Total	**2,623,314**	**2,619,248**	**-4,065.80**

Source: Own calculation based on the LFS data

Table 11. Descriptive statistics on the net changes in employment by clusters

	Number of clusters[1]	Average	Minimum	Maximum	Total change[2]
Q1	138	570	-9,515	11,380	78,622
Q2	139	-684	-16,103	14,107	-95,031
Q3	141	-41	-10,305	10,841	-5,779
Q4	141	29	-9,354	9,838	4,100
Total	559	-43	-16,103	14,107	-18,087

Source: Own calculation based on the LFS data

[1]The maximum number of clusters was 144 for each quarter. The lower number of clusters is due to the fact that in some cases the original cluster was too small to estimate the changes, so we joined individuals into that cluster to a bigger cluster until the minimum number of units per cluster was 20. This was typically done by grouping different age groups and keeping other characteristics constant.

[2]Total change is calculated as the product of the number of clusters and the average change in the cluster.

Table 12. Job loss and job finding models (estimates from the LFS applied to SILC data)

	Job loss model		Job finding model	
	coef	se	coef	se
Age group 15-19	omitted		omitted	
20-24	-0.448***	(0.160)	0.486***	(0.069)
25-29	-0.678***	(0.159)	0.851***	(0.078)
30-34	-0.867***	(0.162)	0.824***	(0.084)
35-39	-0.875***	(0.161)	0.804***	(0.083)
40-44	-0.855***	(0.160)	0.766***	(0.083)
45-49	-0.943***	(0.163)	0.564***	(0.087)
50-54	-0.923***	(0.163)	0.520***	(0.084)
55-59	-0.774***	(0.163)	0.383***	(0.085)
60-64	-0.510***	(0.167)	0.150*	(0.091)
Female	-0.235***	(0.040)	-0.330***	(0.035)
Education - Primary	omitted		omitted	
Secondary	-0.074	(0.051)	0.223***	(0.041)
Tertiary	-0.183***	(0.066)	0.386***	(0.063)
Marital status - Single	omitted		omitted	
Married	-0.104**	(0.051)	0.215***	(0.051)
Widowed	0.123	(0.117)	0.372***	(0.132)
Divorced	0.051	(0.076)	0.231***	(0.085)
Number of children (aged 0 to 4)	0.099***	(0.026)	-0.043	(0.027)
Number of children (aged 5 to 9)	0.021	(0.025)	-0.033	(0.025)
Household size	-0.033**	(0.015)	0.039***	(0.013)
Region/settlement: Belgrade urban	omitted		omitted	
Belgrade rural	-0.035	(0.083)	0.109	(0.088)
South-East urban	0.023	(0.067)	0.005	(0.063)
South-East rural	-0.174**	(0.070)	0.298***	(0.065)
Vojvodina urban	-0.018	(0.063)	0.231***	(0.065)
Vojvodina rural	-0.060	(0.072)	0.108	(0.069)
West urban	-0.013	(0.068)	-0.071	(0.066)
West rural	-0.182***	(0.068)	0.171***	(0.063)
Employment status - Formal self-employed	omitted		n/a	
Formal wage employed - permanent	0.049	(0.074)		
Formal wage employed - temporary	0.626***	(0.080)		
Informal self-employed	0.629***	(0.085)		
Informal wage employed	0.976***	(0.085)		
Part time job	0.302***	(0.060)	n/a	
Employment sector - Agriculture	omitted		n/a	
Industry (sectors B, C, D, E)	0.117	(0.080)		

	Job loss model		Job finding model	
	coef	se	coef	se
Construction	0.409***	(0.088)		
Trade	0.102	(0.087)		
Transport	0.278***	(0.100)		
Food and accommodation	0.503***	(0.101)		
IT finance, real estate (sectors J, K, L)	0.102	(0.115)		
Professional services (Sectors M and N)	0.145	(0.104)		
State sector (sectors O, P and Q)	-0.366***	(0.097)		
Arts, entertainment and recreation	0.334***	(0.128)		
Other services (Sectors S, T, U)	0.042	(0.089)		
Constant	-1.009***	(0.187)	-1.781***	(0.087)
Observations	14,833		8,971	

Standard errors in parentheses; *** $p<0.01$, ** $p<0.05$, * $p<0.1$

Table 13. Number of people who have lost/gained labour income

Income losses (employed)			Income gains (unemployed/inactive)		
Income loss	Number of people	Share	Income gain	Number of people	Share
0%	2,473,693	88.5	0%	938,962.02	75.4
25%	184,891.60	6.6	25%	167,928.98	13.5
50%	82,136.27	2.9	50%	77,808.75	6.3
75%	38,887.43	1.4	75%	41,255.73	3.3
100%	16,103.35	0.6	100%	19,073.42	1.5
	2,795,711			1,245,029	

Source: Own calculation based on the simulations in the SILC data

Table 14. Descriptive statistics on the net changes in employment by clusters

	Change in income (employed)		Simulated loss/gain (average)	
Status in 2019	Yearly earnings in 2019[1]	Yearly earnings in 2020	Income loss in 2020	Imputed earnings in 2020
Non-employed				171,528
Employed	547,008	577,607	165,222	.
Population size	2,795,711	2,795,711	322,019	306,067

Source: Own calculation based on the simulations in the SILC data

[1]Includes total yearly earnings calculated by summing variables py010n and py050n from harmonized SILC dataset

Table 15. Uprate factors for pensions and other social transfers

Pensions	Uprate factor	Notes
Old age pensions	1.052448	
Private pensions	1.052448	no info - used old age pensions rate
Survivors pensions	1.056386	
Disability pensions	1.049189	
Benefits		
Unemployment benefits	1.09396	
Sickness benefits	1.091432	
Education benefits	1.04141	no info - used MSA rate
Housing benefits	1.04141	no info - used CA rate

Table 16. Changes in the at risk of poverty (AROP) rate (population: all)

	2019	2020 (simulated data) Change in the AROP rate (in percentage points)			
		before COVID 19 transfers	Change after UCT (in p.p.)	Change after TPS (in p.p.)	Change after UCT and TPS (in p.p.)
	1	2	3	4	5
Total	**21.7%**	**0.4**	**1.3**	**0.2**	**1.5**
Age group					
0-17	24.2%	0.1	1.2	0.2	0.6
18-24	23.6%	0.9	1.2	0.2	1.2
25-54	19.6%	0.5	1.3	0.1	1.6
55-64	23.0%	1.3	1.9	0.4	3.2
65+	22.0%	-0.4	1.2	0.5	1.1
Gender					
Men	21.3%	0.5	1.4	0.2	1.6
Women	22.1%	0.3	1.3	0.3	1.4
Region					
Belgrade	11.3%	0.6	0.5	0.0	1.0
Vojvodina	19.7%	0.5	1.1	0.2	1.5
West Serbia	27.1%	0.1	1.8	0.3	1.3
South-East Serbia	29.3%	0.7	2.1	0.4	2.3
Settlement					
Urban	16.1%	0.2	0.8	0.1	0.4
Rural	30.7%	1.4	2.3	0.4	3.3

Notes: Own calculation based on the 2019 SILC data and simulated 2020 data. Poverty rate calculated a share of individuals below the poverty threshold. Poverty threshold is set at 60% of the median equvalized net disposable income after each phase of the formation of the disposable income.

Chapter 6
Response of the Economic System to COVID-19:
A Review of Evaluation Methods Applied in Assessing Economic Policies' Effects in Serbia

Kosovka Ognjenović
https://orcid.org/0000-0002-3768-8860
Institute of Economic Sciences, Belgrade, Serbia

Mihajlo Đukić
https://orcid.org/0000-0001-5677-330X
Institute of Economic Sciences, Belgrade, Serbia

ABSTRACT

This chapter specifically deals with the response of the Serbian economy and society to the challenges of the crisis, focusing on research conducted from 2020 to 2022. Bearing in mind that economic systems remember and learn from both good and bad experiences, it was interesting to see if, in this relatively short time, in terms of data availability, the analyses that were done offered results not only from a short-term but also from a long-term perspective. After the analysis of the available studies and the systematization of the main findings, it was concluded that the majority of these studies provide only analyses of the short-term effects without applying rigorous evaluation methods. This is an important limit in terms of their reliability, but also with regards to usefulness of the obtained results for the policy making purposes.

DOI: 10.4018/978-1-6684-7285-9.ch006

INTRODUCTION

The COVID-19 pandemic posed new challenges ahead of policymakers regarding developing adequate policy responses aimed at reducing its socioeconomic, political, health, and other adverse effects. Limited financial resources and a highly complex and rapidly changing environment, including the necessity to react promptly, represent critical issues for the successful implementation of anti-COVID-19 measures. The economic effects of the measures undertaken have been extensively analyzed in scientific papers. Although the number of studies related to the impacts of health measures is far more represented in the applied scientific literature, there is no shortage of analyses dealing with the evaluation of non-health measures. What is a characteristic of these studies and what gives them additional value is that the effects are rarely observed at the level of only one country, which is also understandable since the pandemic has been a global phenomenon. Some of the papers dealing with the policy response to the pandemic considering experiences from many countries/areas include Deb et al. (2021), Podolsky et al. (2022), Rezapour et al. (2021), etc.

The Serbian government reacted very quickly and prepared a series of measures to protect the population from the risk of contagion, allow the economy to function, and protect vital sectors, including certain categories of particularly affected citizens (Fiscal Council of the Republic of Serbia, 2020). Economic demand was also boosted by the payment of a one-time financial assistance to citizens and by measures to grant minimum wages to private sector employees (Marjanović & Đukić, 2020). The Serbian economy faced a major challenge. Even before the pandemic, the risk of poverty rate indicated that more than a quarter of the population was vulnerable (Ognjenović & Pavlović, 2021), and apart from high income inequality, which exceeded 37 Gini points, the Serbian labor market was characterized by the persistent practice of gender wage inequality (Ognjenović, 2021a). Although many businesses, especially micro, small and medium enterprises (MSME) and entrepreneurs, faced great uncertainty, most maintained pre-pandemic employment levels, helped by government measures (Adžić & Al-Mansour, 2021; Beraha & Đuričin, 2020; Vladisavljević et al., 2021). At the end of 2020, the decline in real gross domestic product (GDP) was only 1% (Fiscal Council of the Republic of Serbia, 2022a), and the labor market was characterized by an increase in inactivity (Ognjenović, 2022).

Economic impact assessments based on a counterfactual approach were rare in the pre-pandemic period. Some of them include an assessment of the impact of active labor market measures (Ognjenović, 2007). There are almost no studies dealing with the cost-effectiveness of economic and social policies in Serbia; at least they were not identified in the period when the impact of the COVID-19 pandemic was concerned. The purpose of this chapter is to analyze the methodological approaches that were used to evaluate the effectiveness of the policy measures applied in Serbia to minimize the negative socioeconomic impact of the pandemic. Thus, the main research questions include the evaluation approaches used to address specific evaluation issues, attempting to provide answers to whether the evaluation methods used were successful in answering the evaluation questions raised, where the financial support came from, who the evaluators were, and what the limitations of the evaluations conducted were.

This research contributes to the existing literature in at least three aspects. First, to the best of the authors' knowledge, it provides the first review of the applied evaluations of economic measures against COVID-19 in Serbia. Second, it provides a review of the applied evaluations focusing on the gaps between previously developed evaluation methods and the methods applied in this very specific context characterized by the necessity to combine data from different sources, including intervention logic limitations referring to providing support to the economic activity, on the one hand, and preventing social interactions, on the other. Third, obtained results could provide important contribution to the literature analyzing complex relationships between systems thinking and evaluation of system-level interventions including evaluations of interventions with system-wide effects (De Savigny & Adam, 2009). Having large implications for developing systems thinking in various fields, particularly in building effective health systems, this type of summaries could help understanding the challenges evaluators face when considering underlying assumptions about selecting and using specific methods (Gates, 2016).

The following two sections provide a policy background, summarize the economic measures implemented to mitigate the impact of COVID-19 in Serbia, and present the relevant literature, which includes some key evaluation studies and research conducted globally and in Serbia, focusing on different regional and national contexts. As mentioned in this section and given the multiple impacts that COVID-19 has had on the local economy, the focus is on socioeconomic impacts, including labor market, economic activity, and other non-health impacts. The fourth section describes the methods and data sources used for the analysis. Finally, the fifth and sixth sections present key research findings and discussions. The final section concludes the chapter and provides recommendations for policymakers and evaluation practitioners.

INSTITUTIONAL FRAMEWORK

Policy Background

Policy evaluation is of great importance in developing effective public policies. In line with evaluation theory, its importance lies in providing at least two important feedbacks to the policymakers. The first one refers to accountability which deals with resource management and providing evidence of whether the funds, staff, and properties are used as envisaged in public policy documents and intervention plans (Sawer, 1990). Accountability is a very important precondition for building responsibility, also having important role in effective policymaking *per se*, as policymakers would behave differently if knowing that their work will be monitored and assessed according to a certain criterion. The second important feedback is related to learning since learning from previous mistakes helps policymakers to improve their approach and avoid mistakes in policy implementation when conducting similar interventions in the future. This is also very important for developing learning organization culture and institutional memory. Proactive learning requires developing sophisticated views on how learning processes take place and where the eventual bottlenecks might occur (OECD, 2001). Both evaluation dimensions should be considered when designing a proper evaluation approach and objectives, methodology, as well as reporting evaluation findings to the interested stakeholders. COVID-19 environment represents a great challenge for policymakers to use their knowledge and assess the limitations of existing approaches when implementing large-scale interventions without the possibility of spending a lot of time planning. Additionally, it represents a unique opportunity for evaluators to apply different evaluation methods, provide important policy feedback and participate in developing capacities of the policymakers for dealing with similar external shocks that could take place in the future.

This chapter provides a systematic and critical review of the evaluations and research analyses aimed at analyzing research approaches and findings that focused on the effects of policy interventions developed to deal with the negative consequences of the COVID-19 pandemic. It is primarily based on the concept of mapping existing evaluations of economic and social policies implemented during the COVID-19 pandemic in Serbia. By elaborating on the known theories and evaluation methods, this chapter provides a model for estimating gaps in conducted evaluations of measures aimed at reducing the adverse effects of COVID-19 on the Serbian economy and society.

Developed systems for policy monitoring and evaluation represent important preconditions for effective public policy implementation in any field. There are at least two important groups of actors that should be considered when designing such a system that works to promote social improvement. On one side of the system, there are researchers who find sources and verification of the quality of research in papers published in prestigious scientific journals. However, journals are not a suitable channel for communicating knowledge to other stakeholders, given that they are mostly intended to be used by the scientific community. They need knowledge that can be easily translated into concrete activities, knowledge that is freed from the vocabulary of science and has concrete purpose (Zubović et al., 2020). The institutional framework for conducting evaluations in Serbia has been formalized through the adoption of the Law on Planning System in 2018 (Official Gazette of the Republic of Serbia No. 30). The Law on Planning System regulates the planning system of the Republic of Serbia stipulating regular monitoring, *ex-ante* and *ex-post* impact assessments, and other types of public policy documents evaluation. Guidelines for conducting policy evaluations have been provided by the Republic Secretariat for Public Policies representing an important methodological basis not only for the government but also for the other non-government institutions including individual researchers. Although envisaged as mandatory under the national planning system framework, evaluation practice in Serbia is still underdeveloped facing serious challenges in terms of lack of financial resources and a low level of understanding about its importance. However, research and civil society organizations have solid experience in applying evaluation methods performing complementary activities to the ones performed by state institutions. They also have larger capacities and flexibility to conduct evaluations when the need arises, as with the COVID-19 pandemic. Therefore, within this research, the focus will be on the assessments conducted by the research and civil society institutions and other non-research institutions aimed at assessing government response to the COVID-19 issues.

Since public administration reform conducted by the Serbian government strongly relies on the European Union (EU) public policy foundations, experiences in evaluating anti-COVID-19 economic policies applied within the EU support packages will be particularly taken into consideration. Unlike in Serbia, policy evaluation has been the critical concept of the EU public policy management over the last decades. Generous support aimed at reducing pandemic impact provided through EU funds has been assessed within many evaluation studies (Kamberović et al., 2020). In this chapter, there have been analyzed critical reviews, analyses, and impact evaluations of independent researchers and research institutions, as well as assessments produced by the EU authorities, member states' institutions, and other international organizations. Such analysis provided us with comparative research practices including evaluation questions and methods applied in the EU and Serbia as a candidate for accession.

Anti-COVID-19 Program in Serbia

This chapter focuses on the evaluations based on the idea to assess the effects of the economic measures implemented with the aim of reducing adverse shocks of the pandemic on economic activity. Therefore, it will be briefly outlined measures from the economic program of the Serbian government conducted over the period 2020-2021.

As per Fiscal Council calculations, Serbia provided a generous support package that was around 55% higher if compared to other Central and Eastern European (CEE) countries. The total financial injections have been estimated at 10.4% of the gross domestic product, out of which 4.8% was direct support to the enterprises (Fiscal Council of the Republic of Serbia, 2022a). Two thirds of financial support was reimbursed during the first pandemic year. The most important and financially demanding support was non-selective support to the economic entities reimbursed per employee with the only condition of retaining at least 90% of employees compared to the pre-crisis period. Financial support included reimbursement of the minimum wage for the three months period (March-May 2020) and 60% of the net wage for the following two months. Support to large enterprises was less generous since they were reimbursed 50% of the net wage per employee. In order to support consumption, the government introduced different types of one-time financial support for all citizens and targeted financial injections to the youth and retired population. Most evaluation studies regarding the impact of implemented policies were conducted during the first pandemic year (Ognjenović & Đukić, 2022).

Other support measures included loans for maintaining working capital and preserving liquidity, as well as measures aimed at avoiding serious financial difficulties, including tax payment deferrals and moratoriums on debt payments (Institute of Economic Sciences, 2022a). Research conducted by the Institute of Economic Sciences (2022b) shows that the generous government measures helped overcoming pandemic impact on poverty and income inequality in Serbia during the first pandemic year (Marjanović & Đukić, 2020). Simulation of the indicators of poverty and income inequality shows that poverty and inequality were even slightly improved in 2020 compared to 2019. The risk of poverty rate decreased to 20.2% in 2020, 1.5 percentage points lower than in 2019. However, it could be noticed the deterioration of the between-group inequality as the decrease of poverty was lower for the groups that had the lowest risk of poverty rates before the pandemic (youth and low-educated). The worsening of the situation during the pandemic is not surprising since Serbia is among the European countries with the highest rates of risk of poverty and income inequality (Ognjenović & Pavlović, 2021), whereas the model of economic development, including the labor market, strongly relies on foreign investments (Zdravković et al., 2017) which could be highly volatile in pandemic periods.

LITERATURE REVIEW

'State of the art' in analyzing evaluations of anti-COVID-19 interventions could be roughly classified under the two groups of studies. The first group refers to systematic reviews and cross-country analyses of the anti-COVID-19 policy response. The second group involves evaluation studies and papers focused on national approaches, mainly stimulus packages aimed at reducing negative non-health impacts in specific countries. Analyzing the results of these studies made us capable of detecting some of the critical issues related to the design of the evaluation studies, applied methodologies, as well as some of the most frequently obtained results. It should be noted that, despite the rising popularity of using meta evaluations and systematic review methods in social sciences, there is relatively scarce scientific evidence on evaluating overall policy responses in a crisis context. Most of the review studies analyzed partial impacts of the policy response on specific group or social phenomena, for example, children in several crises periods (Tirivayi et al., 2020), health during the financial crisis in Europe (Mladovsky, 2012), foreclosures during the global economic crisis in 2008-2009 (Immergluck, 2013), etc.

Applied methodologies were different with regards to the research objectives. Large number of studies aimed at identifying the effects of government measures on economic activity, usually through stock market returns, since they represent a solid proxy for market expectations regarding economic activity (Siddik, 2020; Ashraf et al., 2020, etc.). Another group of studies was aimed at assessing the cost-effectiveness of the applied measures, including cost-utility and cost-benefit studies, or impact on specific economic sectors (Wang et al., 2020; Shapovalova et al., 2021; Huang et al., 2020, etc.). As noted by Rezapour et al. (2021), most of the studies used cost-effectiveness method intending to capture the economic and financial effects of imposed protection measures.

Short time horizons ever since the realization of the imposed measures and lack of data were important limitations of the conducted evaluation studies. Aiming to provide a systematic review of the economic evaluations of COVID-19 interventions, Podolsky et al. (2022) analyzed the results of 70 articles to assess how the authors addressed non-health impacts and distributional concerns under their research approach. More than half of the studies were pre-print papers including at least one non-health impact analysis. Results of the study confirmed that most of the studies reported cost-effective interventions in the field of treatment, public information campaigns, quarantines, canceling public events, and physical distancing. However, they found that there were many studies facing difficulties capturing non-health impacts, mainly due to a lack of data available to quantify them. This is partly due to the relatively short time horizon since the onset of the crisis. More precise estimates of the non-health impacts are expected to be conducted as soon as more data, particularly ones related to education, environment, and economic indicators, become available. One of the significant contributions of the study that should be noted refers to the importance of capturing non-health impacts since otherwise, local policies could result in considerable misallocation of healthcare resources as well as lower effectiveness of interventions aimed at supporting the most vulnerable population groups. Another important result refers to the effects that the announcement of the measures to be imposed had on reducing panic and pessimism in the market. Deb et al. (2021) empirically examined the impact of fiscal policy measures applied in 52 countries from January 1st to December 31st, 2020. Following Jordà (2005) methodology and focusing on the effects of fiscal announcements on daily exchange rates and stock prices, as well as OECD economic activity tracker, they quantified the economic impacts of anti-pandemic measures on the main economic indicators. Authors confirmed favorable and statistically significant effects of the anti-COVID-19 measures, particularly in terms of reducing supply side shocks. The significant contribution of their study refers to the importance of considering country characteristics such as the share of public debt and the pandemic cycle. The effects of the applied measures were, however, different depending on the size and dynamics of the stimulus packages. In that context, important research contributions were provided by Siddik (2020), who aimed to identify the determinants of the stimulation packages in 168 countries. Siddik (2020) applied the Euclidean distance formula and developed a new index that confirmed that median age, hospital beds *per capita*, and total cases of COVID-19 pandemic, GDP, health expenditure, and COVID-19 government stringency index are the most important determinants of economic stimulus. Similarly, to other authors, a repeated study that would include a longer time span and more indicators would provide more robust results. In many studies, authors used pooled panel ordinary least square regression model to examine the direct impact of government actions. These studies often had limitations of the short study sample and the lack of individual country data to develop control variables; thus, the authors used country fixed effects dummy variables (Ashraf, 2020; Alfaro et al., 2020, etc.).

METHODOLOGICAL APPROACH

Concept and Related Evaluation Theories

A context that includes the social, political, and economic environment under the conditions of a crisis such as the pandemic triggered by COVID-19 may be more important than the mere fact of whether the government's interventions are such that the impact of the interventions can be assessed by applying an evaluation based on one of the well-known theoretical approaches. Nevertheless, it is important to know what theory-based assessment approaches can be used in such situations.

Most evaluation studies are based on a theoretical approach that focuses on the effects of intervention (White, 2009). Therefore, it is logical that evaluations based on a known theoretical approach contain two key components: conceptual and empirical (Leeuw, 2012; Angrist & Pischke, 2008). Conceptually, any evaluation based on a theoretical approach includes a description of the intervention behind the economic policy. Crucial to this type of evaluation, however, is whether it includes an 'evaluation question' that describes why the evaluation is being conducted and what the expected impact is. An evaluation based on a theoretical approach empirically tests and evaluates the theory to show whether the policies implemented led to the observed outcomes and the extent to which they were responsible for those outcomes.

Theory-based impact evaluation focuses on intervention theories elaborated by White (2009). It is primarily premised on the theory of change, which is based on a logical model that explains how 'things' should work to 'effect change', as Leeuw (2012) further elaborates. Counterfactual impact evaluation focuses on how much of the change was brought about by the intervention, i.e., what portion of the change, in terms of numbers, was due to the intervention, with the difference being a pre- and post-intervention state. For example, a valuable synthesis of evaluation methods that use the counterfactual history is provided by Angrist and Pischke (2008).

By explaining the known theories and evaluation methods in the following, the authors of this chapter try to establish a relationship between conducted impact evaluations and applied evaluation approaches in the study of the impact of the COVID-19 pandemic on the Serbian economic system.

Method

The authors used a two-step approach aimed at systematizing and generalizing studies to determine whether they were based on evaluation theories:

- As a first step, the authors applied the scientific review method-a meta-analysis approach-to select a sample of policy documents that met inclusion criteria.
- In the second step, the authors created a model, i.e., a schematic diagram, to estimate the gap between the known and the implemented evaluation approaches and to suggest which evaluation types should be applied depending on the implemented policies.

Using the scientific review method, as a first step, the authors estimated the gap between the types of evaluations that were conducted and that should be conducted depending on the policy measure implemented. The authors researched the latest available documents on conducted evaluations in Serbia and systematized them in terms of the main areas of expertise as follows: those focusing on labor market participants, economically disadvantaged groups, evaluation of the overall impact on the economy, evaluation of the impact on companies and other business entities, and those considering the impact on the economic and social well-being of citizens. Only slightly more than half of these studies use some assessment methods to determine the impact of current health crises. Most often, these are combinations of *ad hoc* surveys and expert opinion on the impact of an intervention applied to specific categories of citizens or businesses, while sometimes questionnaires collect general opinions on the government's response to the COVID-19 pandemic. Sophisticated evaluation methods and techniques are rarely considered evaluation methods in Serbia. Moreover, most policy evaluation reports are written based on the analysis of secondary statistical data comparing the situation before and during the COVID-19 pandemic, leaving no room for conclusions based on causal relationships.

Sample

The authors made an inventory of the assessments conducted of economic policy responses to the current crisis and, based on the documents collected, made three types of generalizations: (a) critical views, (b) reports or research papers (studies), and (c) impact assessments. The search process ended with more than 50 documents collected. The next step was to exclude comments and assessments related only to health or environmental impacts, conference papers, working papers, duplicates, etc. Review of the bulk of the documents finally yielded 31 studies that were included in the final analysis. Thus, the sample size is n=31. The authors applied the Preferred Reporting Items for Systematic Reviews and Meta-Analyses (PRISMA) approach, which generally involves three steps. The process begins with broader identification, followed by screening, and ends with selection of a sample of studies (Hansen et al., 2022).

When meta-analysis is used to obtain a systematic review (or classification) of studies, it is necessary to have a critical number of reports and scientific papers, i.e., assessments of the impact of the COVID-19 pandemic on the economy and society. It is recommended that at least three sources of data be used to create an inventory of the studies conducted (Hansen et al., 2022).

How many responses from citizens or companies did the studies reviewed have? And how many in-depth interviews with various stakeholders did the studies involve? These questions may be crucial for the relevance of the selected studies and the reliability of the results. The data in the published studies include a total sample of 5,061 businesses and entrepreneurs, 6,710 individuals, and more than 130 stakeholders who participated in the qualitative data collection process.

The following search engines were used in this chapter: Google Scholar, Scopus, the websites of the Serbian government and related state agencies, employers' associations, and universities, as well as the websites of international, i.e., multilateral, organizations such as the World Bank, the OECD, the United Nations (UN), etc. All collected evaluation reports, studies and research papers cover the period from January 2020 to December 2022.

RESULTS

Descriptive Analysis

The main characteristics of the selected evaluation studies are shown in Table 1. Most of the studies included in the analysis were completed in 2020 (16 studies in total), and a smaller number were scientific papers (3), taking into account that publication of papers in peer-reviewed journals takes more time. Of all the cited papers (9), 25.8% (8 documents) are articles in peer-reviewed journals, and one chapter was published in the collection of papers. None of these articles were published in 2022. Of the two articles published exclusively in journals on the Web of Science, one was from a business journal. Most of the collected papers refer to reports, and only one of them can be categorized as research studies with a clearly defined research methodology.

Most of these studies focus, on the one hand, on the assessment of the impact of the COVID-19 pandemic and, on the other hand, on the measures taken by the government to combat the negative consequences of the pandemic. Therefore, it was to be expected that an above-average number of these studies would deal with impact assessments on the economy and significantly fewer with assessments of the impact on the labor market or with assessments of the negative impact on socially disadvantaged groups. It is also important to note that many of these studies do not only deal with one-sided impact assessments, but look at the impact from several sides, e.g., by combining the impact on the overall economy and on businesses.

Only a few evaluation reports involve stakeholders. These are reports in which qualitative evaluations were conducted (9 reports). In most cases, the evaluators are from academic institutions (one-third of the studies) and civil society organizations (one-quarter of the studies). Funding of evaluations by international organizations (38.7%) and the Serbian government (25.8%) is therefore more widespread. The EU is also an important factor in the structure of providers of financial support (12.9%). At least indirectly, the government is involved as the second most important source of funding, as some of the evaluation work was carried out by research institutions directly supported by the Ministry of Science.

Key Findings

A typical evaluation question is whether the COVID-19 pandemic or government actions against COVID-19 have affected the economy, business activity, or welfare of citizens. The main findings are presented in Table 2. As can be seen, all the selected studies asked an evaluation question prior to analysis. Only two reports used a public barometer to collect the attitudes of citizens and businesses, asking their opinions about the negative impact of the pandemic. Considering the different categories of citizens and businesses for which the impact of the pandemic or of the measures taken by the government is assessed, several groups of expected assessment effects are distinguished: on the overall economy (n=7 or 22.6% of the total), on the business sector only (n=9 or 29.0% of the total), on economically disadvantaged populations (n=9 or 29.0% of the total), and on multiple effects (n=5 or 16.1% of the total), while one of the evaluation studies looks at the lives of citizens in the context of the COVID-19 pandemic from the local perspective.

The evaluation approach used implies the use of different techniques, which can be mainly described as those that can quickly produce results. Due to the limited freedom of movement and the maintenance of physical distance in the first year of the pandemic, the surveys were mainly online, using the technique of computer-assisted telephone interviewing (CATI). The Zoom platform is a mainly used communication channel for one-on-one interviews. These techniques proved to be effective, so they continued to be used in subsequent years (see Table 2) and could even be described as a good legacy of the pandemic.

This part of the analysis will yield narrow results because the selected studies rarely include elasticities or marginal effects from the estimation of the regression models. Only one research study in the sample (n=31) uses regression analysis (Vladisavljević et al., 2021), and one of the papers published in a scientific journal uses multivariate factor analysis (Adžić & Al-Mansour, 2021). Another report uses simulation methods without disclosing the inputs that are part of the simulation process (Fiscal Council of the Republic of Serbia, 2022b). This study has the potential to show the evaluation effects of the impact that economic policies can have on beneficiaries.

Table 1. Characteristics of evaluation studies

Characteristics	No. of Studies (%)
Year of publication	
2020	16 (51.6%)
2021	13 (41.9%)
2022	2 (6.5%)
Type of study	
Scientific paper	9 (29.0%)
Research report	1 (3.2%)
Report	21 (67.7%)
Connection with public policy	
Labor market participants	6 out of 31 (19.4%)
Socially disadvantaged groups	6 out of 31 (19.4%)
Overall impact on the economy (economic growth, public debt, sectoral vulnerability, etc.)	10 out of 31 (32.3%)
Impact on enterprises and other business entities	17 out of 31 (54.8%)
Impact on the economic and social well-being of citizens	8 out of 31 (25.8%)
Stakeholders	
Included	9 (29.0%)
Not included	22 (71.0%)
Evaluators	
Academia	10 (32.3%)
Independent assessment of official staff	6 (19.4%)
Civil sector	8 (25.8%)
International organizations	4 (12.9%)
Other	3 (9.7%)
Funding	
Government	8 (25.8%)
Serbian Science Fund	1 (3.2%)
Employers' association	1 (3.2%)
European Union	4 (12.9%)
International organizations	12 (38.7%)
No funding	3 (9.7%)
Unknown	2 (6.5%)
Total	n=31

Source: The authors based on the collected studies.

Table 2. Collected studies by author, year of publication, evaluation approach, and evaluation question

Author (Year)	Evaluation Approach	Evaluation Question
Adžić & Al-Mansour (2021)	Online survey questionnaire for companies (n=40)	Impact on competitiveness
AmCham et al. (2020)	Survey questionnaire in Belgrade only and descriptive analysis of statistical data (situation analysis)	Attitudes about the impact on citizens' well-being
Beraha & Đuričin (2020)	Online survey questionnaire for companies (n=50)	Impact on the operation of medium-sized enterprises
Bodroža & Lazić (2021)	Descriptive analysis of statistical data (situation analysis)	Impact on the gross domestic product, public debt, fiscal balance, and unemployment
CEVES (2020)	Survey questionnaire (n=994) and semi-structured interviews with company representatives and principal stakeholders (n=55)	Impact on the MSMEs sector activities and employment
Đurović (2020)	Descriptive analysis of statistical data (situation analysis) and Public Barometer and Business Opinion surveys	Attitudes about the short-term impacts of government's measures
EAPN (2020)	Descriptive analysis of statistical data (situation analysis)	Impact of anti-poverty government's measures
Fiscal Council (2020)	Descriptive analysis of statistical data (situation analysis)	Impact of economic measures of support to the business sector and citizens on the state budget
Fiscal Council (2022a)	Descriptive analysis of statistical data (situation analysis)	Impact of economic measures of support to the business sector and citizens on the state budget
Fiscal Council (2022b)	Descriptive analysis of statistical data (situation analyses) and simulation analysis	Impact of economic measures on economically deprived population groups
Gajić (2020)	Descriptive analysis of statistical data (situation analysis)	Impact of measures on economic growth and budget deficit
Lazarević-Moravčević & Kamenković (2021)	Descriptive analysis of statistical data (situation analysis)	Diverse impacts depending on the economic sector and company size
Kamberović et al. (2020)	Descriptive analysis of statistical data (situation analysis) and survey questionnaire (n=80)	The long-term impact of economic measures based on sustainable business models
Martin (2020)	Descriptive analysis of statistical data (situation analysis)	Impact of monetary and fiscal policy on economic activity
NALED (2021)	Descriptive analysis of statistical data (situation analyses) and survey questionnaire (n=253)	Impact of economic measures on the business sector
Nikolić & Filipović (2021)	Short-term evaluation of economic measures response (situation analysis)	Impact on public debt, unemployment and economic growth, as well as the indebtedness of companies
Ognjenović (2021b)	Descriptive analysis of statistical data (situation analysis)	Impact on labor market functioning
Pajvančić et al. (2020)	Descriptive analysis of statistical data (situation analysis) and semi-structured interviews with female entrepreneurs and socially disadvantaged women (n=84)	Impact of economic measures from a gender perspective
Paunović & Aničić (2021)	Survey questionnaire (n=689)	Impact on MSMEs and innovation response-demand, logistics, organization of business activities, collection of trade receivables and access to financing
Pejin Stokić (2020)	Descriptive analysis of statistical data (situation analysis)	Impact of government's measures on the economy

continues on following page

Table 2. Continued

Author (Year)	Evaluation Approach	Evaluation Question
SeCons (2020a)	Survey questionnaire (CATI) for employed citizens old 18 years and over (n=1,603)	Impact of government's measures on the labor market with a focus on gender perspective
SeCons (2020b)	Survey questionnaire (CATI) for citizens old 18 years and over (n=1,925)	Impact on gender inequalities and livelihoods of men and women
SeCons (2021)	Qualitative assessments based on individual interviews with citizens, health care providers, members of local crisis headquarters, and civil society organizations (n=37)	Local institutions' responses to the COVID-19 pandemic
SIPRU (2020)	Descriptive analysis of statistical data (situation analysis) and survey questionnaire for civil society organizations (n=54)	Impact on economically deprived population groups and groups at risk
Sovilj & Stojković Zlatanović (2021)	Descriptive assessment of the support to the economy, private sector, and citizens, including legal perspective	Impact of lockdown, trade, and job restrictions measures on the economy and labor
Srdić (2021)	Online survey questionnaire (n=200)	Impact on women entrepreneurs
Udovički & Medić (2021)	Survey questionnaire (CATI) for companies (n=1,082) and entrepreneurs (n=547) and semi-structured interviews with company representatives and principal stakeholders (n=40)	Impact on employment with a focus on deprived population groups
UNDP (2020)	Evaluation based on available official statistics, government documents, surveys conducted by UN agencies, as well as research conducted by civil society organizations	Assessment of the socioeconomic impact through an economic and human development perspective
UNECE (2021)	Survey questionnaire with export-oriented MSMEs (n=726)	Impact on trade and structural transformation with a focus on export companies
Serbian Association of Employers (2020)	Online survey questionnaire with companies (n=400)	Impact on the business sector activities with a focus on specific vulnerable sectors
Vladisavljević et al. (2021)	Survey questionnaire (n=3,044) and regression analysis	Impact on the labor market and vulnerable groups

Source: The authors based on the collected studies.

Figure 1 shows one way to assess the gap in the assessment process used, using a schematic diagram and the data taken from the studies reviewed. Impact assessments are classified according to methods and techniques commonly used in meta-analytic studies (see, for example, Hansen et al. (2022) and Podolsky et al. (2022)).

As emphasized in evaluation methods, any well-intentioned evaluation, economic or otherwise, should begin with an evaluation question. The results of the review of studies presented in the systematic Table 2 show that only two studies, instead of analyzing the impact of the pandemic or governmental actions, exclusively monitored public opinion using a tool called barometer (AmCham et al., 2020; Đurović, 2020). As can be seen, both studies examined opinions and sought to provide early findings on the impact of COVID-19 pandemic. With regard to the other important component of gap assessment, i.e., the application of the assessment methodology, three main characteristics of the studies reviewed were highlighted. Data collection uses surveys, interviews, and secondary data sources, especially when a macroeconomic approach is taken. The methods used to derive the effects can be systematized into qualitative, dealing mainly with narrative findings, and quantitative, such as multivariate regression and factor analysis. The latter is rarely used. Although the impact of interventions on economically disadvantaged groups, at-risk groups, women entrepreneurs, economically disadvantaged local communities, etc., has been studied, none of the observed studies used a counterfactual evaluation approach. These main reasons show how large the gap is between the methods identified and the actual need for impact assessments.

Figure 1. The gap in the evaluation approaches applied in the collected studies

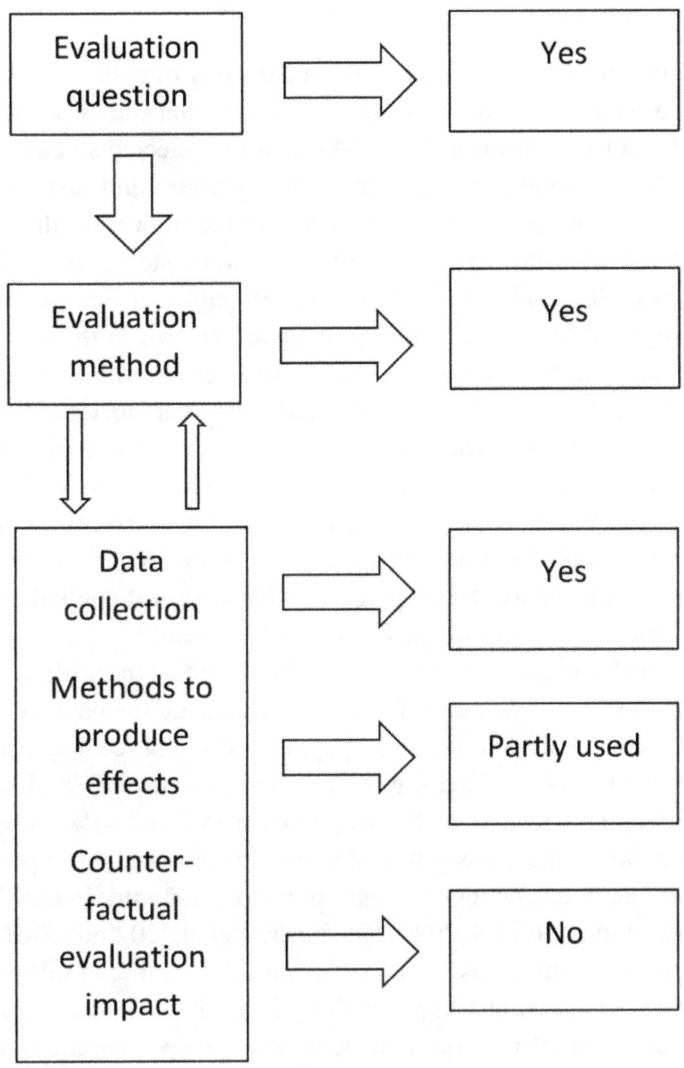

Sensitivity Analysis

A pre-screening of fourteen reports, studies, and scientific papers from various research areas was conducted. The review concluded that the selected documents were a combination of situation analyses and assessments. It also identified a significant gap between the need and application of evaluation techniques and methodologies to support the evidence-based decision-making process in the important areas of economic and social policy in Serbia. State aid to citizens and companies implemented in 2020-2021 was an immediate and direct financial response to the current health crisis. However, a significant portion of the government's actions were not adequately followed up, suggesting that no *interim* or *ex-post* evaluations were planned after the implementation of the policies.

DISCUSSION

Considering that economic impact assessments rely on the study of elasticity and causality rather than pairwise correlations, no papers were found in the 31 studies that relied on the cost-effectiveness of interventions against COVID-19. However, 17 or 54.8% of the selected studies assess quantitative impacts, mainly limited to short-term impacts or attitudes of businesses and citizens about the extent of the negative impact of the crisis, with short-term anticipation of the impact for the next period.

The crisis did not affect the deeper decline of the Serbian economy in 2020, with GDP declining by only 1% in real terms in 2020 (Fiscal Council of the Republic of Serbia, 2020), and employment remaining almost unchanged at a rate of 49.1% (Ognjenović, 2021b). At the same time, the response of the global economy to the pandemic was much more robust in 2020 (Marjanović & Đukić, 2020; wiiw, 2021). However, the impact of the crisis was not felt equally by all categories of labor market participants or economic sectors. Without the government's intervention, the contraction of the economy would undoubtedly have been even more pronounced and would have continued in 2021. The Fiscal Council of the Republic of Serbia (2022a) estimates that a similar impact of the economic measures, on which 5.4 billion euros were spent, would have been achieved even with 3.5 billion euros. This is a part of the analysis where the economic evaluation would give a reliable answer if applied at the level of individual measures to cope with the consequences of the COVID-19 pandemic.

The studies dealing with the impact on the business sector examine the positive effects of the measures or the negative effects of the crisis. The most effective measures are the payment of three minimum net wages to workers in MSMEs and the deferral of the payment of social security contributions and payroll taxes, as well as the payment of 60% of the minimum net wages twice (NALED, 2021). The negative effects are often manifested in the form of lost sales (between 18% and 44%) or cost increases (for up to 64% of firms). In addition, 30% of businesses were unable to operate and 45% operated at reduced capacity, with the most affected activities being passenger transport and work in travel agencies, HORECA, personal services and education (CEVES, 2020; Kamberović et al., 2020; Udovički & Medić, 2021). In addition, 76% of women-owned small businesses experienced a decline in sales during the first year of the pandemic (Srdić, 2021). The situation was particularly difficult for exporting companies. The UNECE study (2021) estimates that 17% of export-oriented businesses ceased operations and only 31% managed to regain their pre-pandemic revenues by the last quarter of 2020. It is also estimated that 78% of these companies survived thanks to the government's tax and contribution deferral measures. As estimated in the survey of the Serbian Association of Employers (2020), companies were exposed to additional costs for the purchase of disinfectants and protective equipment, then the activities of the hospitality industry, part of the retail sector and transportation were suspended, and about 45% of companies would face difficulties because they do not have access to sources of financing for recovery. The impact of the crisis is also reflected in the increase in gender inequalities. However, both men and women were affected, in terms of employment sector, occupation or the fact that working from home became a new reality for those jobs for which it was possible to organize it. Therefore, it is not surprising that gender inequalities were also manifested in differences in working hours (Vladisavljević et al., 2021). Pajvančić et al. (2020) noted a lack of targeted interventions for vulnerable populations.

CONCLUSION WITH RECOMMENDATIONS, LIMITATIONS, AND FURTHER RESEARCH

In summary, some specific recommendations for the further development of policy evaluation approaches need to be derived. The existing gap in systematic evaluation methods for assessing the impact of public policies on the state of the economy and society, including actors (e.g., firms, small businesses, or professions), regions (including underdeveloped or rural regions), and groups of people (e.g., economically and socially disadvantaged groups) should be addressed by conducting timely and comprehensive evaluations with clear conclusions drawn from the magnitude of the estimated impacts or quantitative effects of the policies implemented. There are also many other underexploited methods, already applied in Serbia within different policy fields, that could provide important insights into the effectiveness of the applied policy measures such as cost-benefit analyses (Jenkins et al., 2022) or the analyses of crowding-out effects of applied policy measures that assess changes in human behavior or consumption patterns (Vladisavljević et al., 2023). In addition, this chapter highlighted the limitations of using evaluation studies to support policy responses to 'black swans' in Serbia and suggested important recommendations for improvement. Considering the most important findings of the performed analysis and presented discussion, several recommendations would help close the observed gap, probably resulting from underdeveloped evaluation culture and insufficient institutional awareness on the potential benefits which conducting rigorous evaluations might have on achieving desired policy outcomes. First, demand for rigorous impact evaluations should be strengthened since conducting high-quality evaluations is often motivated by the policymakers' need to improve their policy approach and create systems that can change and learn from previous experiences. This is, however, closely related to the evaluation culture and awareness about the concrete evidence that could help achieve better social outcomes in the long-term. Second, applying rigorous evaluation studies requires well-developed research and evaluation capacities. This is a long-term process that assumes political will to implement evidence-based policy-making concepts and should also be promoted by top-level government officials. Third, available and easily accessible data is a necessary prerequisite for any research, including high-quality evaluations. Serbia has well developed statistical infrastructure that has been significantly improved over the previous period. However, there is no statistical infrastructure that could serve all potential needs for data. At the institutional level is, therefore, necessary to invest further efforts in developing learning organizations that continuously monitor the progress of the existing policies through the set of previously established indicators and initiate additional data collection in the case of specific needs occur, such as the COVID-19 pandemic circumstances.

ACKNOWLEDGMENT

The research presented in this chapter was funded by the Ministry of Science, Technological Development and Innovation of the Republic of Serbia under contract number 451-03-47/2023-01/200005.

REFERENCES

Adžić, S., & Al-Mansour, J. (2021). The negative impact of COVID-19 on firms: Insights from Serbia. *Eastern European Economics*, *59*(5), 472–486. doi:10.1080/00128775.2021.1953387

Alfaro, L., Chari, A., Greenland, A. N., & Schott, P. K. (2020). *Aggregate and Firm-level Stock Returns During Pandemics, in Real Time* (National Bureau of Economic Research Working Paper No. 26950). https://www.nber.org/system/files/working_papers/w26950/w26950.pdf

AmCham Serbia, Economics Institute & Faculty of Medicine of Belgrade University. (2020). *Studija o mogućim pravcima adekvatnog odgovora zdravstvenog sektora izazovima epidemije COVID-19 u periodu 2020-2021* [A Study on the Possible Directions of an Adequate Response of the Health Sector to the Challenges of the COVID-19 Epidemic in the Period 2020-2021]. https://www.amcham.rs/upload/documents/AmCham%20COVID%20studija%20Final.pdf

Angrist, J. D., & Pischke, J.-S. (2009). *Mostly harmless econometrics: an empiricist's companion*. Princeton University Press. doi:10.1515/9781400829828

Ashraf, B. N. (2020). Stock markets' reaction to COVID-19: Cases or fatalities*? Research in International Business and Finance*, *54*, 101249. doi:10.1016/j.ribaf.2020.101249 PMID:34170989

Beraha, I., & Đuričin, S. (2020). The impact of COVID-19 crisis on medium-sized enterprises in Serbia. *Economic Analysis*, *53*(1), 14–27. https://www.library.ien.bg.ac.rs/index.php/ea/article/view/1264/1083

Bodroža, D., & Lazić, M. (2021). Economic impact of COVID-19 pandemic on Western Balkan countries. *Economic Analysis*, *54*(2), 30–40. doi:10.28934/ea.21.54.2.pp30-40

Center for Advanced Economic Studies. (2020). *The COVID-Crisis and Serbia's SMEs: Assessment of Impact and Outline of Future Scenarios*. https://ceves.org.rs/wp-content/uploads/2020/11/WB-Covid-19_-Report-final.pdf

De Savigny, D., & Adam, T. (Eds.). (2009). *Systems Thinking for Health Systems Strengthening*. World Health Organization. https://apps.who.int/iris/bitstream/handle/10665/44204/9789241563895_eng.pdf?sequence=1&isAllowed=y

Deb, P., Furceri, D., Ostry, J. D., Tawk, N., & Yang, N. (2021). *The Effects of Fiscal Measures during COVID-19* (International Monetary Fund Working Paper No. 262). https://www.imf.org/-/media/Files/Publications/WP/2021/English/wpiea2021262-print-pdf.ashx

Đurović, G. (2020). *Analysis of Economic and Social Consequences of the COVID-19 Pandemic in Western Balkans Administrations*. ReSPA. https://www.respaweb.eu/download/doc/Analysis+of+economic+and+social+consequences+of+the+Covid-19+pandemic+in+Western+Balkans+administrations.pdf/e97586d30e7b07fa0dd3f00ebf84dcc8.pdf

Fiscal Council of the Republic of Serbia. (2020). *Ocena antikriznog programa ekonomskih mera* [Evaluation of the Anti-Crisis Program of Economic Measures]. https://www.fiskalnisavet.rs/doc/ocene-i-misljenja/2020/FS-Ocena_antikriznog_programa_ekonomskih_mera.pdf

Fiscal Council of the Republic of Serbia. (2022a). *Antikrizne budžetske mere tokom pandemije COVID-19: troškovi, rezultati i pouke* [Anti-crisis Budgetary Measures During the COVID-19 Pandemic: Costs, Results and Lessons Learned]. https://www.fiskalnisavet.rs/doc/analize-stavovi-predlozi/2022/FS_Antikrizne_budzetske_mere_tokom_pandemije.pdf

Fiscal Council of the Republic of Serbia. (2022b). *Predlog mera socijalne i poreske politike za smanjenje nejednakosti i rizika od siromaštva u Republici Srbiji* [Proposed Social and Tax Policy Measures for Reducing Inequality and Poverty in the Republic of Serbia]. https://www.fiskalnisavet.rs/doc/analize-stavovi-predlozi/2022/FS_Predlog%20mera%20socijalne%20i%20poreske%20politike.pdf

Gajić, M. (2020). *The Impact of the COVID-19 on the Serbian Economy in 2020*. Friedrich Naumann Foundation for Freedom. https://shop.freiheit.org/download/P2@1040/398508/Final____LIFE_WITH_THE_Covid19_Serbia.pdf

Gates, E. F. (2016). Making sense of the emerging conversation in evaluation about systems thinking and complexity science. *Evaluation and Program Planning*, *59*, 62–73. doi:10.1016/j.evalprogplan.2016.08.004 PMID:27591941

Hansen, C., Steinmetz, H., & Block, J. (2022). How to conduct a meta-analysis in eight steps: A practical guide. *Management Review Quarterly*, *72*(1), 1–19. doi:10.100711301-021-00247-4

Huang, A., Makridis, C., Baker, M., Medeiros, M., & Guo, Z. (2020). Understanding the impact of COVID-19 intervention policies on the hospitality labor market. *International Journal of Hospitality Management*, *91*, 102660. doi:10.1016/j.ijhm.2020.102660 PMID:32904433

Immergluck, D. (2013). Too little, too late, and too timid: The federal response to the foreclosure crisis at the five-year mark. *Housing Policy Debate*, *23*(1), 199–232. doi:10.1080/10511482.2012.749933

Institute of Economic Sciences. (2022a). *Institutional Response to COVID-19 Pandemic Effects on the Labour Market – Lessons Learned and Future Perspectives*. Policy Brief [Research Project INEQ RS COVID-19 No. 7552225]. https://www.ien.bg.ac.rs/data/images/proj_ineqrs/policy_brief_institutional_response_co_covid19_pandemic_effects_labour_market.pdf

Institute of Economic Sciences. (2022b). *Effects of the COVID-19 Pandemic on Poverty and Income Inequality in Serbia and Cost-Benefit Analysis of the Implemented Income Support Measures*. Policy Brief [Research Project INEQ RS COVID-19 No. 7552225]. https://www.ien.bg.ac.rs/data/images/proj_ineqrs/policy_brief_effects_covid19_pandemic_on_poverty_income_inequality_inserbia.pdf

Jenkins, G. P., Olubamiro, O. C., & Miklyaev, M. (2022). *Cost-Benefit Analysis of Tax Incentives in Serbia* (Development Discussion Papers No. 2022-01, JDI Executive Programs Support to Enhancing the Quality of the Tax System in Serbia: Analysis of Tax Incentives in Serbia). https://cri-world.com/publications/qed_dp_4579.pdf

Jordà, Ò. (2005). Estimation and inference of impulse responses by local projections. *The American Economic Review*, *95*(1), 161–182. doi:10.1257/0002828053828518

Kamberović, S., Mitrović, S., & Behrens, A. (2020). *GAP Analysis - Opportunities for Economic Empowerment through Sustainable Business Models in the Aftermath of the COVID-19 Pandemic in the Republic of Serbia*. Organization for Security and Co-operation in Europe Mission to Serbia. https://www.osce.org/files/f/documents/9/1/479678_0.pdf

Lazarević-Moravčević, M., & Kamenković, S. (2021). The impact of the COVID-19 crisis on the Serbian economy - consequences and recovery. *Economic Analysis*, *54*(2), 41–54. doi:10.28934/ea.21.54.2.pp41-54

Leeuw, L. F. (2012). Linking theory-based evaluation and contribution analysis: Three problems and a few solutions. *Evaluation*, *18*(3), 348–363. doi:10.1177/1356389012452051

Marjanović, D., & Đukić, M. (2020). Ekonomske mere za ublažavanje posledica COVID-19 [Economic measures to mitigate the consequences of COVID-19]. In P. Mitić & D. Marjanović (Eds.), *Black swan in the world economy 2020* (pp. 91–105). Institute of Economic Sciences.

Martin, V. (2020). The response of the monetary and fiscal policies on COVID-19 in Serbia. *Bankarstvo*, *49*(2), 92–114. doi:10.5937/bankarstvo2002070M

Mladovsky, P., Srivastava, D., Cylus, J., Karanikolos, M., Evetovits, T., Thomson, S., & McKee, M. (2012). *Health Policy Responses to the Financial Crisis in Europe* (World Health Organization Policy Summary No. 5). https://www.euro.who.int/__data/assets/pdf_file/0009/170865/e96643.pdf

National Alliance for Local Economic Development. (2021). *Efekti ekonomskih mera za ublažavanje negativnih posledica COVID-19 na privredu* [Effects of Economic Measures to Mitigate the Negative Consequences of COVID-19 on the Economy]. https://naled.rs/htdocs/Files/06839/Analiza_efekata_ekonomskih_mera_za_ublazavanje_negativnih_posledica_COVID-19_na_privredu.pdf

Nikolić, I., & Filipović, S. (2021). Comparative analysis of the economic policy response to COVID-19 in the EU-27 and Republic of Serbia. *Ekonomika preduzeća*, *7-8*, 460-475. doi:10.5937/EKOPRE2108460N

Ognjenović, K. (2007). The use of propensity score-matching methods in evaluation of active labour market programs in Serbia. *Economic Annals*, *52*(172), 21–53. doi:10.2298/EKA0772021O

Ognjenović, K. (2021a). Gender wage gap in Serbia: Inheritance and sources of the wage gap. *Stanovnistvo*, *59*(2), 123–141. doi:10.2298/STNV2102123O

Ognjenović, K. (2021b). Prilagođavanje tržišta rada u Srbiji novim uslovima izazvanim COVID-19 krizom [Adjustment of the labour market in Serbia in the conditions of the crisis caused by the COVID-19 pandemic]. In A. Bradić-Martinović, I. Beraha, & S. Đuričin (Eds.), *Poslovanje u periodu pandemije – izazovi i šanse* [Pursuing business during the pandemic – challenges and chances] (pp. 18–37). Institute of Economic Sciences.

Ognjenović, K. (2022). Multiple job-holding and informality: may pandemic change their course? In S. Redžepagić, A. Portugal Durate, M. Siničáková & D. Bodroža (Eds.), Economic and financial implications of COVID-19 crisis (pp. 200-218). Balkan Institute of Science and Innovation, Université Côte d'Azur, Nice.

Ognjenović, K., & Đukić, M. (2022). A review into evaluation methods applied in assessing economic and social policies' responses to COVID-19 in Serbia: 2020-2021 [Paper presentation]. PROFEEDBACK Evaluation of Public Policy Responses to 'Black Swans' Conference, Budapest, Hungary.

Ognjenović, K., & Pavlović, D. (2021). *SILC u Republici Srbiji: metodološki okvir i analiza izabranih pokazatelja siromaštva i nejednakosti* [SILC in the Republic of Serbia: methodological framework and analysis of poverty and inequality indicators]. Institute of Economic Sciences.

Organisation for Economic Cooperation and Development. (2001). *Evaluation feedback for effective learning and accountability*. OECD.

Pajvančić, M., Petrušić, N., Nikolin, S., Vladisavljević, A., & Baćanović, V. (2020). *Rodna analiza odgovora na COVID-19 u Republici Srbiji* [Gender Analysis of the Response to COVID-19 in the Republic of Serbia]. Organization for Security and Co-operation in Europe Mission to Serbia & Women's Platform for the Development of Serbia 2014-2020. https://www.osce.org/files/f/documents/3/9/459391.pdf

Paunović, B., & Aničić, Z. (2021). Impact of the COVID-19 crisis on SMEs and possible innovation responses. *Ekonomika preduzeća, 3-4*, 169-184. doi:10.5937/EKOPRE2103169P

Pejin Stokić, Lj. (2020). *COVID-19: Impact on the Serbian Economy and the Government Recovery Package* (ESPN Flash Report No. 45). https://ec.europa.eu/social/BlobServlet?docId=22945&langId=en

Podolsky, M. I., Present, I., Neumann, P. J., & Kim, D. D. (2022). A systematic review of economic evaluations of COVID-19 interventions: Considerations of non-health impacts and distributional issues. *Value in Health, 25*(8), 1298–1306. doi:10.1016/j.jval.2022.02.003 PMID:35398012

Republic Secretariat for Public Policies. (2018). *Law on the Planning System of the Republic of Serbia.* Official Gazette RS No. 30.

Rezapour, A., Souresrafil, A., Peighambari, M. M., Heidarali, M., & Tashakori-Miyanroudi, M. (2021). Economic evaluation of programs against COVID-19: A systematic review. *International Journal of Surgery, 85*, 10–18. doi:10.1016/j.ijsu.2020.11.015 PMID:33227532

Sawer, B. J. (1992). *Evaluating for accountability: a practical guide for the inexperienced evaluator.* Oregon State University Extension Service.

SeCons. (2020a). *Impact of COVID-19 Pandemic and Measures for its Prevention on Employment and Working Conditions of Women and Men in Serbia*. https://secons.net/wp-content/uploads/2022/01/133-publication.pdf

SeCons. (2020b). *Consequences of COVID-19 on Women's and Men's Economic Empowerment*. https://secons.net/wp-content/uploads/2022/01/134-publication.pdf

SeCons. (2021). *Ocena efikasnosti institucionalnog odgovora na pandemiju COVID-19 u multietničkim zajednicama u Sandžaku/jugozapadnoj Srbiji i na jugu Srbije* [Evaluation of the Effectiveness of the Institutional Response to the COVID-19 Pandemic in Multi-Ethnic Communities in Sandžak/Southwest Serbia and in the South of Serbia]. https://secons.net/wp-content/uploads/2021/12/131-publication.pdf

Serbian Association of Employers. (2020). *Uticaj pandemije COVID-19 na aktivnosti preduzeća u Srbiji: izazovi i očekivanja* [The Impact of the COVID-19 Pandemic on the Activities of Companies in Serbia: Challenges and Expectations]. https://drive.google.com/file/d/1oxElSqLaUB8mibQV06xMZW77L5hrTurA/view

Shapovalova, A. O., Ivanov, Y. B., Tyschenko, V. F., & Karpova, V. V. (2021). Assessment of the effectiveness of anti-COVID tax support for innovation activities of small and medium-sized enterprises in OECD countries. *Journal of Tax Reform*, *7*(1), 68–86. doi:10.15826/jtr.2021.7.1.091

Siddik, M. N. A. (2020). Economic stimulus for COVID-19 pandemic and its determinants: Evidence from cross-country analysis. *Heliyon*, *6*(12), e05634. doi:10.1016/j.heliyon.2020.e05634 PMID:33319096

Social Inclusion and Poverty Reduction Unit. (2020). *Impact of the COVID-19 on Vulnerable Groups and Groups at Risk – Causes, Outcomes and Recommendations*. https://socijalnoukljucivanje.gov.rs/wp-content/uploads/2020/12/Posledice_Kovid_19_na_polozaj_osetljivih_grupa_i_grupa_u_riziku_ENG.pdf

Sovilj, R., & Stojković Zlatanović, S. (2021). Tackling the impact of the COVID-19 pandemic in economy and labour – a case study of Serbia regulation. *Medicine. Law & Society*, *14*(2), 301–320. doi:10.18690/mls.14.2.301-320.2021

Srdić, M. (2021). *Podrška preduzetnicima tokom pandemije COVID-19* [Support for Entrepreneurs during the COVID-19 Pandemic]. Organization for Security and Co-operation in Europe Mission to Serbia. https://www.osce.org/files/f/documents/2/7/513133.pdf

The European Anti-Poverty Network Serbia. (2020). *COVID Crisis and Poverty: Poverty Watch Serbia 2020*. https://www.eapn.eu/wp-content/uploads/2020/09/EAPN-EAPN-Serbia-Poverty-Watch-2020_ENG-4698.pdf

Tirivayi, N., Richardson, D., Gavrilovic, M., Groppo, V., Kajula, L., Valli, E., & Viola, F. (2020). *A Rapid Review of Economic Policy and Social Protection Responses to Health and Economic Crises and Their Effects on Children: Lessons for the COVID-19 Pandemic Response* (UNICEF Office of Research – Innocenti Working Paper No. 02). https://www.unicef-irc.org/publications/pdf/WP2020-02.pdf

Udovički, K., & Medić, P. (2021). *Impact of the COVID-19 Crisis on Employment: Focus on Vulnerable Categories*. Social Inclusion and Poverty Reduction Unit & CEVES. https://socijalnoukljucivanje.gov.rs/wp-content/uploads/2021/01/Uticaj_COVID-19_krize_na_zaposlenost.pdf

United Nations Development Programme Serbia. (2020). *COVID-19: Socio-Economic Impact Assessment*. https://serbia.un.org/sites/default/files/2020-09/seia_report%20%281%29.pdf

United Nations Economic Commission for Europe. (2021). *The Impact of COVID-19 on Trade and Structural Transformation in Serbia – Evidence from UNECE's Survey of Micro, Small and Medium Enterprises*. https://unece.org/sites/default/files/2021-06/Impact_COVID-19_Serbia-Eng.pdf

Vladisavljević, M., Lebedinski, L., Banović, J., & Vukmirović, V. (2021). *Effects of the COVID-19 Pandemic on the Labour Market and the Position of Vulnerable Groups in Serbia*. Institute of Economic Sciences. (Research Project INEQ RS COVID-19 No. 7552225)

Vladisavljević, M., Zubović, J., Jovanovic, O., & Đukić, M. (2023). Crowding-out effect of tobacco consumption in Serbia. *Tobacco Control.* Advance online publication. doi:10.1136/tc-2022-057727 PMID:36990684

Wang, Y., Luangasanatip, N., Pan-Ngum, W., Isaranuwatchai, W., Prawjaeng, J., Saralamba, S., Painter, C., Briones, J. V., & Teerawattananon, Y. (2022). Assessing the cost-effectiveness of COVID-19 vaccines in a low incidence and low mortality setting: The case of Thailand at start of the pandemic. *The European Journal of Health Economics*, 1–14. doi:10.100710198-022-01505-2 PMID:35951243

White, H. (2009). Theory-based impact evaluation: Principles and practice. *Journal of Development Effectiveness, 1*(3), 271–284. doi:10.1080/19439340903114628

Wiener Institut für Internationale Wirtschaftsvergleiche. (2021). *Getting Stronger after COVID-19: Nearshoring Potential in the Western Balkans* (wiiw Research Report No. 453). https://wiiw.ac.at/getting-stronger-after-covid-19-nearshoring-potential-in-the-western-balkans-dlp-5814.pdf

Zdravković, A., Đukić, M., & Bradić-Martinović, A. (2017). Impact of FDI on unemployment in transition countries: panel cointegration approach. *Industrija, 45*(1), 161-174. http://ebooks.ien.bg.ac.rs/1093/1/0350-03731701161Z.pdf

Zubović, J., Đukić, M., & Jovanović, O. (2020). *Ekonomski aspekti kontrole duvana i empirijski nalazi u Srbiji* [Economic aspects of tobacco control and empirical findings in Serbia]. Institute of Economic Sciences.

ADDITIONAL READING

Borenstein, M., Hedges, L. V., Higgins, J. P. T., & Rothstein, H. R. (2009). *Introduction to meta-analysis.* John Wiley & Sons. doi:10.1002/9780470743386

Dawoud, D. M., & Soliman, K. Y. (2020). Cost-effectiveness of antiviral treatments for pandemics and outbreaks of respiratory illnesses, including COVID-19: A systematic review of published economic evaluations. *Value in Health, 23*(11), 1409–1422. doi:10.1016/j.jval.2020.07.002 PMID:33127010

International Monetary Fund. (2020). *Policy Responses to COVID19: Serbia.* IMF. http://www.imf.org/en/Topics/imf-and-covid19/Policy-Responses-to-COVID-19#S

Kirti, D., Liu, Y., Martinez Peria, S., Mishra, P., & Strasky, J. (2022). *Tracking Economic and Financial Policies during COVID-19: An Announcement-Level Database* (International Monetary Fund Working Paper No. 114). https://www.imf.org/-/media/Files/Publications/WP/2022/English/wpiea2022114-print-pdf.ashx

Radović-Marković, M., & Jovanović, J. (2021). Impact of COVID-19 on working from home in Serbia: possibilities and consequences. In N. Vidaković & I. Lovrinović (Eds.), *Macroeconomic responses to the COVID-19 pandemic* (pp. 319–343). Palgrave Macmillan. doi:10.1007/978-3-030-75444-0_12

Steinmetz, H., & Block, B. (2022). Meta-analytic structural equation modeling (MASEM): New tricks of the trade. *Management Review Quarterly*, *72*(3), 605–626. doi:10.100711301-022-00293-6

Weiss, C. H. (1998). *Evaluation: methods for studying programs and policies* (2nd ed.). Prentice Hall.

KEY TERMS AND DEFINITIONS

Economic Impact: The effects specific phenomena have on the economic situation in the country, including results of external factors (e.g., COVID-19 pandemic) and government measures (e.g., anti-COVID-19 policy).

Economic Policy: A set of programs and economic measures used by governments and aimed at creating a favorable business environment, as well as improving the standard of living of a population. In overall, economic policy aims at improving desired economic outcomes.

Economic System: Stands for a complex set of interactions among different economic subjects.

Evaluation: A process of assessing specific social phenomena or activity based on previously defined objectives and using a methodologically founded set of standards.

Evaluation Method: Tools used by evaluators when conducting evaluations. Selection of evaluation methods depends on the evaluation objectives and often includes application of quantitative and qualitative research techniques.

Meta-Analysis: Statistical analysis of the results provided by several similar studies conducted to generate quantitative evidence on a specific phenomenon.

Systematic Review: Use of systematic methods aimed at gathering and assessing a large number of scientific evidence to provide the empirically based answer on a concrete research question.

Chapter 7
A Systems Theory Approach to Maximizing Student Engagement With Reused Video Content

Timothy P. Lombardo
Purdue University, USA

Audrey Beth Lombardo
Wexner Medical Center, The Ohio State University, USA

ABSTRACT

The COVID-19 pandemic may constitute what Kurt Lewin refers to as an unfreezing moment for higher education institutions. When COVID-19 started to spread worldwide, traditional brick-and-mortar institutions were forced to close due to mandated lockdowns. Concerned with student success, universities rushed to move course content online through recorded lectures and synchronous class sessions via videoconferencing platforms. In transitioning back to normal, evaluating how materials developed during the pandemic can be integrated into the classroom setting is vital. Before the world returns to a post-pandemic environment, the question arises regarding what faculty should do with content created during the pandemic. Should they be archived, never to be used again? Or can they look for ways to integrate them into coursework to enhance learning? One-way schools can help to foster quality online education is by preserving the materials created and lessons learned during the pandemic, especially video content.

DOI: 10.4018/978-1-6684-7285-9.ch007

FROM CHAOS TO COHESION: A SYSTEMS THEORY APPROACH TO MAXIMIZING STUDENT ENGAGEMENT WITH REUSED VIDEO CONTENT

The COVID-19 pandemic represented an upheaval for many, both personally and professionally. Across the globe, industries were forced to re-examine their practices and pivot in ways they had never anticipated. The field of education was particularly hard hit during this transition. The entire system of face-to-face classroom instruction was disrupted for both students and instructors. System theory tells us that a system is comprised of a group of interconnected components that work together to create a functional system (Friedman & Neuman Allen, 2014). Pre-COVID-19, face-to-face instruction was comprised of inputs (students, faculty, and the lessons that they would prepare), throughputs (dissemination of the lectures and the learning that happened in the face-to-face classroom), and outputs (students leaving with the knowledge that they received from lectures in the classroom). Environmental factors, such as inclement weather days or instructor illness, that would disrupt this learning model have always existed. However, these disruptions never came close to the pandemic's massive impact on the traditional educational system. COVID-19 created a significant disruption by not allowing the inputs (students and instructors) to meet face-to-face. In order to address this issue, the throughput (dissemination of educational materials and lectures) needed to be moved to a virtual environment. As a result of this shift, a new output (educational videos and lectures) was created. The authors of this chapter will examine this shift in more detail, focusing on practical uses of how video content created during the pandemic can continue to be leveraged in the higher education classroom. The authors will also provide real-life examples from their current practice and an overview of applicable theories and models.

A 2020 EDUCAUSE survey reported that most faculty and undergraduate students preferred face-to-face instruction over remote modalities (Gierdowski, 2019). When COVID-19 spread worldwide, traditional institutions were closed through mandated lockdowns (Gewertz, 2020; Kamenetz, 2020). By March, over 1 million U.S. students transitioned to remote learning, with all U.S. public school buildings closed by March 25 (Gewertz, 2020). Concerned with student success, universities rushed to move course content online through recorded lectures and synchronous class sessions via videoconferencing platforms (Lowenthal et al., 2020; Pokhrel & Chhetri, 2021). Once seen as secondary services by many organizations, instructional design departments came into high demand to help transition courses from in-person to remote learning modalities. Teleconferencing systems such as Zoom and video content became the primary modality for how instructors interacted with their students (Byrnes et al., 2021). During this time instructors were also tasked with creating large amounts of video content, including pre-recorded lecture content and virtual class session recordings, so students who could not attend class due to illness could remain engaged. While these modalities were seen as potentially inferior in quality, they ensured that educational services for students largely remained uninterrupted. From an instructional design perspective, this rapid transition to online modalities was less than ideal. An average online course can take months to develop and often involves many meetings between instructors and instructional designers to ensure it is pedagogically sound. The emergency response to the pandemic made this due diligence impossible.

Renowned psychologist Kurt Lewin might have considered the pandemic an "unfreezing moment" for education institutions (Lewin, 1997; The Mind Tools Content Team, 2022). Lewin's change model helps individuals and groups transition to a desired state in three steps and is still widely used today. The first step, unfreezing, involves a sense of dissatisfaction with the current state. Second, with the changing step, goals are created, and actions are taken toward those goals. Finally, refreezing occurs when the change becomes permanent, including consolidation and stabilization.

As institutions transition back to normal, evaluating how these video materials and methodologies developed during the pandemic can be integrated into the classroom setting will be essential. Before the world returns to a post-pandemic environment, educators may want to define how they envision the new normal. Many would maintain that education was primed for the dramatic shift of the pandemic (Carnegie Mellon University, 2020). With a push toward normalcy, the question arises regarding what faculty should do with video content created during the pandemic. Should they be archived, never to be used again? Or can we look for ways to integrate them into coursework to enhance learning? Can we, as educators, look for ways to leverage these video content libraries to create better educational experiences for students (inputs) and enhance meaningful classroom learning (throughput)?

The authors will argue that educational organizations should consider integrating the video materials created during the pandemic into their synchronous and asynchronous remote learning courses. This repurposing can lead to enhanced student learning and a better teaching experience for instructors. One helpful method for this integration is the flipped classroom model. In this model, students receive instructional materials, such as asynchronous lecture videos to be viewed before class, to introduce concepts so that class time can be freed for instructor-led and collaborative activities (November & Mull, 2012). Collaborative activities can improve engagement, team-based skills, student autonomy, and creative freedom for faculty (Millard, 2012). While traditionally used primarily in the face-to-face classroom model, the application of the flipped classroom model in both synchronous and asynchronous online settings has also emerged and deserves further exploration. The authors recognize that K-12 and post-secondary face-to-face educational systems were highly disrupted during the COVID pandemic. While there are similarities in the strategies used in both sectors to address this shift, several regulatory rules and procedures are unique to the different educational levels. For example, K-12 institutions often must proactively address issues around recording and circulating videos that include minors (a concern that would have come up in recording virtual class sessions during the pandemic) which does not need to be addressed in higher education. Given these differences, the authors will focus their chapter on post-secondary services.

Synchronous vs. Asynchronous Learning

Online learning has two primary types of courses: synchronous and asynchronous. Synchronous courses are conducted in real-time, with students and instructors participating in live, scheduled sessions. Classroom interaction and engagement are similar to the traditional face-to-face classroom in this model. In comparison, asynchronous courses do not have established meeting times. Instead, students access materials at their convenience. The asynchronous online learning environment may use various digital tools such as discussion boards, blogs, wikis, and video conferencing to facilitate interaction and engagement.

Both approaches have their benefits and drawbacks. Synchronous courses offer the advantage of allowing students the flexibility of remote learning while maintaining traditional real-time interactions with their instructors and other students. These immediate interactions may benefit those who learn best through discussion and collaboration. However, synchronous courses can also be somewhat inflexible, requiring students to be available at specific times, and this can be particularly challenging for students in different time zones. In comparison, asynchronous courses provide more flexibility since students do not have to attend scheduled class sessions and can complete coursework on their schedule. However, asynchronous courses can be less interactive, and students might miss live discussions with peers (Delahunty, 2018; Galikyan & Admiraal, 2019).

Post-Pandemic Possibilities for Video

Educational videos are critical in many instructional modalities (Cheng et al., 2018; Laster-Loftus & Cooper, 2019; Tomei, 2008; Wong et al., 2019). Videos can create and enhance the teaching presence vital to the community of inquiry (CoI) model (Belt & Lowenthal, 2021). CoI was first developed by Garrison et al. (1999) and is commonly used when creating online educational experiences (Garrison, 2011, 2016). The framework is broken into three primary areas: social presence, cognitive presence, and teaching presence. Social presence is focused on the social interactions that the students have in the course. It highlights students' and instructors' collaborative environment (Garrison, 2011, 2016). Cognitive presence addresses how students interact with course content. It involves creating materials and assessments that challenge the students to reflect on and critically connect with the content. Teaching presence covers several topics, including instructional design and organization, facilitating discourse, and direct instruction. These three presences can, and often do, overlap. The point where all three areas connect is where instructors and students will find the optimal educational experience.

As previously mentioned, the COVID-19 pandemic completely disrupted the system of face-to-face education. Students and instructors were no longer permitted to meet and receive educational content in person, meaning the input and throughput of the traditional face-to-face system were impacted and needed to be altered. The three areas of the CoI framework can help address this transition by encouraging educators to think critically about changing their throughput while actively engaging students in the learning process. Using videos in the classroom can enhance teaching presence to improve the course and student satisfaction (Anderson et al., 2001; Garrison, 2011, 2016; Garrison et al., 1999; Garrison et al., 2019). Video content is known to enhance learning. Artz et al. (2022) revealed that students who watched lecture videos did better on aligned assessments than peers who attended live lectures. One potential explanation for this difference in performance was that students could watch video content repeatedly and take additional time to ensure they were fully absorbing the material. In contrast, in class, students only hear content once. Another possible explanation for the difference in student experiences from this study was that students often multitask during live lectures. Students attending synchronous classes must often pay attention to the instructor, take notes, and try to comprehend the presented information.

Numerous studies have concluded that multitasking is rarely suitable in education (Aagaard, 2019; Enz et al., 2021; Kirschner & De Bruyckere, 2017). However, it is still expected in a synchronous session, whether in a lecture hall or listening live online (Enz et al., 2021; Kirschner & De Bruyckere, 2017). Using videos instead of live sessions, students can take in information at their own pace and pause videos as desired. This ability to take breaks and pace information intake has positively impacted student learning. For example, in one study of classes addressing advanced topics, students' time watching video lectures (including rewatching) correlated positively with grades on aligned assessments (Aragon, 2016). The psychological benefits of videos include fostering the dual coding of the information presented. For instance, a video that simultaneously presents material auditorily and visually often helps students retain information longer (Joshua, 2016; Pekdağ, 2010).

The Flipped Classroom

As the name implies, the flipped classroom model involves turning traditional classroom activities into ones done outside class time. In the flipped classroom model, lectures are commonly recorded by the instructor and provided to students before class, and students watch them to prepare for class. When students attend class, the instructor can spend more time on active learning activities and discussions since students have already watched the lecture videos. Because students have already watched the lecture videos, this allows students to ask expert questions and have meaningful discussions with peers (Cheng et al., 2018; Millard, 2012).

The flipped classroom was historically associated with face-to-face learning to make the classroom environment a hybrid experience (Singh & Arya, 2020). However, why can't we also use it in an entirely online environment? COVID-19 completely disrupted the system of education. The traditional inputs of students, instructors, and lectures from the face-to-face system still existed. However, instructors were now being asked to record their lectures, which created a new output of video libraries that did not exist beforehand. The flipped classroom can leverage this new output in the post-pandemic world. Students can now watch videos beforehand and enter classroom sessions with a baseline knowledge of the class content, allowing them time to better engage with the instructor. Students watching videos beforehand and entering classroom sessions with that baseline knowledge allows them to spend more time actively engaged with the material while freeing class time for discussions and other activities. The same approach can be used in a synchronous online course. Besides studying at their own pace, students can use class meetings to ask questions about the material or get help from their classmates outside class. This approach can be particularly beneficial for students with difficulty concentrating in a traditional lecture format.

Flipping the asynchronous online classroom is more complex than synchronous courses. However, it is possible and has the potential to be beneficial for both students and instructors. Students and teachers are not usually logged in simultaneously for live sessions in an asynchronous online course. Despite the flexibility it offers, it presents challenges to student engagement and communication. Since there is no live classroom session in the asynchronous model to hold students accountable for viewing pre-recorded lectures, instructors might want other ways to track engagement, such as offering participation points by having students complete a small quiz after each video or embedding a code somewhere in the video for students to find and enter, to access password-protected course content.

Another way to enhance student discussion and engagement is through team-based learning (TBL): a student-centered instructional strategy that organizes students into teams for collaborative learning. Unlike traditional lecture-based instruction, TBL allows students to engage with course content actively and with each other to promote deep understanding and long-term retention. In a typical TBL classroom, students work together on reading assignments, problem sets, and group projects. The TBL collaborative approach encourages students to learn from each other and helps them develop essential teamwork skills such as communication, negotiation, and conflict resolution. TBL also improves student motivation and attitudes toward learning, effectively promoting student success in the classroom (Michaelsen & Sweet, 2008).

Traditionally, TBL happens in a face-to-face classroom setting and follows a sequence of events involving four elements: groups, accountability, feedback, and assignment design (Michaelsen & Sweet, 2008). The typical TBL sequence includes the instructor strategically creating teams of students that may have different strengths to help groups succeed in their tasks; teams are retained throughout the entire semester. Given the topic and materials, students take individual tests to gauge their knowledge of the subject matter. Then, the group takes the same test while collaborating on the answers. If the groups do not agree, they can talk with the instructor to appeal their grades, showing they believe they are correct. After all, grades are given on the test, and the instructor gives a lecture on the content while addressing the areas students find the most difficult. Next, students work on an active learning exercise in their groups. Finally, the instructor summarizes the covered content (Gopalan & Klann, 2017).

Although traditionally used in the face-to-face classroom, TBL has potential in online courses and the flipped classroom model. For example, students could leverage the accountability element of TBL to ensure everyone watches the videos. Discussion boards could also be designed so that students would watch pre-recorded videos to answer questions and participate in discussion threads. Class preparedness can also be included in the grading rubric and the overall group assessment.

OPPORTUNITIES AND CAVEATS

If any positive outcomes resulted from the COVID-19 pandemic, a long overdue reevaluation of curriculum and instruction would be among them. These lessons must be preserved as education transitions to a post-pandemic new normal. Valuable time and effort were dedicated to developing videos and other digital content; those resources must not be wasted. However, as with any significant shift, several opportunities and caveats should be considered.

Teach Students How to Interact With the Content

Students often benefit from being provided with specific strategies for interacting with course content. While this may seem simplistic or apparent to those of us in instructor or instructional designer roles, it is often overlooked. For example, students can prepare for tests in numerous ways, including creating mind maps, developing outlines, and reviewing class notes. Similarly, students may draft an essay from different perspectives, some using a more methodical approach of brainstorming and outlining, while others jump into writing first and editing afterward. Providing students with options and ideas for interacting with content helps them find approaches that work best for them, making them more likely to remember and understand the material and remain engaged.

Guiding students in using course content can be critical regarding digital materials. Many of us remember doing homework with textbooks but cannot recall being taught to stop reading when our eyes were tired or unable to concentrate. We knew it was time to stop and start again later. However, it appears that it is not the same with digital content. Although binge-watching is a popular pastime, this does not translate well to watching videos for learning. Rather than approach video content like readings in multiple sittings, students might simultaneously try to consume all the videos in an online module. Students trying to consume all the videos in an online module simultaneously may lead to feeling overwhelmed, lacking focus, and inability to retain information. Instructors can guide students to use educational videos effectively by modeling proper viewing habits and offering tips on staying focused. Teaching students to maximize digital resources can improve student engagement and overall outcomes.

Make the Content Manageable

Have you ever caught yourself singing along to a favorite song on the radio? How did you learn the lyrics? For many, it was by listening to the song multiple times until the content finally stuck. Similarly, students benefit from the repetition of course material. Laster-Loftus and Cooper (2019) demonstrated that multiple viewings of instructional videos resulted in higher grades. Other studies have shown that repetition assists in the conversion from short-term to long-term memory (Squire & Dede, 2015; Zhan et al., 2018). One way this can occur is through dual coding, taking advantage of the brain's ability to simultaneously take in two information channels and convert that information to long-term memory (Cuevas & Dawson, 2018; Paivio, 1978). One key to dual coding is to ensure it is limited to two channels. More than that can become overwhelming and exceed cognitive load, the amount of information learners can take in before they can no longer retain knowledge (Sweller, 2020), for example, using too many sensory channels, mismatched text, and visual cues, and adding animations, text, and sounds to a classroom presentation. Cognitive load is often associated with video length; however, that applies to all cases and should not be considered when evaluating content (Altinpulluk et al., 2019; Costley et al., 2021).

Educators can instruct students about the importance of viewing videos multiple times. Authors such as Rose (2009) and Scagnoli et al. (2019) have shown that students often do not fully watch all the videos in an online course. There may be several reasons for this. For example, students may become overwhelmed with an extensive list of videos to watch for a course and give up. This feeling of becoming overwhelmed could be alleviated by providing detailed instructions on using video content and stressing the importance of pacing.

Technology Issues and Disparities

While the popularity and usage of videos for learning have skyrocketed in recent years, the costs associated with these technologies have remained prohibitive. The cost of hardware and software for creating and editing videos has made it difficult for many schools. As a result, faculty members may have to use personal equipment or borrow from colleagues to create videos for their classes. While this may not seem like a significant problem at first, it can lead to severe disparities in the quality of education. Schools with higher technology budgets could have more access to state-of-the-art video tools and provide students with higher-quality resources.

In comparison, institutions without the same level of funding may be forced to settle for lesser-quality solutions. In an era when video is rapidly becoming an essential part of the educational landscape, this lack of support by institutions may become untenable. Universities should consider ways to provide faculty with the resources necessary to produce high-quality videos.

Technology Training for Students

As educational technology becomes increasingly common in schools, ensuring students are adequately trained in using these technologies may become more critical. While students might be comfortable using essential technologies like word processing and email, they may not be as familiar with more complex tools like the learning management system (LMS) and media streaming software. As a result, faculty must provide training on how to use these platforms. This training can be accomplished in several ways, and faculty members can hold workshops or provide such instruction in their syllabi. In addition, institutions must provide support services to answer student questions and help troubleshoot technical problems. Ensuring that students are adequately trained to use educational technologies can help them get the most out of their education and the available resources.

Technology Training for Faculty

Faculty members might also benefit from training to use video software and hardware. With the increasing popularity of online education, many students choose to take courses from home. As a result, faculty members should be able to deliver lectures online, use videoconferencing software such as Skype and Zoom for synchronous sessions, and use other technologies, such as digital video cameras and video editing software. Although, in the past, video editing was highly technical and required significant expertise, it has evolved to become much more user-friendly. Several video editing solutions require little technical knowledge to edit video duration. The ability to edit videos is essential as instructors look to repurpose video content that was created during the pandemic.

There are several considerations instructors might consider when repurposing videos. Although instructors have videos of recorded lectures with lengths that meet higher educational credit hour equivalency requirements, longer videos are not ideal for the cognitive load (Aalioui et al., 2022; Altinpulluk et al., 2019). When videos cannot be shortened, instructors can encourage students to stop watching them when fatigued. To benefit fully from the video content, learners should apply what they have watched almost immediately to ensure that the information goes from short-term to long-term memory (Aalioui et al., 2022). Providing faculty members with the necessary training can ensure that faculty are well prepared for remote learning and that students continue receiving high-quality educational content.

Access to Reliable High-Speed Internet

Dependable high-speed Internet access is often overlooked regarding online education. Internet access is still limited in many rural areas; even in urban areas, service can be unreliable and expensive (Lai & Widmar, 2021). In an age when Internet connectivity is increasingly seen as a basic need, those without reliable high-speed connectivity face a significant educational disadvantage. Uploading and downloading videos can be frustratingly slow, and this can be a considerable barrier for instructors who want to repurpose videos for online courses and students in those courses. Instructors using video content can inform students about connectivity requirements at the beginning of the course and provide links to troubleshooting resources.

ACADEMIC REGULATIONS AND POLICIES

Academic regulations are essential in every educational experience, whether online or face-to-face. These regulations are established by larger governing agencies such as the U.S. Department of Education and individual institutions. They ensure students are held to established standards and learn in fair and consistent environments, creating a level playing field for all students. Academic regulations are fundamental in online learning because they help ensure that all students have the same access to materials and resources as students in a face-to-face classroom setting. Typical regulations include attendance requirements, grading policies, and conduct standards. Instructors typically outline such expectations at the beginning of the semester so students can know them before enrolling in a course. Instructors include many expectations to promote student success. For example, attendance requirements exist to ensure that students are physically present for lectures and discussions, which is essential for learning the material. Another common policy concerns grading policies that ensure students receive fair work evaluations. Conduct standards exist to promote respectful and civil interactions between students, faculty, and staff. These regulations are essential for maintaining a positive learning environment. During the COVID-19 pandemic, many academic regulations and policies were suspended due to the unprecedented nature of the situation (U.S. Department of Education, 2020c). Higher education institutions were intentional in their efforts to distinguish emergency remote learning during COVID from pedagogically sound, fully asynchronous online courses (Members of the National Council for Online Education, 2022). However, recently many of those suspensions have been lifted.

Accessibility

The Americans with Disabilities Act (ADA) of 1990 prohibited discrimination against people with disabilities in employment, education, transportation, and all state and local government activities. The Rehabilitation Act of 1973 covered similar territory but was more limited in scope. Section 504 of the Rehabilitation Act prohibited discrimination against people with disabilities in programs and activities that receive federal financial assistance. Section 508 of the Rehabilitation Act extended the prohibition against discrimination to include electronic and information technology. These two laws allowed people with disabilities to participate fully in the educational process and access information available electronically (U.S. Department of Education, 2020a, 2020b).

While online learning has many benefits, such as increased flexibility and the ability to learn at your own pace, it has brought with it some additional areas of concern regarding accessibility, such as video content. Although videos can convey information effectively, they can also pose difficulties for students with visual impairments or other disabilities. Students who cannot access reliable Internet connections may have trouble streaming video content. As instructors seek to repurpose video content created during the pandemic and create new digital artifacts, these accessibility needs should be kept in mind constantly.

The World Wide Web Consortium's "Web Content Accessibility Guidelines" (WCAG) provide standards for making web content accessible to people with disabilities. There are three levels of compliance: A, A.A., and AAA. Level A sets a minimum level of accessibility but does not achieve broad accessibility for many situations. Although AAA is the highest and most challenging level of compliance, it is often difficult to achieve for most institutions due to available resources and time. Most organizations aim for A.A. compliance, which is more reasonable for organizations and provides a substantial level of accessibility for most users. However, it is essential to note that even A.A. compliance does not guarantee that all users can access the content; it merely sets a minimum standard (Nelson, 2020; W3C Web Accessibility Initiative (WAI), 2018, 2021).

Several considerations must be considered as an institution begins to consider WCAG ratings and guidelines. A specific WCAG rating highly depends on the video type of content. For example, suppose a video only contains speech or other audio. In that case, captions are required for an A rating, and a transcript of the audio information is required for a AAA rating. In comparison, if a video includes visual information (for example, a PowerPoint presentation or pictures) needed to understand the content, an audio description of the visual information is required for an A rating, and a descriptive transcript is required for a AAA rating. For the A.A. rating, either an audio description of the visual information or a descriptive transcript would be sufficient.

Creating video transcripts also depends on where videos are hosted. Using YouTube or Kaltura to host videos allows for auto-generated captions that will produce editable caption files that can then be used to create transcripts. However, other media streaming services do not assist in creating captions or transcripts. Another way to create them is to use a text-to-speech program to generate an audio recording of your video: the text would become the transcript. As a video plays, An instructor could also type out the transcript. While this can be time-consuming, it will allow for creating more accurate transcripts. Whichever method one chooses, creating a transcript can help ensure that everyone can understand a video; as you can see, making course video content accessible can be a difficult and somewhat complex process. However, ensuring all students receive the same level of quality online education is also vital.

Credit Hour Equivalency

Even before the COVID-19 pandemic, many colleges and universities had begun to transition to offering online courses to remain competitive. However, despite this growth, there has always remained some debate around the quality of online education compared to traditional classroom instruction. One primary question among critics is whether online courses can provide the same value in terms of academic credit. Most institutions often address the rigor of an online course using credit hour equivalency. Credit hour equivalency means that a student who takes an online course should expect to receive the same number of credits as a student who takes the same course in a traditional classroom setting. In most higher education institutions, online courses are equivalent to traditional courses in terms of credit hours. Many programs follow the U.S. Department of Education's definition of credit hour equivalency (34 CFR 600.2), which is determined by a student's total time spent on learning, studying, and completing assignments or tasks for a course (The Electronic Code of Federal Regulations, 2022). By using this metric, institutions can ensure that their online courses meet the same level of interaction as their face-to-face offerings.

Due to the format, there are some cases where an online course offering may differ from its face-to-face counterpart. For example, an online course may offer more flexibility regarding credit hours (e.g., providing a three-hour credit option for a course instead of the traditional four-hour option). This flexibility can be a valuable alternative for students who want to lighten their workload or who desire to take courses at a faster pace. As the profile of students pursuing higher education continues to evolve, this flexibility will become an even larger asset for institutions by offering the same academic value as traditional courses while accommodating the desire for flexibility in terms of credit hours.

Credit hour equivalency requirements can also be a consideration to keep in mind when looking at repurposing video content created during the COVID-19 pandemic. This repurposing can result in a significant benefit for instructors looking to meet these requirements. While it is impossible to assess the total video content created during the pandemic accurately, it is fair to assume that faculty members have invested many hours in developing these materials over the years. By integrating these recorded lectures into their courses, instructors enhance their curriculum while addressing content length requirements for higher education credit hour equivalency (Downs, 2020a, 2020b).

Privacy

Privacy concerns are one final caveat that the practitioner should consider when integrating video into an online classroom. Privacy concerns may depend on how the original videos were created during the pandemic. Overall, video recordings tend to fall into two primary buckets. In the first bucket are videos recorded before each class session and provided for students to watch asynchronously. These types of videos are commonly used in the flipped classroom model. From a privacy perspective, there are very few issues that instructors should mind as they repurpose this content. The second bucket includes video content created from synchronous lecture sessions recorded during live sessions with students. There are many benefits from video content created in this fashion, including time efficiency for the instructor and the ability for students to learn from the discussions and questions asked by their peers who have taken the class before them. Although most video conferencing software like Zoom requires attendees to agree to be recorded during a session, that does not always mean they consent to share their interactions with future classes.

However, there are some solutions instructors can utilize so that these hours of classroom videos do not go to waste. One way to address this concern is to take the time to edit these types of videos recorded during the live sessions so that student questions or other interactions are not included. Not including these interactions allows the lectures to remain intact while respecting the students' privacy. Another potential solution is to explicitly ask students for permission to use their recorded classroom discussions and interactions in future sessions. This option helps the instructor to hopefully maintain the integrity of the recording without having to spend a great deal of time and resources on editing while still allowing the students to have a voice in the process. If going with this solution, one thing to remember is that if any student does object to being included in future materials, they should be edited out to respect privacy.

REAL-WORLD EXAMPLE

The primary motivation for this chapter was the practical application of the issue. As an instructional designer and instructor, I continually encounter faculty facing the quandary of what to do with the large amount of video content they created during the COVID-19 pandemic. While many will agree that there is room for improvement in the materials, there is also a desire to continue to build upon what they developed and use it to enhance their courses. I recently encountered a prime example of this with a faculty member with whom I collaborated on a course. This faculty member was teaching an undergraduate statistics course, a complex and highly technical class for many students. Traditionally, this class was always taught in person; however, it was moved into a remote learning format during the pandemic. To accommodate this move and to ensure student learning continued despite the disruption, the instructor worked hard to create several hours of video content.

The instructor offered face-to-face and online courses when the university returned to in-person classes. As we worked together to develop the asynchronous online course, we intentionally tried to find ways to repurpose the videos used in the previous iteration of the course. In a regular online course build, the faculty member creating the online course has to split time between creating meaningful lecture videos and creating the rest of the materials for the course. Typically, the time allotted for these tasks does not allow for the instructor to do both entirely, resulting in work having to be done after the official course build has concluded. However, the videos had already been recorded in this particular case. Reusing these videos that the instructor created during the pandemic allowed them to focus their efforts on the other elements of the course, such as the quizzes, written elements, shell build-outs, and assessments. Some minimal video editing was done, including machine-generated captioning and minor manual caption file cleanup by the instructor. The course included approximately 18 hours of video content across 16 weeks (about three and a half months) of instruction.

By focusing on repurposing the video content rather than creating new versions of the material, we produced a comprehensive course that addressed a very complex technical topic in an asynchronous online format. Faculty and instructional designer time and staff resources were maximized by reusing the materials created during the pandemic rather than creating new material. It also resulted in a more complete course for the students with more resources and materials than would have usually been included. Some may argue that the instructor had too much video in the course, with around 18 hours of lectures. However, if that amount is divided by the length of the semester, it results in approximately 60 minutes of in-class time each week. The rest of the in-class time included readings and other activities to achieve credit hour equivalency.

CONCLUSION

The COVID-19 pandemic disrupted many elements of our world in ways that had never been seen before, including education. While the transition has not been without challenges, it has ultimately resulted in several new and innovative approaches to learning. Models such as the flipped classroom method, once thought to be limited to the face-to-face classroom, are finding new applications in the distance education environment and leading to enhanced classroom experiences for students and instructors. These lessons must not be lost as colleges and universities return to normal. Organizations can continue to examine materials, such as video content and other online learning materials that were created and discover new and innovative ways to integrate them into the classroom. As Kurt Lewin's change model noted, every unfreezing moment eventually has that transition back to refreezing, in which individuals or groups fully embrace the change and make it part of their new way of life. The COVID-19 pandemic is a once-in-a-lifetime moment for education to seize. As we rapidly move closer to this post-pandemic in higher education and online learning, we can shape a new way of life that we genuinely believe should be.

REFERENCES

W3C Web Accessibility Initiative (WAI). (2021). *Planning Audio and Video Media*. https://www.w3.org/WAI/media/av/planning/

Aagaard, J. (2019). Multitasking as distraction: A conceptual analysis of media multitasking research. *Theory & Psychology*, *29*(1), 87–99. doi:10.1177/0959354318815766

Aalioui, L., Gouzi, F., & Tricot, A. (2022). Reducing cognitive load during video lectures in physiology with eye movement modeling and pauses: A randomized controlled study. *Advances in Physiology Education*, *46*(2), 288–296. doi:10.1152/advan.00185.2021 PMID:35297692

Altinpulluk, H., Kilinc, H., Firat, M., & Yumurtaci, O. (2019). The influence of segmented and complete educational videos on the cognitive load, satisfaction, engagement, and academic achievement levels of learners. *Journal of Computers in Education, 7*(2), 155-182. doi:10.1007/s40692-019-00151-7

Anderson, T., Rourke, L., Garrison, D. R., & Archer, W. (2001). Assessing teaching presence in a computer conferencing context. *Journal of Asynchronous Learning Networks*, *5*(2), 1–17.

Aragon, R. (2016). What has an Impact on Grades? Instructor-Made Videos, Communication, and Timing in an Online Statistics Course. *Journal of Humanistic Mathematics*, *6*(2), 84–95. doi:10.5642/jhummath.201602.07

Artz, B., Johnson, M., Robson, D., & Siemers, S. (2022). Live or lecture capture: Evidence from a classroom random control trial. *International Review of Economics Education, 40*, 100240. doi:10.1016/j.iree.2022.100240

Belt, E. S., & Lowenthal, P. R. (2021). Video use in online and blended courses: A qualitative synthesis. *Distance Education, 42*(3), 410–440. doi:10.1080/01587919.2021.1954882

Byrnes, K. G., Kiely, P. A., Dunne, C. P., McDermott, K. W., & Coffey, J. C. (2021). Communication, collaboration and contagion: "Virtualisation" of anatomy during COVID-19. *Clinical Anatomy (New York, N.Y.), 34*(1), 82–89. doi:10.1002/ca.23649 PMID:32648289

Carnegie Mellon University. (2020). *Higher Education Was Already Ripe for Disruption. Then, COVID-19 Happened.* Carnegie Mellon University. https://www.cmu.edu/news/stories/archives/2020/september/higher-education-covid-disruption.html

Cheng, L., Ritzhaupt, A. D., & Antonenko, P. (2018). Effects of the flipped classroom instructional strategy on students' learning outcomes: A meta-analysis. *Educational Technology Research and Development, 67*(4), 793–824. doi:10.100711423-018-9633-7

Costley, J., Fanguy, M., Lange, C., & Baldwin, M. (2021). The effects of video lecture viewing strategies on cognitive load. *Journal of Computing in Higher Education, 33*(1), 19–38. doi:10.100712528-020-09254-y

Cuevas, J., & Dawson, B. L. (2018). A test of two alternative cognitive processing models: Learning styles and dual coding. *Theory and Research in Education, 16*(1), 40–64. doi:10.1177/1477878517731450

Delahunty, J. (2018). Connecting to learn, learning to connect: Thinking together in asynchronous forum discussion. *Linguistics and Education, 46*, 12–22. doi:10.1016/j.linged.2018.05.003

Downs, L. R. (2020a). *New Regulations Review #2: Credit Hour, Title IV Eligibility, & What Exactly is a Week?* https://wcet.wiche.edu/frontiers/2020/04/06/new-regulations-review-2/

Downs, L. R. (2020b). *Surprise! Newly Released Final Regulations.* https://wcet.wiche.edu/frontiers/2020/08/26/newly-released-final-regulations/

Enz, S., Hall, A. C. G., & Williams, K. K. (2021). The Myth of Multitasking and What It Means for Future Pharmacists. *American Journal of Pharmaceutical Education, 85*(10), 8267–1020. doi:10.5688/ajpe8267 PMID:34965913

Friedman, B. D., & Neuman Allen, K. (2014). Systems Theory. In J. R. Brandell (Ed.), *Essentials of clinical social work* (1st ed., pp. 3–20). SAGE. doi:10.4135/9781483398266.n2

Galikyan, I., & Admiraal, W. (2019). Students' engagement in asynchronous online discussion: The relationship between cognitive presence, learner prominence, and academic performance. *The Internet and Higher Education, 43*, 100692. doi:10.1016/j.iheduc.2019.100692

Garrison, D. R. (2011). A Framework for Research and Practice (2nd ed.). Routledge.

Garrison, D. R. (2016). E-Learning in the 21st Century: A Community of Inquiry Framework for Research and Practice (3rd ed.). Routledge.

Garrison, D. R., Anderson, T., & Archer, W. (1999). Critical Inquiry in a Text-Based Environment: Computer Conferencing in Higher Education. *The Internet and Higher Education*, *2*(2), 87–105. doi:10.1016/S1096-7516(00)00016-6

GarrisonD. R.Cleveland-InnesM.VaughanN. (2019). *CoI Publications.* https://coi.athabascau.ca/publications/

Gewertz, C. (2020). *The Coronavirus Spring: The Historic Closing of U.S. Schools.* https://www.edweek.org/leadership/the-coronavirus-spring-the-historic-closing-of-u-s-schools-a-timeline/2020/07

Gierdowski, D. C. (2019). *2019 Study of Undergraduate Students and Information Technology.* https://library.educause.edu/resources/2019/10/2019-study-of-undergraduate-students-and-information-technology

Gopalan, C., & Klann, M. C. (2017). The effect of flipped teaching combined with modified team-based learning on student performance in physiology. *Advances in Physiology Education*, *41*(3), 363–367. doi:10.1152/advan.00179.2016 PMID:28679573

Joshua, C. (2016). An analysis of current evidence supporting two alternate learning models: learning styles and dual coding. *Journal of Educational Sciences & Psychology, 6*(68).

Kamenetz, A. (2020). *6 Ways Universities Are Responding To Coronavirus.* https://www.npr.org/2020/03/06/812462913/6-ways-universities-are-responding-to-coronavirus

Kirschner, P. A., & De Bruyckere, P. (2017). The myths of the digital native and the multitasker. *Teaching and Teacher Education*, *67*, 135–142. doi:10.1016/j.tate.2017.06.001

Lai, J., & Widmar, N. O. (2021). Revisiting the Digital Divide in the COVID-19 Era. *Applied Economic Perspectives and Policy*, *43*(1), 458–464. doi:10.1002/aepp.13104 PMID:33230409

Laster-Loftus, A., & Cooper, L. (2019). The Role of Instructional Videos in Psychology Students' Course Success. *The Journal of Educators Online*, *16*. Advance online publication. doi:10.9743/JEO.2019.16.2.5

Lewin, K. (1997). *Resolving social conflicts; &, Field theory in social science.* American Psychological Association. doi:10.1037/10269-000

Lowenthal, P., Borup, J., West, R., & Archambault, L. (2020). Thinking Beyond Zoom: Using Asynchronous Video to Maintain Connection and Engagement During the COVID-19 Pandemic. *Journal of Technology and Teacher Education*, *28*(2), 383.

Members of the National Council for Online Education. (2022). *Emergency Remote Instruction Is Not Quality Online Learning.* https://www.insidehighered.com/views/2022/02/03/remote-instruction-and-online-learning-arent-same-thing-opinion

Michaelsen, L. K., & Sweet, M. (2008). The essential elements of team-based learning. *New Directions for Teaching and Learning*, *2008*(116), 7–27. doi:10.1002/tl.330

Millard, E. (2012). 5 Reasons FLIPPED Classrooms Work. *University Business*, *15*(11), 26–29.

Nelson, G. (2020). *WCAG 2.1 Guidelines Explained with Examples.* https://medium.com/c2-group/wcag-2-1-guidelines-explained-with-examples-5c7c5d8b69eb

November, A., & Mull, B. (2012). *Flipped Learning: A response to 5 Criticisms -What's Flipped Learning?* https://novemberlearning.com/article/flipped-learning-a-response-to-five-common-criticisms-article/

Paivio, A. (1978). *A Dual Coding Approach to Perception and Cognition.* Psychology Press.

Pekdağ, B. (2010). Alternative methods in learning chemistry: Learning with animation, simulation, video and multimedia. *Journal of Turkish Science Education*, *7*(2), 79–110.

Pokhrel, S., & Chhetri, R. (2021). A Literature Review on Impact of COVID-19 Pandemic on Teaching and Learning. *Higher Education for the Future*, *8*(1), 133–141. doi:10.1177/2347631120983481

Rose, K. K. (2009). Student perceptions of the use of instructor-made videos in online and face-to-face classes. *Journal of Online Learning and Teaching*, *5*(3), 487–495.

Scagnoli, N. I., Choo, J., & Tian, J. (2019). Students' insights on the use of video lectures in online classes: Students' insights on video lectures. *British Journal of Educational Technology*, *50*(1), 399–414. doi:10.1111/bjet.12572

Singh, S., & Arya, A. (2020). A hybrid flipped-classroom approach for online teaching of biochemistry in developing countries during Covid-19 crisis. *Biochemistry and Molecular Biology Education*, *48*(5), 502–503. doi:10.1002/bmb.21418 PMID:32790218

Squire, L. R., & Dede, A. J. O. (2015). Conscious and unconscious memory systems. *Cold Spring Harbor Perspectives in Biology*, *7*(3), a021667–a021667. doi:10.1101/cshperspect.a021667 PMID:25731765

Sweller, J. (2020). Cognitive load theory and educational technology. *Educational Technology Research and Development*, *68*(1), 1–16. doi:10.100711423-019-09701-3

The Electronic Code of Federal Regulations. (2022). *eCFR: 34 CFR Part 600 -- Institutional Eligibility Under the Higher Education Act of 1965, as Amended.* https://www.ecfr.gov/current/title-34/subtitle-B/chapter-VI/part-600

The Mind Tools Content Team. (2022). *Lewin's Change Management Model: Understanding the Three Stages of Change.* https://www.mindtools.com/ajm911e/lewins-change-management-model

Tomei, L. A. (2008). *Online and distance learning concepts, methodologies, tools, and applications.* Information Science Reference. doi:10.4018/978-1-59904-935-9

U.S. Department of Education. (2020a). *Accessibility Statement.* https://www2.ed.gov/notices/accessibility/index.html

U.S. Department of Education. (2020b). *Protecting Students With Disabilities.* https://www2.ed.gov/about/offices/list/ocr/504faq.html

U.S. Department of Education. (2020c). *UPDATED Guidance for interruptions of study related to Coronavirus (COVID-19) | Knowledge Center.* https://fsapartners.ed.gov/knowledge-center/library/electronic-announcements/2020-05-15/updated-guidance-interruptions-study-related-coronavirus-covid-19-updated-june-16-2020

W3C Web Accessibility Initiative (WAI). (2018). *Web Content Accessibility Guidelines (WCAG) 2.1.* https://www.w3.org/TR/WCAG21/

Wong, G., Apthorpe, H. C., Ruiz, K., & Nanayakkara, S. (2019). An innovative educational approach in using instructional videos to teach dental local anaesthetic skills. *European Journal of Dental Education, 23*(1), 28–34. doi:10.1111/eje.12382 PMID:30069994

Zhan, L., Guo, D., Chen, G., & Yang, J. (2018). Effects of Repetition Learning on Associative Recognition Over Time: Role of the Hippocampus and Prefrontal Cortex. *Frontiers in Human Neuroscience, 12*, 277–277. doi:10.3389/fnhum.2018.00277 PMID:30050418

Chapter 8
Systems Thinking Through Mass Customization in Response to COVID-19 in the Higher Education Context

Daniel Gutmore
Seton Hall University, USA

Al Galloway
Seton Hall University, USA

ABSTRACT

This chapter about the systemic changes that took place at one small university during the pandemic tells the story of what happened in March 2020 and how the department was positioned to address the needs of the students. The Corona-19 virus was the event that closed schools and forced the college's remote response for teaching and communication. A short survey was issued to students about their experience. The survey yielded a positive response about the department's readiness and response to the students. Yet, when looked at holistically and systemically, another paradigm emerges that resembles the mass customization concept evolving in the manufacturing and service industries for greater than 60 years. Mass customization as a systems thinking paradigm is the theoretical frame of this chapter. The mass customization paradigm is adapted from the manufacturing industry. This paradigm addresses the decision to customize products based upon customer needs.

INTRODUCTION

On March 6, 2020, Stanford University and Touro College became the first institutions in the United States to announce transitions to completely online instruction because of the COVID-19 pandemic. Three days later, six colleges and universities in California, Connecticut, and Massachusetts followed. The next day, 78 institutions joined them. By the end of that week, just over half of all degree-granting

DOI: 10.4018/978-1-6684-7285-9.ch008

private, non-profit, and public 4-year institutions in the country announced transitions to online learning practices. By the end of March, nearly 1,400 institutions determined online delivery of coursework for the foreseeable future (Marsicano et al., 2020). In the follow-up analysis of how higher education institutions responded, many of the theoretical frameworks focused on isomorphic applications (DiMaggio & Powell, 1983). Clearly much may be derived from such an application. In better understanding university responses, however, a system thinking application may add a nuanced level of insight and complement the extant isomorphic analysis. To contextualize the discussion, we will use an evaluation of one university's response to the pandemic as a template for the application of systems thinking.

The University, under discussion for this chapter, began online instruction in March 2020 during the Corona-19 shutdown. Marsicano et al. (2020) offer a useful conceptual framework that guided much of the analysis that followed higher education's responses. It is the neo-institutional theory of isomorphism that has a history of application to higher education departmental behavior (DiMaggio & Powell, 1983). The concept advances three typologies; mimetic, coercive, and normative behaviors. Coercive isomorphism may be evident when the governors of various states required all institutions to shift to remote learning, resulting in potential mimetic behavior as higher education institutions examined how their counterparts responded. Mimetic is the common, almost identical approach institutions engaged in, even though they were each left to their own responses. Normative behavior was less observed because of the immediacy of the responses, with no organizational group present to influence the institution. Another common analytical framework was offered by Hodges et al. (2020) in their study of "emergency remote teaching." They distinguished the strategic development of online learning and what they characterize when responding to the emergency conditions the Covid-19 pandemic produced. Often overlooked in the previous analysis is the connection of systems thinking and its impact on the various responses, particularly as they differ from the application of isomorphic applications.

In reviewing the extant literature on what higher education institutions did in response to the pandemic, most studies make no connections to the application of systems thinking in their analysis. Among the approaches are those outlined by Senge (1990) in his treatise, *The Fifth Discipline.* Senge may not have started the systems thinking trend, but his work accelerated it. In his book, Senge provides some excellent examples of compensating feedback, application of systems thinking to terrorism, and systems archetypes. Senge (1990, p136 also provides a generic definition: "Systems thinking is a discipline for seeing wholes. It is a framework for seeing interrelationships rather than things, for seeing patterns of change rather than 'snapshots.'" It is also a set of specific tools and techniques, originating in two threads: in "feedback" concepts of cybernetics and in "servo-mechanism" engineering theory." Another approach is provided by Ballé (1996), in the book *Managing With Systems Thinking: Making Dynamics Work for You in Business Decision Making.* Ballé's text focuses on applying systems concepts to the workplace. He observes that a typical management reaction to an issue is a myopic, short -term solution instead of a long-term systemic analysis, and that therefore many "solved" problems recur. Checkland's (1981), *Systems Thinking, Systems Practice* book, originally published in 1981, distinguishes "hard" systems thinking (for which problems may be formulated by making choices among alternatives to achieve an end) from "soft" systems thinking, such as human activity or social systems, which are poorly structured and often harder to deal with. Although all these ideas contribute to connections to how systems thinking may have contributed applications taken by higher education institutions to the pandemic, we advance the most appropriate is mass customization. Thus, the systems thinking guiding this discussion is the idea of the *Mass Customization* approach developed by Seddon (2008), which depends on the choice of a proper service operations design (Da Silveira et al., 2001; Moon et al., 2010). In addition, when considering how to design services from which customers can 'pull value'– i.e., get what they want – Seddon (2008) believes that it is critical to shift from thinking of the system as one that pulls things together at the rate of customer demand to thinking of the system as one that offers services together in response to the variety of customer demands, called 'systems-thinking'. Developed since 2003 by Seddon, the terminology of 'systems-thinking' used throughout this chapter to describe the service delivery system has emerged from the translation to the service sector of lean manufacturing principles (Seddon & Brand, 2008), incorporating Deming's (1982) intervention theory together with Checkland's Soft Systems Methodology (Wheeler, 2000). According to Jaaron and Backhouse (2013), the employment of the systems-thinking approach for service operations design has a significant impact on service-offering organizations which provide real-time mass-customized service. Although mass customization of student services is essential, as analyzed above, little attention was paid by earlier empirical studies to models and tools that can operationalize mass customization in universities, particularly as they responded to the vagaries of the pandemic. If we consider students as the customers, then their satisfaction with the university's response would be demonstrative of the systems thinking application of mass customization. To further enhance the understanding and application in a university setting, one university's experiences may offer illustrative insights.

LITERATURE REVIEW

Defining Mass Customization

Mass customization is the process of delivering market goods and services that are modified to satisfy a specific customer's needs. The key is to view it basically as a process for aligning an organization with its customers' needs. That is, "mass customization is [...] about moving toward [stated] goals by developing a set of organizational capabilities that will, over time, supplement and enrich an existing business" (Salvador et al., 2009, p71).

Using a systems approach to mass customization encompasses cutting-edge management methods and tools that give companies the ability to produce customized, affordable, and high-quality services. The key dimensions of this framework are customer sensitivity, process amenability, competitive environment, and organizational readiness (Hart, 1995).

The Corona-19 virus event made Hart's vision of readiness apparent as professors could deliver course material and teach remotely despite the hastened response to change. We also found that focusing on the needs of the customer (student) the hybrid design of a delivery system (classroom) morphed into teaching methods that were on-demand and included additional individualized student meetings and classroom accommodations.

Neo-institutional Theory

Marsicano et al. (2020) offer a useful conceptual framework that will guide the analysis. It is the neo-institutional theory of isomorphism that has a history of application to higher education department behavior (DiMaggio & Powell, 1983). The concept advances three typologies: mimetic, coercive, and normative behaviors. Coercive isomorphism may be evident when the governor of New Jersey required all institutions to shift to remote learning, resulting in potential mimetic behavior as higher education institutions examined how their counterparts responded. Another useful analytical framework is offered by Hodges et al. (2020) in their study of "emergency remote teaching" They distinguished strategic development of online learning and what they characterize responding to the emergency conditions the Covid-19 pandemic produced.

Categorical Aspects

Extant literature falls into three categories. First is descriptive analysis of what action institutions of higher education engaged in responding to the pandemic and the resulting shutdown of on-campus instruction. Second is the exploration of the relative advantages and disadvantages of various approaches to remote learning. Third are the attempts to derive meaning and insight into events and actions and their implications for the future of higher education. Within the context of this study, the focus is primarily on descriptive analysis of what institutions did during the pandemic crisis and then determining what policy implications emerged and what the implications might be for future practice in the higher education environment. In terms of descriptive analysis of the various responses, Marsicano et al. (2020) provide a comprehensive analysis of the unfolding of higher education's responses from the beginning of the pandemic, with Stanford and Touro responding first in March 2020. They were followed by over 1400 institutions closing by the end of that month. They confirmed their perspective of bounded rationality and isomorphism by noting the similarity in responses throughout the higher education community. Those responses were mimetic in orientation, with all the institutions using various forms of remote learning.

Bao (2020) presented a case study of Peking University's responses. Noting the history of Chinese Universities in providing online learning, the results were traumatic because of the suddenness of the transition and the lack of faculty experience and preparation to implement the transition. Prokopenko and Berezhna (2020) explored the responses in Ukraine institutions of higher education. They found similar levels of responses as described regarding China and the United States (advancing the isomorphic nature of the responses) and added a survey component of seeking feedback from students and teachers regarding their experiences. The identified issues related to technical preparedness, with not all institutions equally prepared and psychological dissonance because of confinement during instructional delivery. Prokopenko and Berezhna (2020) concluded that online learning was not perceived as inferior to in-seat instruction, with positive effects on student-teacher relations.

Hodges et al. (2020) in examining responses distinguished traditional concepts of online learning and what they characterized as "emergency remote teaching (ERT) (page 1)" They rejected comparing the relative merits of online learning with in-seat and instead advanced a different conceptual framework, which more precisely identified the situation: emergency remote learning. They focused their analysis on the context, input, and process elements that occurred during the pandemic and the quick move to ERT. Thus, they purport that understanding how the crisis affected the product (learning) is not the most relevant. Instead, understanding the effects of the customization are.

ONE UNIVERSITY'S RESPONSE

The focus was on the students and faculty in an educational leadership department and their perceptions of the sudden convergence to remote learning. Concern was raised by faculty in the department regarding how the process unfolded in terms of students' and professors' perspectives. Two faculty members were commissioned to engage in the study and then share the results accordingly. The faculty hypothesized that the department under consideration was strategically positioned to accommodate the changes brought on by the pandemic because of the history of online instruction, along with the training and experience of its faculty in providing remote instruction along with a perceived disposition toward mass customization. The university had been offering online instruction for 20 years and consistently and continually trained faculty in online delivery techniques and processes. In promoting student satisfaction, as related to mass customization, the department within the university consistently surveyed students relative to their satisfaction with online instruction.

Ninety-one (91) students were identified who met the criteria of having been directly impacted by the shutdown of the university and the provision of only remote learning in response to the pandemics mandated shut down protocols. Using the concept of emergency remote learning (Hodges et al., 2020), a short survey was constructed, which avoided comparing the relative merits of online versus in-seat instruction while only examining the students' perception of the department's implementation of emergency remote learning. (Faculty were also identified based on their participation in converting to emergency remote learning and interviewed regarding their perceptions but the results are not discussed in this chapter). This method provided a purposive sample that included graduate students enrolled in courses offered by the department during the spring semester of 2020. This set of students was enrolled in master's, education specialist, and doctoral programs. These students were identified because they were directly affected by the move to switch from in-seat to online course delivery (coercive) brought on by the Corona-19 virus.

Results From the Response

Ninety-one (91) students received a three-question survey through the Survey Monkey platform with the opportunity for open-ended comments in a fourth question. Specific demographic information was requested. Forty-two (42) students responded. Specific details of the responses are provided in the Appendix. In terms of the first question, "my experience with exclusive remote/online learning provided a meaningful educational experience," the results are presented in Figure 1.

In response to the second question, "my instructor attempted to accommodate my personal needs in the transition to remote learning," the results are indicated in Figure 2.

In response to the third question, "I would categorize my transition to remote learning as meaningful and accommodating," we show the results are indicated in Figure 3.

There were also opportunities for respondents to comment about their transition to remote learning. Thirty-six (36) students commented on their experiences. *Ascribe Verbatim Analysis* software was used to analyze the responses and three thematic conclusions emerged. Although efforts were made in the instructions to respondents regarding the avoidance of comparing remote and or online learning with in-seat instruction, students regularly showed a preference for in-seat instruction. A few also commented on the difference in discussions through screens versus in person or the lack of comprehensive communication when online. There was also an appreciation for the personal accommodation faculty provided (concerning the application of mass customization) in transitioning to remote learning. Last, the theme of consistent rigor both in content and course expectations emerged.

The department deemed that action research (survey) narrowly focused on the students' satisfaction would yield information for further changes and/or redesigning the course delivery methods during the Covid-19 restrictions. Running courses remotely resulted in customization of the courses, as they were delivered online and via remote access methods. The methods included adding real-time meetings while teaching both asynchronously and synchronously based on the material, the professors' availability, and our student's needs for learning.

Student Responses

A summary of the survey that was presented when the university moved to ERT shows that an evaluation of the process including context, input, and process elements was considered when implementing ERT in a short amount of time, which is critical according to Moore et al. (2002c). In our case study, the implementation time occurred while the semester was in progress during spring 2020. Here is a summary of some of the students' responses capturing the overall mood.

Online learning has been challenging for our students, who indicated a preference for in-seat instruction. The students enjoy in person classes for the social aspect but would not mind strictly online provided there was a web-based (Zoom) option to meet as a class. The students also prefer a mix of online and an in-seat instruction, but do not like the student interaction components when online.

The experience of getting a master's degree was challenging, but the instructors were very helpful. Some professors did not skip a beat going from classroom to virtual, while others uploaded videos of themselves lecturing and posted them. Most of the students appreciate in person interaction more because online learning is long and not as interactive as in person. Using a mix of materials and having the professors accessible, makes learning informative and interactive.

For a complete list of all the responses, see the Appendix. These responses were most significant to the mood of the students (customers) and their attitudes to the rollout of online courses, either synchronous or asynchronous modalities, which occurred without choice for the students.

A high level of customer satisfaction was found in the student survey. The respondents consistently showed that the university/department considered their needs and attempted to accommodate those needs.

Looking at how mass customization intersected in response to the challenges of the pandemic, several observations are relevant. First, customer satisfaction seems evident in the responses of participants. In fact, satisfaction was quite high, which indicatives successful mass customization alignment. Additionally, satisfaction was not achieved by compromising on rigor of the instructional delivery. Students consistently commented on the high level of rigor throughout the shutdown period. This balance between accommodation and rigor is also aligned with the principles of mass customization.

DISCUSSION

The department under review conducted a short evaluation of the effect on their students. This was a Soft Systems research study. Checkland's (2000) hierarchal approach to systems thinking labeled Soft Systems Methodology (SSM) places a lens on the intersection of mass customization and this isomorphic event. Checkland honed a system thinking approach over three decades and has been referenced for his SSM studies since the 1970s. Building a purposeful model of an SSM paradigm or stated differently, creating a pictorial view of this event at this school in 2020, we conjure a systemic picture of the students at the center of our event. One concerns of the department under review included trying to make sense of the Corona-19 events from the perspective of the client (student learner). The intended outcome of the short evaluation was to understand how students were affected by the changes brought on by the pandemic. We found an alignment with Checkland's hierarchy of soft systems methods wherein we discover a positive outcome for the learner. This is more than a mere discovery it was a welcomed yet sought-after outcome of the department's evaluation of the Covid-19 event on their students.

Another lens used to enlighten the results of the departments survey of their students is the holistic approach taken by Macadam and Packham (1989). They explain this phenomenon while pointing to Checkland's (1972, 1981) hierarchy of problem-solving methodologies. Checkland's outline of a soft system methodology (SSM) hierarchy, when explored, moves upward from reductionism to holism. At the upper limit of his soft systems hierarchy, Checkland focuses on the question, "how can I improve the situation?" It is the same question that inspired the department to survey their students. The department's problem in this case study was the response to the Covid-19 move to remote instruction. This problem inspired the question posed by the department's survey of the students who were affected by moving to remote learning.

CONCLUSIONS AND LOOKING FORWARD

The evaluation of the students' satisfaction was an action research study by the department, which could be enhanced. In this study, the students responded with a high level of customer satisfaction. They regularly indicated that the university/department considered their needs and attempted to accommodate those needs. However, this analysis of the question (what can I do to improve the situation?) begs more detailed information from the student's perspective.

We found that more information should be sought to provide a deeper understanding of the customization that was implemented from the student's perspective. Perhaps open-ended questions or a qualitative study would enhance the findings. For example, the professors revealed that students who taught in low-income communities experienced higher levels of stress during the transition. This development required more frequent and increased time spent with the students. Given the flexibility of the technology, software, and willingness of the professors to engage with students, the authors found the students could talk through their issues via contact with their professors. Yet, we are lacking complete understanding of the effect on the students from their perspective.

Additional revelations emerged in this study. First, satisfaction was achieved without compromising the rigor of instructional delivery. Students consistently commented on the high level of rigor throughout the shutdown period. This balance between accommodation and rigor is also aligned to the principles of mass customization (Hart, 1995; Jaaron & Backhouse, 2013). Regarding this context, understanding precisely the mass customization that occurred can be further defined and delineated from the perspective of students.

Customization of course delivery was essential at the onslaught of the Covid-19 pandemic. We found customization to be productive while meeting the needs of the learner (student) at a critical and pivotal moment. We learned that combining technology with the requirements of distancing during Covid-19 crisis may have provided a lab for how to extend the in-person classroom to more participants. Extending its impact on the transformation of educational design can/may be the most important takeaway from our assessment (Eri, 2013).

In the wake of our short survey, we may have discovered an educational design/method that meets the needs of learners without borders. Customizing for the sake of reaching students may be the wave of the future for higher education institutions. Possibly it will stand as a solid design lab (Eri, 2013) to apply to situational changes, whether driven by pandemic forces or internal forces, wherein change is required for students who cannot attend class sessions in-person.

REFERENCES

Ansmann, M., & Seyfried, M. (2022). Isomorphism and organizational performance: Evidence from quality management in higher education. *Quality Assurance in Education, 30*(1), 135–149. doi:10.1108/QAE-07-2021-0114

Ballé, M. (1996). *Managing with systems thinking: Making dynamics work for you in business decision making*. McGraw Hill.

Cardona Mejía, L. M., Pardo del Val, M., & Dasí Coscollar, M. S. À. (2020). The institutional isomorphism in the context of organizational changes in higher education institutions. *International Journal of Research in Education and Science*, *6*(1), 61–73. doi:10.46328/ijres.v6i1.639

Checkland, P. (2000). Soft systems methodology: A thirty year retrospective. *Systems Research and Behavioral Science*, *17*(S1), S11–S58. doi:10.1002/1099-1743(200011)17:1+<::AID-SRES374>3.0.CO;2-O

Da Silveira, G., Borenstein, D., & Fogliatto, F. S. (2001). Mass customization: Literature review and research directions. *International Journal of Production Economics*, *72*(1), 1–13. doi:10.1016/S0925-5273(00)00079-7

Dacin, M. T. (1997). Isomorphism in context: The power and prescription of institutional norms. *Academy of Management Journal*, *40*(1), 46–81. doi:10.2307/257020

Deming, W. E. (1982). *Quality, productivity, and competitive position*. Massachusetts Inst Technology.

DiMaggio, P. J., & Powell, W. W. (1983). The iron cage revisited: Institutional isomorphism and collective rationality in organizational fields. *American Sociological Review*, *48*(2), 147–160. doi:10.2307/2095101

Eri, T. (2013). The best way to conduct intervention research: Methodological considerations. *Quality & Quantity*, *47*(5), 2459–2472. doi:10.100711135-012-9664-9

Gutmore, D., & Galloway, A. (2021). Case study of a higher education's department's responses to the Covid 19 pandemic implications for policy and practice. *Journal of Education & Social Policy*, *8*(3), 60–66. doi:10.30845/jesp.v8n3p8

Hart, C. W. L. (1995). Mass customization: Conceptual underpinnings, opportunities and limits. *International Journal of Service Industry Management*, *6*(2), 36–45. doi:10.1108/09564239510084932

Hodges, C. B., Moore, S., Lockee, B. B., Trust, T., & Bond, M. A. (2020). *The difference between emergency remote teaching and online learning*. EDUCAUSE Review. https://er.educause.edu/articles/2020/3/the-difference-between-emergency-remote-teaching-and-online-learning

Jaaron, A. A. M., & Backhouse, C. J. (2013). Systems thinking for service delivery design: A real time mass customisation model. *IFAC Proceedings Volumes, 46*(9), 228-233. 10.3182/20130619-3-RU-3018.00053

Macadam, R. D., & Packham, R. G. (1989). A case study in the use of soft systems methodology: Restructuring an academic organisation to facilitate the education of systems agriculturalists. *Agricultural Systems, 30*(4), 351–367. doi:10.1016/0308-521X(89)90004-8

Marsicano, C., Felten, K., Toledo, L., & Buitendorp, M. (2020). *Tracking campus responses to the COVID-19 Pandemic*. American Political Science Association Preprints. doi:10.33774/apsa-2020-3wvrl

Marsicano, C. R., Barnshaw, J., & Letukas, L. (2020). Crisis and change: How COVID-19 exacerbated institutional inequality and how institutions are responding. *New Directions for Institutional Research, 2020*(187/188), 7–30. doi:10.1002/ir.20344

Moon, S. K., Shu, J., Simpson, T. W., & Kumara, S. R. (2010). A module-based service model for mass customization: Service family design. *IIE Transactions, 43*(3), 153–163. doi:10.1080/07408171003705383

Moritz, A., & Markus, S. (2021). Isomorphism and organizational performance: Evidence from quality management in higher education. *Quality Assurance in Education, 30*(1), 135–149. doi:10.1108/QAE-07-2021-0114

Salvador, F., De Holan, P. M., & Piller, F. (2009). Cracking the code of mass customization. *MIT Sloan Management Review, 50*(3), 71–78.

Seddon, J. (2008). *Systems thinking in the public sector*. Triarchy Press.

Seddon, J., & Brand, C. (2008). Debate: Systems thinking and public sector performance. *Public Money & Management, 28*(1), 7–9. doi:10.1111/j.1467-9302.2008.00611.x

Wheeler, F. P., & Checkland, P. (2000). Review of systems thinking, systems practice: Includes a 30-year retrospective. *The Journal of the Operational Research Society, 51*(5), 647–648. doi:10.2307/254200

APPENDIX

Student Response to conversion to all digital learning
Q4 Please indicate/describe your experience to exclusive online learning.

Answered	39
Sipped	1
Rsponses	**Rsponse Date**
It was engaging.	Ag 05 2020 09:45 AM
Outstanding experience.	Ag 03 2020 11:46 PM
I find the online learning to be more time and cost effective. It allows me to complete school while serving in a demanding career.	Ag 03 2020 10:57 PM
I feel that I was able to learn a lot more in person. A lot of time is spent figuring out issues with technology (no one's fault) and I feel we were unable to do activities to practice skills.	Jl 18 2020 01:31 PM
It has been challenging academically and emotionally. The interaction and ability to converse with professors are missed and minimally it would be helpful to have opportunities to converse with the professor as we do in Dr. Tienken's class.	Jl 12 2020 10:19 PM
Online learning is conceivable however I feel that all should have some form of interaction via a Zoom or other platform because some material is better comprehended when discussions are had.	Jl 11 2020 11:55 AM
Weekend class to meet 6x all together went online. Not the same, tech issues hindered class.	Jl 11 2020 08:46 AM
It was a good experience and simple transition. I was thankful my health and education weren't compromised.	Jl 09 2020 01:54 AM
It went well.	Jl 06 2020 02:25 PM
I enjoy in person classes for the social aspect but would not mind strictly online provided there was a zoom option to meet as a class.	Jul 06 2020 08:29 AM
It's difficult to sit on a computer every day to give and get feedback for discussions. It's difficult to word questions for clarity electronically. Communication is not consistent.	Jul 05 2020 10:39 PM
Not enough time to address changes to online classes.	Jl 05 2020 07:54 PM
I prefer the mix of online and an in-seat instruction. Online learning through blackboard was effectively used to communicate with professors and classmates	Jl 05 2020 07:53 PM
My program was already exclusively online.	Jl 05 2020 07:38 AM
I learn best when interacting with my professor & peers in a formal class setting. I managed well in a hybrid setting but teaching & learning online has been greatly difficult for me.	Jl 05 2020 12:02 AM
The directions were clear. Assignment deadlines were also clear. Did not like the student interaction portion. Getting a master's degree, it would have been better to really have a discussion for topics, not typed up responses that could take days to view.	Jl 04 2020 08:27 PM
Excellent	Jl 04 2020 08:07 PM
Zoom sessions with my Cohort was comfortable as it is a small class size. I have been able to easily communicate with my professors to address any questions or concerns. The course of study has been varied in approach and materials accessible.	Jl 04 2020 04:24 PM
It was still very informative and interactive. We were still able to speak and give our opinions on questions asked or statements that were made.	Jn 30 2020 08:25 AM
At the graduate level, there is very little (perhaps nothing) that requires in-seat learning. Videos can be watched, reading can be assigned, and research and writing can be accomplished without in-person instruction. In fact, I've learned in this program that the entire concept of "instruction" can be re-imagined in a number of more constructive ways - including online learning.	Jn 25 2020 01:39 PM
The experience was challenging especially with class in session; however, the instructors were very helpful.	Jn 24 2020 05:40 PM

continues on following page

Table continued

Some professors did not skip a beat going from classroom to virtual. Classes met online and it was great. Others just uploaded videos of themselves lecturing and posted them. This was not a good thing. No time to ask questions or interact. Also, some professors need to learn how to use technology!	Jn 24 2020 11:33 AM
It would depend on the course. Some professors took into consideration the attempt to work, attend class, and tend to children simultaneously. Others continued as normal.	Jn 24 2020 10:05 AM
As an "old school "student, I appreciate the in-person interaction more. I understand it is not feasible at this juncture, but that is my preference.	Jn 23 2020 03:00 PM
I definitely prefer learning in the classroom but Dr. G did a great job at communicating with us and providing quick feedback throughout the online course.	Jn 23 2020 01:44 PM
Due to being part of a cohort with weekend learning, online learning is long and not as interactive as in person.	Jn 22 2020 01:42 PM
My program was all online, so there were no changes for me.	Jn 22 2020 01:10 PM
Online learning becomes frustrating when the professors do not set up Blackboard correctly.	Jn 21 2020 01:05 PM
No issues, the transition was smooth, and the instruction was top of the line.	Jn 21 2020 08:28 AM
Dr. G & Dr. C created an excellent online learning experience.	Jn 21 2020 12:27 AM
I feel that I received the same quality instruction in my Ethics Course. There was a strong understanding of what we, as teachers are experiencing right now. The professors were more than understanding and fair to our situation. The online course of Leadership and Management was not as understanding. There was not much consideration taken for our situation. The grading was not the same between both professors and the course is a source of stress during this time.	Jn 20 2020 09:37 PM
I have to take three classes this summer (one additional that I need to qualify for the program) and having none of them be in seat is terrible. I feel like the whole point of the cohort is to meet in person and have a special synergy as a group. Even though the school year is over, having to balance my children's remote learning with my own crisis teaching for the past six weeks has been stressful. I don't foresee things being much easier over the summer, since my children will not be attending camps or leaving the house. In addition, we lost my mother-in-law to COVID19, and my father-in-law is now living with us, so I no longer have a home office or dedicated workspace. During the school year, I could at least stay in my classroom and do work in a quiet place at the end of the day or during lunch, but that is not an option either. I think the pacing of the classes and the amount of work assigned needs to be reconsidered given the circumstances. I feel that the live, in seat class experience has always been, dare I say, fun? And right now, there is not much fun at all. Maybe this is more information than you need but thank you for the opportunity to share.	Jn 20 2020 04:44 PM
The professors attempt to keep the schedule of our weekend meetings was done with good intentions but was not realistic. An online learning session long than 90 minutes is not optimal. We met for about 3 hours or more at a pop. Yikes.	Jn 20 2020 04:21 PM
You don't build relationships with professors. Zooming and online learning is not collaboration. I enjoyed getting out of my house and going somewhere else to meet people.	Jn 20 2020 03:49 PM
It does not even come close to the experience of being in seat in a classroom. The professors did their best to adjust their teaching styles, but that is a lot of time to be sitting in front of a computer screen over the course of a weekend. We appreciate the efforts to keep our cohort on track for our August graduation!!	Jn 20 2020 02:01 PM
Seamless.	Jn 20 2020 12:58 PM
This is an executive cohort program so online learning was a component of the program prior to the inception of all online learning. My professors were accommodating and everyone, professors and students, seemed to do the best they could under the circumstances that we were all unexpectedly thrust into. I do miss meeting with my cohort since we have a great chemistry, but I do feel as though it made me personally more aware of how my own children feel missing their teachers, classrooms, and friends. I did my best to think outside of the box with my own university program's movement to all online to how I could view it from a larger perspective - which lead me right back to my own family. It helped me to assist them better than I could have, had I not been a student during this time.	Jn 20 2020 11:19 AM
It's very difficult to sit in front of a computer for hours on end listening to someone talk, especially after being on the computer all week. The in class was a nice break and had great interactions with cohorts and professors. The 100% online is nowhere near as effective.	Jn 20 2020 10:43 AM
The transition to online learning was quite seamless. I think SHUs hybrid style of instruction has helped prepare us for these changes.	Jn 20 2020 10:41 AM
Good experience	Jn 04 2020 09:39 PM

Q5 Demographic Information - check all that apply

Answer	Responses	Percentages
Male	12	30%
Female	28	70%
Totals by Gender	40	
Black	3	8.11%
White	28	75.68%
Hispanic	4	10.81%
Asian	1	2.70%
Native American	0	0.00%
Other	1	2.70%
Totals by Ethnicity	38	
Master's Degree	20	52.63%
Education Specialist Degree	16	42.11%
Doctoral Degree	2	5.26%
Totals by Degree	37	
Respondents to demographic Questions = 41 Skipped = 1		

Figure 1. Responses of students to emergency remote learning

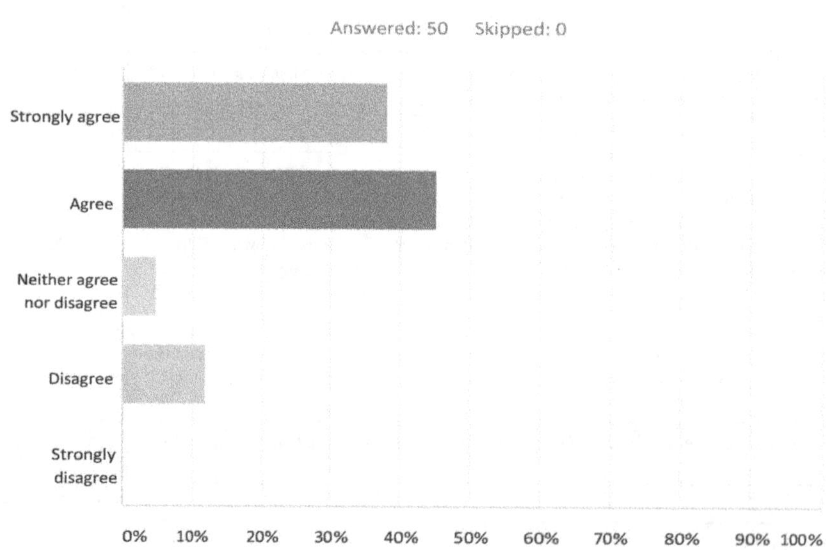

Figure 2. Responses of students to personal accommodations made by professors

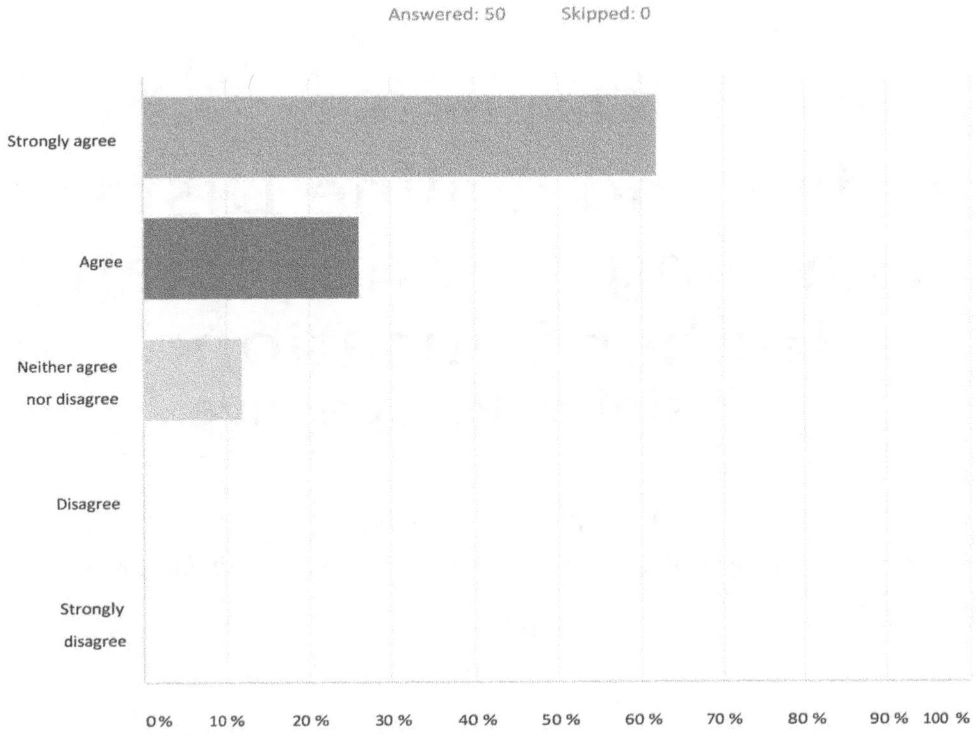

Figure 3. Responses of students regarding quality of remote learning

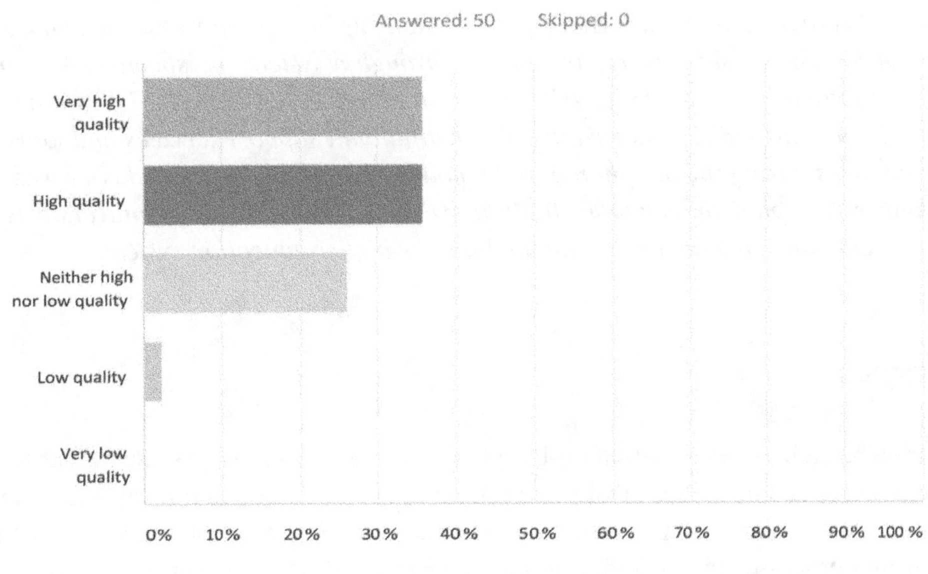

Chapter 9
Blended and Online Distance Learning in Indonesian Higher Education:
A Systems Perspective

Dewi Wahyu Mustikasari
University of Technology Sydney, Australia & UIN Salatiga, Indonesia

Keith Heggart
University of Technology Sydney, Australia

ABSTRACT

This study investigated the external factors that influenced teachers' design processes during the adoption of blended learning (before COVID-19) and online distance learning (during COVID-19) in higher education settings. This study sought to understand the relationship between university leaders, educators, and other actors inside and outside university contexts through a systems perspective. A multiple case study design was employed, and eight Indonesian university English teachers (IUETs) were voluntarily recruited to participate. The findings showed that the institutional support services and tools influence teacher design practices during the adoption of both modes of learning. Activity theory was used as a conceptual framework to explore the common pitfalls that disconnected the roles of university leadership, institutional policy and units, and actors within the broader higher education system.

INTRODUCTION

Blended learning has been chosen as an institution-based approach to accommodate students with technology-integrated learning in some universities involved in this study. The Indonesian government has mandated that universities shift to an online distance learning mode, a terminology used for online learning to deliver teaching and learning processes remotely during the pandemic in 2020. Later in this chapter, an online distance learning approach will also be called emergency remote learning (Green et al., 2020).

DOI: 10.4018/978-1-6684-7285-9.ch009

This chapter discusses the interconnected roles of actors in the university ecosystem and how these roles either supported or hindered the adoption of blended and online distance learning for English language teachers. It seeks to find common ground between activity theory's focus on human activity as a socially situated phenomenon (Engestrom, 1999) and the holistic and self-organising nature of systems thinking (Arnold & Wade, 2015). One feature of systems thinking is that systems possess a continuous feedback loop (Monat & Gannon, 2015). This feature can be used to improve each system element and, thus, the system as a whole. However, a clearer understanding of the actors themselves and their interactions is required to use the feedback loop. Activity Theory (Engeström, 2001) provides a framework to understand each system element and how they interact comprehensively. It is possible to visualise connections between each element in a complex system and recognise obstacles and solutions experienced by the element and those between the elements in the system. Adopting different learning modalities in a university setting, especially during a crisis such as COVID-19, relies on a complex interaction between various elements; there is more to consider than teachers' pedagogical decisions! By examining this complex system through the lens of Activity Theory, it is possible to arrive at a deeper understanding of how these elements influenced each other and ultimately determined the success or failure of blended or online learning at specific institutions.

Several rationales of this study are described as follows. Blended and online learning have been practised for at least two decades in Western countries, but adoption has been slower in developing countries (Farooq et al., 2012; Priyatno, 2017). Universities and other higher education institutions in Indonesia have recently started experimenting and trialling blended learning modalities in diverse disciplines (Dewi, 2014; Dwiyogo, 2018; Irawan et al., 2017; Kristanto, 2017; Mulyadi et al., 2020; Priyatno, 2017; Sulisworo et al., 2016). These trials were made more challenging and urgent due to the unprecedented global health crisis caused by COVID-19 in 2020-2022. Indonesian teachers (like their counterparts in other countries) were required to move much of their teaching from blended or face-to-face to entirely online to ensure that instruction continued while also seeking to minimise the opportunity for the virus to spread.

The study described in this chapter shows that this change in modality relied upon more than just teachers' pedagogical decisions or learning designs. While that was an important part, other factors either assisted or hindered the success of the shift in modality. This paper highlights the different elements of university organisation and administration and their effects on the quality of the teaching and learning experience for teachers and students. In particular, misalignment between institutional leaders and educators, the efficacy of different support forms, and individual practitioners' resilience were exposed during the pandemic. While many universities and higher education institutions are returning to normality, essential lessons to be learned from this experience might have enduring consequences on how Indonesian higher education institutions are structured and how the different elements within these systems relate.

In summary, one key consequence is how educators design learning experiences for their students and how the relationship mediates this between educators, students, colleagues and their access to various resources within the university system. Research in other, mainly Westernised, countries has explored how teachers design online learning activities for their students in a manner that includes a consideration of the affordances of various technologies (Agostinho, 2011; Asensio-Pérez et al., 2017; Bennett et al., 2015). However, contextual factors are generally less considered, especially in higher education settings. In addition, there has been less empirical research about the design processes of educators from non-western countries. For example, there is a lack of understanding of the design processes and influences upon IUETs, specifically, how environmental and contextual factors shape their design choices. This study investigated how these external contextual factors influenced the adoption of blended learning modalities before COVID-19 and online distance learning modalities during COVID-19. It aimed to understand the access and nature of institutional support services and tools, particularly the technological tools provided to the teachers during the design processes of both modalities and especially the way that these support services and tools worked - and sometimes did not -within the university system. This study also considers the role of socio-economic factors and, specifically, the technological affordances and financial costs of the technologies employed in the IUETs' learning designs. In order to determine how these connections are drawn, this study used the Activity Theory (Engeström, 2001) - which is presented in the discussion section.

LITERATURE REVIEW

The use of technology in order to assist or improve learning outcomes has a significant and lengthy history. A complete discussion of the history of educational technology is beyond the scope of this chapter, but what is particularly pertinent is a consideration of the modality in which this technology is employed. Various terms have currency in these current debates, including online, flipped, hybrid and hyflex learning. The term *blended learning* has widespread appeal. The phrase itself is imprecise, as it can be used to relate to any use of technology for learning (Knussen, 2020), or it can be used to describe a mix of both in-person and asynchronous learning supported by technological tools, such as learning management systems (LMSs) (Priyatno, 2017; Moreno-López et al., 2017; Young, 2008; Thang et al., 2013). This latter definition is applied in this case. Blended learning can be contrasted with the term *online learning*, which is usually used to refer to a mix of synchronous and asynchronous learning activities, again usually supported by technological tools, but one that exists solely in the online space; that is, there is no in-person component (Comas-Quinn et al., 2012; Farooq et al., 2012; Liyanagunawardena et al., 2014; Sun, 2014, 2021). The existing research, as it pertains to language instruction (which is the focus of this study), is discussed below.

Learning Designs in Higher Education

Literature from multiple disciplines has identified the external factors influencing teachers' design practices in developed countries. These factors include colleague support, engagement with literature, participation in workshops and seminars, conferences, and institutional support services (such as teaching and learning units) (Agostinho et al., 2018). Some authors (Agostinho, 2011; Asensio-Pérez et al., 2017; Bennett et al., 2015; Sun, 2021) have reported that educational technologies (such as learning management systems and video conference tools) have a role to play in supporting learning designs. Conversely, a non-technology tool such as a set curriculum is also a factor that contributes to teacher design practices (Bennett et al., 2011). Other authors such as Bennett et al. (2011), Buus and Georgsen (2018), Jordan and Elsden-Clifton (2016), Kim (2015), Lakkala et al. (2015), Sun and Goodyear (2020) and Sun (2021) point out that team design (i.e., learning experiences devised by a team including students, teachers, learning designers, educational designers, technological developers) is an essential factor in design practices. In addition, Asensio-Pérez et al. (2017) and Ma et al. (2018) highlight that professional development programs are essential in supporting teachers' design work.

Blended Language Learning

Studies on blended language learning in higher education settings have revealed external factors influencing student learning experiences. Some authors (Priyatno, 2017; Moreno-López et al., 2017) point out that technology has enhanced language learning. Others note that there are challenges involved in using technology. This can include the technological affordances of a particular institution's technologies: for example, the limited capacity of servers (Young, 2008) or poor internet connectivity (Thang et al., 2013).

Most research into blended language instruction focuses on the impact or evaluation of interventions in course designs. The results have been varied. For example, Priyatno (2017) investigated the effects of implementing Schoology. He reported that this e-learning platform had fostered autonomy, support in processing content, and improved learning material selection by students in an English for Academic Purposes (EAP) course at an Indonesian university.

Moreno-López et al. (2017) examined several pedagogical models: 1) a traditional face-to-face classroom, 2) face-to-face classes with a community-based learning component, 3) face-to-face courses with an online telecollaborative language learning component, and 4) study abroad. They found no overall difference in student learning between the different models but did note significant differences in students' perceptions of the value of the models. Students believed their language acquisition and curiosity about intercultural awareness were improved by moving beyond traditional (i.e., in-person-only) instructional models.

Other research has highlighted the challenges related to implementation. Young (2008) investigated the impact of a redesigned intensive Spanish course. Despite the technical problems associated with the capacity of the server experienced during the implementation of the course, teachers facilitated learner-centred, communicative and task-based learning. As a result, students who received a redesigned course format with a 2-day per week schedule could achieve similar learning outcomes to students who received the primary course in a 3-day per week format. Thang et al. (2013) indicated that a challenge while implementing blended English courses in a Malaysian university was poor internet connectivity, leading to technical and pedagogical problems.

Online Language Learning

Similarly, published studies on online language learning in higher education contexts have indicated that external contextual factors have influenced student learning experiences. Several researchers such as Comas-Quinn et al. (2012), Farooq et al. (2012). Similarly, Liyanagunawardena et al. (2014) and Sun (2014, 2021) have investigated technologies for enhancing language learning, fostering student-teacher relationships and advancing effective teaching practice.

Studies have revealed that teachers have used various technologies to improve students' language acquisition. For example, Ogan et al. (2009) testified that students in an experimental group significantly improved their intercultural perspectives compared to the control group. The experimental group received a pause-predict-ponder technique after using a tool called an Intercultural Competence Attention-focusing Tutor in French online classes. Similarly, Farooq et al. (2012) reported that Computer-Mediated Communication (CMC) in a master's program in English Language Teaching (ELT) in distance education significantly improved students' language learning. Other authors, such as Comas-Quinn et al. (2012), stated that students valued interacting in virtual forums to increase their language skills and personal interaction with tutors in Spanish courses.

Limited studies have indicated that technology-enhanced language learning has promoted effective teaching practices. For instance, Sun (2021) highlighted that teachers, educational designers, technological developers and researchers have benefitted from Computer Assisted Language Learning (CALL) tasks at the classroom level for online Chinese courses.

Conversely, studies have also indicated challenges with implementing technologies in online learning modes for language learners. For example, Sun (2014) reported that students had experienced difficulties in managing their motivation, learning and socialisation in a paired or small-group activity that uses online voice/ video tools integrated into LMS in fully online Chinese classes. Authors like Liyanagunawardena et al. (2014) indicated that students experienced challenges regarding low-level technological knowledge and local language content areas in distance language education that aided Information and Communications Technology (ICT) in Sri Lanka.

Studies have also highlighted that a broader institutional context and environmental factors have influenced technology used for online learning approaches. Liyanagunawardena et al. (2014) reported some barriers regarding institutional contexts to access computers, the internet, and institutional support services. Meanwhile, Natalia (2017) pointed out that university leaders were reluctant to adopt an institutional policy, invest in digital infrastructure, and fully reaccredit online learning methods.

Research Gaps in Blended and Online Language Learning Designs

Most of the studies on language education have focused on investigating contextual factors that influence student learning experiences using technological aids. These studies also focused on how to develop effective teacher-student relationships in online language modes. Limited studies have examined external contextual factors that support teachers' design work in blended and online learning courses. For these reasons, this study aimed to understand the external factors, including institutional support services and tools and socio-economics factors that influence teacher design practices. The study was undertaken to analyse the relationship between educators, students, colleagues and other resources at the institutional level.

METHODOLOGY

Conceptual Framework

In order to examine the relationships between the different elements that influenced the learning designs of the IUETs, this study adopted an Activity Theory conceptual framework (Engeström, 2001). In this framework, the six elements of an activity system represent functionality to contribute to or constrain the complex interaction within a community or a workplace. The elements are subjects (someone who manages his/ her activities), objects (observed problem), mediating artefacts (tools), rules (conventions), community (subject and other actors in an activity), and division of labour (distribution of tasks) (Engeström, 2001).

This framework focuses on the way that different elements within the systems dynamically influence each other, and hence is a suitable approach for examining the interconnection between teaching practices, technology, the role of external groups such as teaching and learning or ICT units, the directives of the university leadership, and students. The elements of Activity Theory can be used to examine the external factors (Rules, Mediating Artefacts, Community, and Division of Labour) within the workplaces and broader contexts (i.e., environmental factors) presented in the individual case studies. These elements in the Activity Theory triangle represent the interconnectedness of each external factor to support or constrain the IUETs as the subject in designing for blended or online distance language learning as the object in the activity system. Access to support tools, governmental and institutional policies and procedures are some of the conventions existing in the Rules, and indeed whether these conventions are provided or not by other actors in the Community (i.e., university leaders, partners, colleagues, academic/ educational and IT centre staff and students). How do other actors distribute tasks and responsibilities? How do other actors facilitate the English teachers by organising physical support tools? These questions show that each external factor plays a vital role in the activity system. Activity theory facilitates a better understanding of how these elements interlink to support or constrain the English teachers during their design processes (see Figure 1).

Research Design

A case study methodology was adopted to obtain in-depth information on this phenomenon. The current phenomenon is an example of a representative case in real-life settings (Yin, 2018). This study is a multiple-study design that represents multiple bounded systems of the examined phenomenon (Creswell & Poth, 2018) and employs multiple units of analysis within individual cases (Yin, 2018). Eight IUETs and their teaching and learning practices were examined in eight different Indonesian universities. The participants were experienced and full-time teachers working in different public and private universities in five of the six Java Island provinces. Four female and four male teachers aged between 35 and 55 years old and varied teaching experience (5 to 29 years) were recruited. Due to the restrictions on travel caused by COVID-19, it was impossible to conduct fieldwork in Indonesia; thus, online research methods were undertaken digitally and remotely in Australia. Data collection techniques were employed, including online interviews, observations, digital participants' diaries, and digital document analyses adapted from the E-Interview Research Framework (Salmons, 2014). Deductive and inductive strategies were undertaken to analyse the data based on the Data Analysis Spiral (Creswell & Poth, 2018). An electronic data store was created to store digital recordings (i.e., audio-video recordings) and documents (i.e., transcripts, observation notes, copies of policies/ regulations, and participants' diaries) for the data management system. The participating teachers and universities were de-identified. The reading and memo-ing phase was carried out by reading the dataset multiple times, making notes on themes, and outlining analytical thinking and framework based on the Activity Theory (Engeström, 2001) of the individual cases. The coding process was undertaken by developing an analytical framework. It included defining initial codes, creating a list of code categories and descriptions, applying initial codes to individual cases, reducing codes to themes, and adding emerging codes. The interpretation of the data phase started with the development of case summaries and cross-case analysis. The representation of the data phase was demonstrated by presenting a thick summary of the individual cases and cross-case analysis in the findings section.

FINDINGS

Analysis of the data demonstrated interconnections between the elements. This included connections between university leaders' facilitation and directives and actors' (in this case, teachers') understanding of their roles and responsibilities in designing learning experiences. It demonstrated the agency of the university leaders and other actors within the university. It also illustrated a secondary, voluntary support system from colleagues and professional communities outside the university (often to address a perceived lack of resources from the institution).

Adoption of Blended Learning Designs

Before COVID-19, some universities were already exploring adopting blended learning models. This was stimulated because of commitments from the university leaders to deliver a more modern approach to teaching and learning and supported by teachers who sought to accommodate these directives in their classrooms and learning designs. Recognising a significant disparity between the different universities' approaches to technology-integrated learning is vital. Some universities were forward-thinking and had implemented plans and infrastructure to support this change in modality, whereas the other universities still needed to guarantee these commitments. Besides the institutional commitments, the IUETs' interest and ability to deliver blended learning designs also played a significant role in adopting these modalities.

Institutional support services and technology tools influenced the adoption of the blended learning designs and the design processes of the IUETs. In universities where teachers were supported by their institution in the adoption of blended modalities, these teachers (hereafter referred to as 'supported teachers') received training and professional development, access to technologies such as LMSs to deliver online learning and the opportunity to gain funding via competitive grants to develop online content for the university's LMS. Much of this was organised at a high level within the university: for example, at one university, the Academic/ Education Institute, an institutional unit at the whole-university level, organised teacher training and competitive grants. In addition, supported teachers could also use hands-on support from IT Centre staff who maintained the LMS and supported teachers when they experienced technical issues. These different elements in the university's learning provision system were interconnected and supported teacher design practices. The actors (i.e., university leaders, teachers and support staff) in the internal community of the university recognised their roles and responsibilities in supporting the teachers in designing learning experiences and thus providing the best outcomes for students. Thus, in these cases where support was provided, teachers felt more satisfied regarding the technological affordances and financial costs of designing learning compared to the unsupported teachers.

The 'unsupported' teachers also received assistance. The support they did receive was located externally to their institutions and was often something they had to seek out and leverage existing relationships to benefit from personally. For example, the unsupported teachers received volunteer support from professional communities outside the universities to shape their technological knowledge on freely available technologies: Edmodo, Google Classroom and Kahoot. In other words, the unsupported teachers needed more access to institutional support services and tools. The actors in the internal community had yet to recognise their roles and responsibilities or were not provided with the resources they required to achieve this, and this demonstrates a lack of alignment and interconnectedness between the university leaders' directions and the practices of the actors within the university.

Another factor that impacted the adoption of blended learning modalities was student access to the internet. At most institutions, it was a (sometimes unspoken) expectation that students would bring their own device (BYOD) and cover the financial costs regarding internet data packages. This has been described as a 'hidden' policy of BYOD. While this covert approach is expected to cause significant issues regarding access and equity, this was not the case (at least, pre-COVID). A possible explanation for this might be that most universities provide computers in the library and wireless internet access;

thus, students might use those services rather than having to fund their technology use. This was another example of the interconnection between institutional policies and the physical and digital infrastructure.

The teachers at universities had all received teacher training. However, this training and their knowledge mainly focused on pedagogical content knowledge (i.e., how to teach about their particular subjects) and technological knowledge (how to use various systems for primarily administrative purposes, such as the LMS). While they recognised that tools like the LMS could offer pedagogical affordances to support online learning, their use was limited to providing online quizzes and online content resources in the LMS. Most teachers reported that online discussion forums could have been more effectively conducted. They considered discussion to be more effective in the classroom setting. This situation can be related to the fact that these teachers needed more technological pedagogical knowledge to design practical online discussions, learning activities and online assessments because the offered teacher training was focused on understanding technological knowledge.

Shifting to Online Distance Designs

COVID-19 caused significant disruption in the higher education sector in Indonesia. Principally, this meant that all universities and teachers who worked within them were required to adopt online distance learning modalities throughout the pandemic. This research examined the teachers' reflections (taken from the interviews, participants' teaching diaries, and teaching and curriculum documents) and researchers' observations based on lesson observations and examination of documents and other course materials. While the disruption caused by the pandemic was much more remarkable, the role of supportive leadership continued to influence teachers' design processes significantly. Different institutional units demonstrated continuous support for the teachers during the pandemic at the universities designated as supportive. At other universities, teachers had to rely on voluntary support systems (i.e., colleagues and professional communities), which remained prominent during the pandemic. The actors in the various activity systems were the key influences on the teachers' design practices. The following paragraphs illustrate these interrelationships.

As the pandemic developed, teachers at all universities were mandated to teach remotely (i.e., online) to minimise the spread of the virus. Although governmental policy accommodated several strategies to deliver online learning, different universities could employ different strategies to meet the requirements of the governmental policy. The universities committed to technology-enhanced learning (i.e., the supportive universities) had minimal disruptions to accommodate the teachers and students in this new modality. They generally deployed some combination of the following supports:

1. More access to technological tools such as video conferencing tools, e.g., Zoom and Microsoft Teams, and software or application to create video content for teaching, e.g., Panopto and PowerPoint 365, and
2. Internet package data arrangements for staff and students so they could access course materials online.

The other universities (i.e., those that provided limited or no access to support tools) needed to prepare to handle online learning. For example, one university needed to prepare to facilitate online learning, and its first response was to require teachers and students to take leave until something could be arranged. Other universities in this category did not provide video conferencing tools, although they could manage to provide internet data packages. This was not acceptable to many of the teachers; considering teacher presence is an integral part of online distance learning, many teachers felt that some form of video conferencing tool needed to be added. Therefore, through networks and professional communities, they source free versions. The unsupported teachers used freely available video conferences such as the free Zoom version and Google Meet.

Regardless of whether their institutions were supportive, all teachers' design processes were influenced by the problems caused by the pandemic. It is important to note that many of these problems were external to the university or the traditional teacher-student paradigm. Instead, these problems were caused by environmental factors, including technological affordances and financial costs. Nevertheless, teachers were required to design – and often redesign – learning experiences to overcome these challenges.

One example relates to the mix of synchronous and asynchronous video sessions. Initially, teachers sought to provide live synchronous video sessions. However, the teachers had to adapt these plans and provide more asynchronous sessions because delivering synchronous sessions consumed many internet data, which students could not afford. This situation led to the quality of teaching being questioned because less teacher presence was provided and concerns from teachers who felt that students needed to be more engaged in the course as fully as they should have been.

Other environmental factors were internet connectivity and device ownership. Both teachers and students experienced low internet connectivity during emergency remote learning. The teachers experienced difficulties designing learning activities at the classroom level because many students needed devices such as laptops. Instead, they accessed much of the course through mobile phones. The smaller screen size and limited functionality was a problem teachers struggled to overcome – and one that had not been foreseen in the move to online learning. Another frustration was that unsupported teachers were still required to deliver online classes via the university's LMS for record-keeping. However, the LMS went wrong such as could not be accessed or delayed receiving real-time information during the online teaching delivery. The low bandwidth capacity caused this problem. In order to address this, teachers shifted to other technologies, such as freely available e-learning platforms (e.g., Google Classroom) and social media applications (e.g., WhatsApp and Telegram).

DISCUSSION

The findings discussed above, both pre-COVID and during the pandemic, illustrate some common pitfalls relating to the interconnectivity between university leadership, institutional policy and units, and actors within systems. This section also discusses design practices through the lens of Activity Theory.

Common Pitfalls of the System Thinking for the Adoptions of Blended and Online Distance Learning Designs

Limited Access in Unsupported Universities

The individual cases demonstrated that limited university directives or efforts to think systematically about learning and teaching provision across entire university contexts have meant that many IUETs needed support to change their teaching and learning modality. The majority of the universities in this study were unprepared to facilitate institutional support services and tools for online learning; this meant that there was limited access to training in the effective use of tools received by the unsupported teachers: this situation led to frustration on behalf of many IUETs as they sought to design and implement online learning.

An effective learning designs was a source of frustration for the teachers because they needed access to suitable institutional technologies (i.e., the LMS and video conferencing tools). In addition, some teachers had yet to be given the chance to learn to use these tools. This caused detrimental effects for the unsupported teachers who were required to spend more time researching other tools – including leveraging non-institutional networks and offerings – so that they could deliver online classes. In these cases, university directives, despite claims of technologically enhanced teaching and learning, did not work as intended to link the actors in the institutional units to support the teachers who needed the support before the pandemic in 2020. This situation was ongoing – and indeed, even exacerbated during the pandemic; there was limited or no inter-unit coordination to overcome the obstacles in the teaching and learning process. For example, university leaders failed to resolve the bandwidth issues, maintain LMSs and provide professional development programs for the teachers and IT Centre Staff. This study has similar findings as in other studies by Liyanagunawardena et al. (2014) and Natalia (2017). They argued that institutional contexts and policies were barriers to adopting online modalities for language teaching. However, colleagues have a pivotal role in this situation by providing ad hoc teacher training (i.e., non-institutional) and hands-on support in tackling the technical issues for the unsupported teachers.

Mismatch Between University's Expectations and Teachers' Preferences

Tensions between policy and practices during the adoption of online learning approaches could have been avoided. Several universities tried to accommodate online learning delivery by using LMSs; however, this was used more for administrative purposes (such as record-keeping) than to maintain or improve pedagogy. For example, the teachers were required to implement an online discussion forum. The university leaders needed to ensure that the teachers had undertaken their duty to teach online, which was one of the measures they used to prove that. However, this directive led to significant frustrations on behalf of the teachers. The teacher's pedagogical intent was being ignored. They had to use limited features of the LMS (i.e., delay in receiving real-time information and interactivity issues); therefore, the teachers needed help to engage students during online discussions.

The university leaders placed more value on the administration of teaching and learning rather than pedagogical practices; thus, the teacher's creativity could have been improved. The relationship between the university leaders and teachers regarding the institutional support services and tools was interconnected but dysfunctional. It influenced how the teachers had to provide online learning at the classroom level.

No Alignment of the University's Directives

Limited university leaders' facilitation and directions regarding the university culture of the covert policy of BYOD have impacted teachers' design work and influenced student learning experiences. It has caused difficulties for teachers in delivering online learning; for example, no clear institutional policy addressed the necessity of BYOD. These teachers experienced difficulties making decisions about learning activities that required them to use technologies at the classroom level; keeping in mind the notion that technologies should be prepared at every classroom level (Knussen, 2020), teachers were forced to adapt their original designs to take into account the limited access that's students had to devices during the pandemic. Student learning experiences could not be guaranteed because students did not necessarily have the same – or any - devices to access course contents. Building teacher-student relationships was challenging because the university directives and instructions did not interconnect or align.

Accidental Pedagogies

In many cases, teacher design practices could be considered 'accidental pedagogies' because the university focused on digital infrastructure – but this focus does not necessarily imply that teachers are competent to teach online (McGee & Carmean, 2012). A minority of universities had made institutional commitments to provide blended and online learning. In this case, they provided institutional support services and tools for online and blended learning. In contrast, unsupported teachers did not receive sufficient institutional support services and tools, but they reported similar pedagogical content knowledge to those teachers who received better institutional support. This means that the institutional support services and tools – in and of themselves - did not guarantee that teachers develop effective learning designs. The explanation for this was related to the teacher training provided: it only addressed technological knowledge rather than pedagogical knowledge. The teachers were trained to use the technologies regarding the conceptions of digital technologies for online teaching; that is, they learned how to use the features of technology, not how to integrate technological pedagogical content knowledge when they delivered online classes. Therefore, the teachers reported having obstacles to engaging students during online learning. The teacher training only influenced teachers understanding of pedagogical content knowledge and technological knowledge; thus, they faced difficulties regarding pedagogy to deliver online learning activities and assessments.

Activity Theory: A Lens to Understand Institutional Support Services and Tools

The contextual factors that influenced the teachers' design processes can be conceived through the lens of Activity Theory. Limited university leaders' directives caused several disruptions in learning designs. This includes the teachers' solutions to their design problems.

Common External Factors Led to Disruptions of Teacher Design Practices

The connections between each element of the Activity Theory triangle can be used to visualise the relationship between university leaders, teachers, students, and other resources at the university level. These relationships were captured during the blended and online distance learning design processes undertaken by the teachers before and during COVID-19. The triangle of the Activity Theory illustrates the disruptions and solutions (see Figure 2).

The majority of the universities illustrated some form of disconnection between the actors at the university level. The university leaders needed to provide more institutional policies regarding the support tools (Rules) to support adopting the blended and online distance language learning approaches (Object). Therefore, the unsupported teachers (Subject) received limited access to institutional support services and tools. The university leaders also needed to accommodate more of an inter-unit connectivity line to share tasks and responsibilities (Division of Labour) among the actors, particularly IT Centre staff (Community), in assisting the teachers in implementing the online learning sessions. Some of the unsupported teachers reported that the IT Centre could not manage technical issues because of their insufficient technical knowledge to maintain the university's LMS. However, this disruption was slightly different from IT Centre Staff side. A limited digital infrastructure development budget was also reported to influence the university leaders' facilitations and directives (Division of Labour). Low-cost budgets to add the bandwidth capacity and to accommodate internet data packages were indicated as the causes that tended to affect the direction of the university leaders' facilitation during online teaching and learning implementations. The inter-disconnectedness can be seen in the Activity Theory triangle (see Figure 2); the disruptions were visualised into two lines of interlink disruptions:

1. Rules-Community-Division of Labour

This type of disruption captured the obstacles regarding the Rules that could not be effectively solved by the actors in the Community, i.e., the university leaders and IT Centre Staff, because of limited leadership and shared responsibilities in assisting the teachers in designing learning.

2. Rules-Object

Limited access to institutional support services and tools has influenced the teachers' designs for blended and online distance language learning.

The teachers tended to find more personal than institutional solutions to their obstacles. In general, these teachers could resolve their problems with support from the other actors (Community) such as partners, colleagues, students, students' parents and professional communities outside the universities in designing for learning by adopting the blended and online distance learning course (Object). These teachers found different actors to sort out their obstacles in learning designs. For example, the teachers found developing the content-area focus and managing the technical issues challenging. They could resolve the problems with support from partners, colleagues and professional communities outside the university. This solution is illustrated by the arrow that connects the Community and Object in the activity system. These teachers resolved their obstacles individually. This means that limited institutional solutions were offered.

Current and Future Challenges

Activity Theory provides a tool to explore how universities can systematically address current and future challenges. This section presents several current and future challenges concerning teacher learning designs, studies on internal factors that influence learning designs, external factors that play a significant role in supporting learning designs, a wide range of different institutional support levels to assist learning designs, and professional development program in improving teachers' capacity to design learning.

Teachers Creating Learning Designs is Essential for Student Achievement in Higher Education

The individual cases demonstrated that institutional and ad hoc teacher training focused on using technologies for teaching and learning. Therefore, the teachers found frustrations and difficulties in designing learning because they needed sufficient knowledge to design learnings at the classroom level. Although the teachers benefited from non-institutional technologies, i.e., freely available e-learning platforms, video conferencing tools, and social media applications, when the institutional technologies could not be accessed, the teachers' teacher-student relationship took time to manage. As previously discussed, these teachers need help engaging students in online discussions and delivering online assessments for speaking and listening skills. Developing learning activities at the classroom level was also difficult to organise because there was no guarantee that students would bring their own devices; no institutional policymakers about this regulation to conduct online learning. This interconnectedness policy and practices have created difficulties in building teacher-student relationships. Students' achievement would be difficult to obtain. Thus, improving teachers' capacity to design learning is essential to build teacher-student relationships and support students to achieve learning outcomes.

Traditional Models Have Really Only Examined Internal Factors

Many studies have examined internal factors to support teachers in designing learning. Some studies have tried to shed light on teachers' conception as designers, although most multidisciplinary studies were undertaken in developed countries. Markauskaite and Goodyear (2009) have investigated teachers' design thinking. Kali et al. (2011) have revealed teachers' design practices and cognition. Bennett et al. (2017) have reported university teachers' design processes that showed similar design routines as learning designers. An up-to-date study on teachers as designers in developing countries, i.e., Indonesia, have been conducted by Mustikasari and Heggart (2022). They supported that English teachers have undertaken similar design phases as learning designers. For example, these teachers must understand how an unknown event may turn into a known one requiring them to undertake a trial-and-error approach while adopting emergency remote learning. In general, these studies on the internal factors that influence design work have acknowledged that teachers should be accommodated with institutional support services such as instructional design support and tools to improve teachers' capability to design learning because these teachers did not receive support from learning or educational designers.

Pandemic Brought Into Sharp Relief That External Factors Can Play a Role in Design Work

This study indicated that institutional support services and tools have significantly influenced teacher design practices. Access to support tools was essential for effective learning designs (Bennett et al., 2017; Knussen, 2020); however, this institutional support was only guaranteed in some of the individual cases in this study. Another area for improvement was environmental factors. Although the majority of previous studies undertaken by Agostinho (2011), Asensio-Pérez et al. (2017), Bennett et al. (2015) and Thang et al. (2013) have reported that technological affordance and internet connectivity have influenced design work; this study indicated other environmental factors, such as financial costing and device ownership issues, that notably influence teachers and students during the implementation of online learning in emergency remote learning.

Institutional Policies on a Wide Range of Different Levels of Assistance

Universities must provide support at various levels to assist teachers in planning for these external factors. Although the supported teachers have received sufficient institutional services and tools, they experienced obstacles during emergency remote learning. However, the supportive university leaders of the individual cases have highlighted several strategies to support teachers in designing for learning. These strategies involved different levels of assistance by interconnecting the institutional units at the university level. The Academic/ Educational unit is responsible for training and professional development for teaching and non-teaching staff. The IT Centre establishes and maintains LMS and provides hands-on support when teachers experience technical issues. The university leaders are responsible for endorsing institutional policies that interlink the units' roles and other regulations supporting teacher learning designs, such as providing a competitive grant that uses the school's budget to develop online classes in the LMS. These universities have shown a better way to build relationships between university leaders and teaching and non-teaching staff to accommodate teacher learning designs and student learning.

Figure 1. Conceptualisation of Activity Theory Model for the Study
(Based on Engeström, 2001)

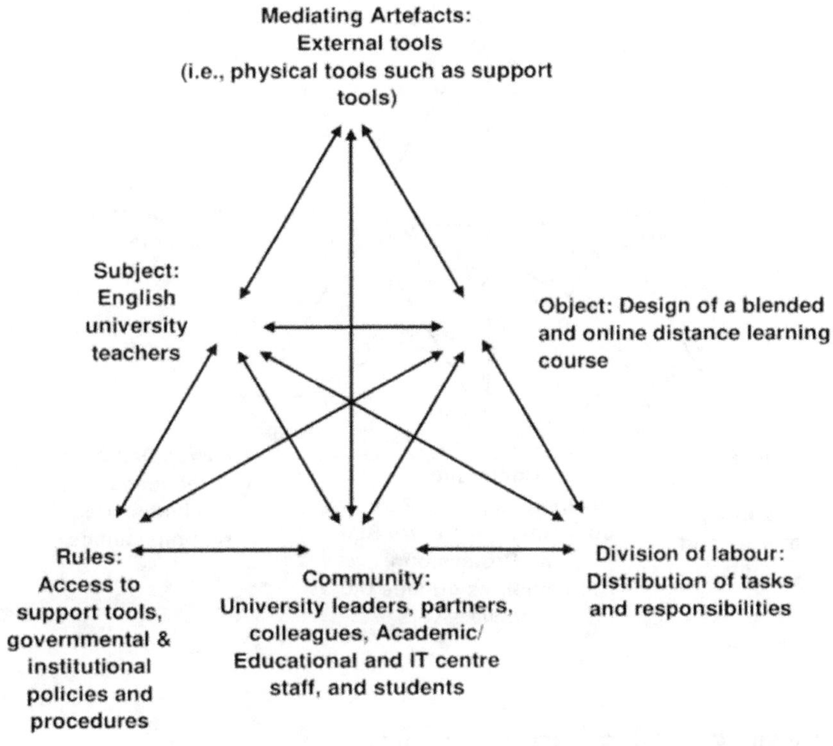

Professional Development Program to Improve Teacher Learning Designs

Future professional development programs should accommodate teachers' capability in designing learning. Some previous studies have acknowledged that professional development improves teacher design practices (Asensio-Pérez et al., 2017; Ma et al., 2018). As previously discussed, teachers have similar design routines to learning designers. Moreover, the fact that there is no role for learning designers in Indonesian higher education contexts means that teachers are responsible for designing everything themselves. Therefore, some alternatives for professional development programs can be arranged by adapting programs created for learning designers. Heggart and Dickson-Deane (2021) have formulated a microcredential program for learning designers. Heggart (2021) has also described a framework for teachers to use to assist them in thinking of themselves as learning designers. However, and perhaps most importantly, while ongoing training for teachers is essential, it is only one part of the broader system; in order to effectively manage crises like COVID-19 or even a shift in teaching modality, each element in the system, and the relationships between them, must be considered and adopted.

Figure 2. Disruptions in the Design Process of the Blended and Online Distance Learning Designs
(Based on Engeström, 2001)

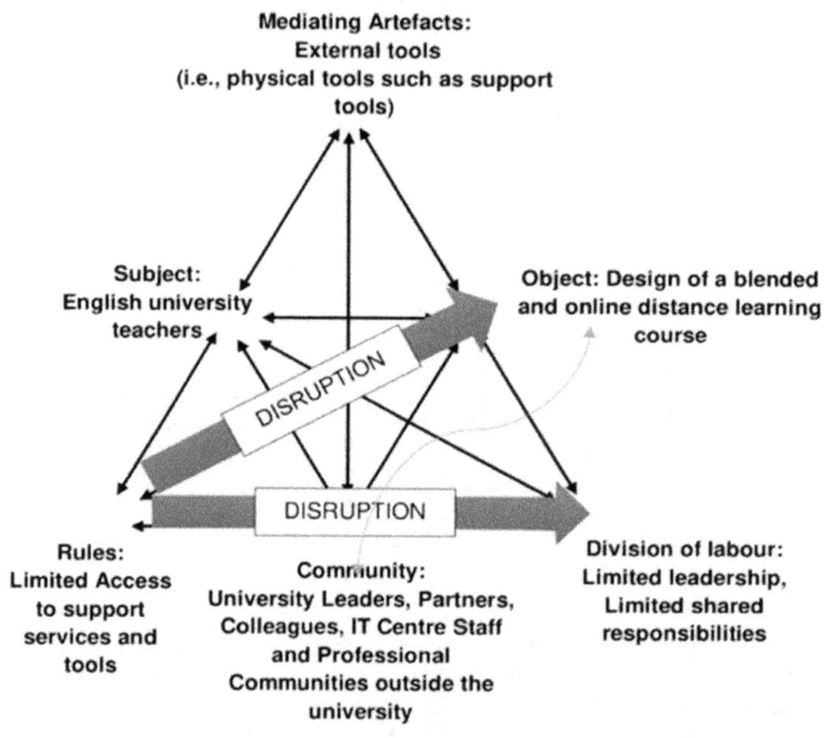

CONCLUSION AND FUTURE IMPLICATIONS

The study has provided a better understanding of the interconnected relationship between university leaders' facilitations and directions and actors' understanding of their roles and responsibility in supporting teacher design practices. In the above analysis, some actors within a system or community can work at cross-purposes with each other. This can infiltrate understanding of the different purposes of educational technology, the way it is used, and ultimately, the learning designs developed and implemented by teaching staff. The most notable example is the different conceptions of using the LMS between teachers (for teaching purposes) and university staff (for administrative purposes).

This study also has contributed to the current knowledge of Activity Theory concerning the different understanding of actors in the activity system who support teacher design practices for blended and online distance learning in developing countries. This opens up opportunities to develop a model or framework as a guide for supporting teacher design practices regarding institutional support services and tools.

REFERENCES

Agostinho, S. (2011). The use of a visual learning design representation to support the design process of teaching in higher education. *Australasian Journal of Educational Technology*, *27*(6), 961–978. doi:10.14742/ajet.923

Agostinho, S., Lockyer, L., & Bennett, S. (2018). Identifying the characteristics of support Australian university teachers use in their design work: Implications for the learning design field. *Australasian Journal of Educational Technology*, *34*(2), 1–15. doi:10.14742/ajet.3776

Arnold, R. D., & Wade, J. P. (2015). A definition of systems thinking: A systems approach. *Procedia Computer Science*, *44*, 669–678. doi:10.1016/j.procs.2015.03.050

Asensio-Pérez, J. I., Dimitriadis, Y., Pozzi, F., Hernández-Leo, D., Prieto, L. P., Persico, D., & Villagrá-Sobrino, S. L. (2017). Towards teaching as design: Exploring the interplay between full-lifecycle learning design tooling and teacher professional development. *Computers & Education*, *114*, 92–116. doi:10.1016/j.compedu.2017.06.011

Bennett, S., Agostinho, S., & Lockyer, L. (2015). Technology tools to support learning design: Implications derived from an investigation of university teachers' design practices. *Computers & Education*, *81*, 211–220. doi:10.1016/j.compedu.2014.10.016

Bennett, S., Agostinho, S., & Lockyer, L. (2017). The process of designing for learning: Understanding university teachers' design work. *Educational Technology Research and Development*, *65*(1), 125–145. doi:10.100711423-016-9469-y

Bennett, S., Thomas, L., Agostinho, S., Lockyer, L., Jones, J., & Harper, B. (2011). Understanding the design context for Australian university teachers: Implications for the future of learning design. *Learning, Media and Technology*, *36*(2), 151–167. doi:10.1080/17439884.2011.553622

Buus, L., & Georgsen, M. (2018). A learning design methodology for developing short learning programmes in further and continuing education. *Journal of Interactive Media in Education*, *1*(8), 1–10. doi:10.5334/jime.469

Comas-Quinn, A., de los Arcos, B., & Mardomingo, R. (2012). Virtual learning environments (VLEs) for distance language learning: Shifting tutor roles in a contested space for interaction. *Computer Assisted Language Learning*, *25*(2), 129–143. doi:10.1080/09588221.2011.636055

Creswell, J. W., & Poth, C. N. (2018). *Research design: Qualitative, quantitative, and mixed methods approaches* (4th ed.). Sage Publications.

Dewi, F. (2014). Edmodo: A social learning platform for blended learning classes in higher education. *Research in Education Technology: Pedagogy and Technology Journal, 11*(2), 1-11. https://www.academia.edu/20300729/EDMODO_ASocial_Learning_Platform_for_Blended_Learning_Class_in_Higher_Education

Dwiyogo, W. D. 2018. Developing a blended learning-based method for problem-solving in capability learning. *Turkish Online Journal of Educational Technology-TOJET, 17*(1),51-61. https://files.eric.ed.gov/fulltext/EJ1165753.pdf

Engeström, Y. (1999). Activity theory and individual and social transformation. In Y. Engeström, R. Miettinen, & R. L. Punamaki (Eds.), *Perspectives on activity theory* (pp. 19–38). Cambridge University Press. doi:10.1017/CBO9780511812774.003

Engeström, Y. (2001). Expansive learning at work: Toward an activity theoretical reconceptualization. *Journal of Education and Work, 14*(1), 133–156. doi:10.1080/13639080020028747

Farooq, M. U., Al Asmari, A., & Javid, C. Z. (2012). A study of online English language teacher education programmes in distance education context in Pakistan. *English Language Teaching*, *5*(11), 91–103. doi:10.5539/elt.v5n11p91

Green, J. K., Burrow, M. S., & Carvalho, L. (2020). Designing for transition: Supporting teachers and students cope with emergency remote education. *Postdigital Science and Education*, *2*(3), 906–922. doi:10.100742438-020-00185-6

Heggart, K. (2021). Formulated professional identity of learning designers and the role of open education in maintaining that identity. In A. Marcus-Quinn & T. Hourigan (Eds.), *Handbook for online learning contexts: digital, mobile and open* (pp. 21–34). Springer Nature. doi:10.1007/978-3-030-67349-9_3

Heggart, K., & Dickson-Deane, C. (2021). Microcredentials. *Pacific Journal of Technology Enhanced Learning*, *3*(1), 13–14. doi:10.24135/pjtel.v3i1.87

Irawan, V. T., Sutadji, E., & Widiyanti, W. (2017). Blended learning based on schoology: Effort of improvement learning outcome and practicum chance in vocational high school. *Cogent Education*, *4*(1), 1282031. doi:10.1080/2331186X.2017.1282031

Jordan, K., & Elsden-Clifton, J. (2016). Designing meaningful learning environments in a work-integrated-learning course using a DOCC Design. In B. H. Khan (Ed.), *Revolutionizing modern education through meaningful e-learning implementation* (pp. 1–17). IGI Global. doi:10.4018/978-1-5225-0466-5.ch001

Kali, Y., Goodyear, P., & Markauskaite, L. (2011). Researching design practices and design cognition: Contexts, experiences and pedagogical knowledge-in-pieces. *Learning, Media and Technology*, *36*(2), 129–149. doi:10.1080/17439884.2011.553621

Kim, B. (2015). Playing with gameful activities and assessments: Avatars and experience points in a graduate course. *Proceedings of the Computer-Supported Collaborative Learning (CSCL) Conference*. https://repository.isls.org//handle/1/450

Knussen, L. J. (2020). *Investigating early career teachers' design of technology-integrated learning in context* [Doctoral dissertation]. University of Technology Sydney. https://opus.lib.uts.edu.au/handle/10453/144102

Kristanto, A., Mustaji, M., & Mariono, A. (2017). The development of instructional materials e-Learning based onblended learning. *International Education Studies*, *10*(7), 10–17. doi:10.5539/ies.v10n7p10

Lakkala, M., Toom, A., Ilomäki, L., & Muukkonen, H. (2015). Re-designing university courses to support collaborative knowledge creation practices. *Australasian Journal of Educational Technology*, *31*(5), 521–536. doi:10.14742/ajet.2526

Liyanagunawardena, T., Adams, A., Rassool, N., & Williams, S. (2014). Blended learning in distance education: Sri Lankan perspective. *International Journal of Education and Development using ICT, 10*(1). https://www.learntechlib.org/p/147446/

Ma, N., Xin, S., & Du, J. Y. (2018). A peer coaching-based professional development approach to improving the learning participation and learning design skills of in-service teachers. *Journal of Educational Technology & Society*, *21*(2), 291–304. https://www.jstor.org/stable/26388408

Markauskaite, L., & Goodyear, P. (2009). Designing for complex ICT-based learning: Understanding teacher thinking to help improve educational design. *Proceedings of the 26th Annual Conference of the Australasian Society for Computers in Learning in Tertiary Education (ASCILITE) in Same Places, Different Spaces.* http://www.ascilite.org.au/conferences/auckland09/procs/markauskaite.pdf

McGee, P., & Carmean, C. (2012). Accidental pedagogy redux. *Distance Learning, 9*(4), 5. https://www.proquest.com/scholarly-journals/accidental-pedagogy-redux/docview/1353085461/se-2

Monat, J. P., & Gannon, T. F. (2015). What is systems thinking? A review of selected literature plus recommendations. *American Journal of Systems Science*, *4*(1), 11–26. http://article.sapub.org/10.5923.j.ajss.20150401.02.html

Moreno-López, I., Ramos-Sellman, A., Miranda-Aldaco, C., & Gomis Quinto, M. T. (2017). Transforming ways of enhancing foreign language acquisition in the Spanish classroom: Experiential learning approaches. *Foreign Language Annals*, *50*(2), 398–409. doi:10.1111/flan.12267

Mulyadi, D., Arifani, Y., Wijayantingsih, T. D., & Budiastuti, R. E. (2020). Blended learning in English for specific purposes (ESP) instruction: Lecturers' perspectives. *Computer-Assisted Language Learning Electronic Journal, 21*(2), 204-219. http://callej.org/journal/21-2/Mulyadi-Arifani-Wijayantingsih-Budiastuti2020.pdf

Mustikasari, D., & Heggart, K. (2022). *Conceptualising Teachers as Designers in the Higher Education Context: Case Studies from Indonesian English University Teachers* [Pecha Kucha Presentation]. *39th International Conference on Innovation, Practice and Research in the Use of Educational Technologies in Tertiary Education, ASCILITE 2022*, Sydney, NSW, Australia. 10.14742/apubs.2022.49

Natalia, F. (2017). Electronic means of foreign language learning in the system of higher education. *Educational Research Review*, *12*(3), 116–119. doi:10.5897/ERR2016.3062

Ogan, A., Aleven, V., & Jones, C. (2009). Advancing development of intercultural competence through supporting predictions in narrative video. *International Journal of Artificial Intelligence in Education, 19*(3), 267–288. https://www.learntechlib.org/p/69607/

Priyatno, A. (2017). Promoting learner autonomy through Schoology m-learning platform in an EAP class at an Indonesian university. *Teaching English with Technology, 17*(2), 55-76. https://www.ceeol.com/search/article-detail?id=527569

Salmons, J. (2014). *Qualitative online interviews: Strategies, design, and skills*. Sage Publications. doi:10.4135/9781071878880

Sulisworo, D., Agustin, S. P., & Sudarmiyati, E. (2016). Cooperative-blended learning using Moodle as an open source learning platform. *International Journal of Technology Enhanced Learning, 8*(2), 187–198. doi:10.1504/IJTEL.2016.078089

Sun, S. Y. (2014). Learner perspectives on fully online language learning. *Distance Education, 35*(1), 18–42. doi:10.1080/01587919.2014.891428

Sun, S. Y. (2021). Using patterns-based learning design for CALL tasks. *Computer Assisted Language Learning, 34*(8), 1061–1084. doi:10.1080/09588221.2019.1657902

Sun, S. Y., & Goodyear, P. (2020). Social co-configuration in online language learning. *Australasian Journal of Educational Technology, 36*(2), 13–26. doi:10.14742/ajet.5102

Thang, S. M., Mustaffa, R., Wong, F. F., Noor, N. M., Mahmud, N., Latif, H., Aziz, M., & Abd, S. (2013). A Quantitative Inquiry into the effects of blended learning on English language learning: The case of Malaysian undergraduates. *International Education Studies, 6*(6), 1–7. doi:10.5539/ies.v6n6p1

Yin, R. K. (2018). *Case study research and applications: Design and methods* (6th ed.). Sage Publications.

Young, D. J. (2008). An empirical investigation of the effects of blended learning on student outcomes in a redesigned intensive Spanish course. *CALICO Journal, 26*(1), 160–181. doi:10.1558/cj.v26i1.160-181

KEY TERMS AND DEFINITIONS

Blended Learning: Teaching and learning processes that comprise face-to-face and online sessions. A face-to-face session can be delivered in a classroom setting, whereas an online session can be delivered via e-learning, mobile learning, and social media applications.

Distance Education: Teaching and learning processes in the online space that are offered in distance education BA and MA degrees via remote learning coursework programs.

Emergency Remote Learning: Teaching and learning processes that shift from face-to-face or blended learning mode and adopt online learning formats during the pandemic in 2020. It is categorised as emergency learning because the pandemic has caused anxiety for teachers and students to recognise the unknown circumstances of the future ahead. However, this approach must be provided to accommodate students with education.

Learning Designs: Various ways to design technology used for student learning experiences that include sequences of types of activities and interactions at the level of a subject or subject component.

Online Distance Learning: A term in the study used by the Indonesian government for teaching and learning processes that comprise online learning remotely during the pandemic in 2020.

Online Learning: Teaching and learning processes that combine synchronous and asynchronous learning activities supported by technological tools. It is delivered in the online space. Students do not require to attend in person. However, online learning as a part of online courses is not offered in distance education programs.

Chapter 10
Simplified Systems Thinking:
The PPE Approach in Post-Pandemic Online Education

Ericka Kathleen Eppler
Oklahoma State University, USA

Monalisa Dash
Brajarajnagar College, India

Matthew Meador
University of Tennessee at Martin, USA

ABSTRACT

This chapter explores the people, process, and environment (PPE) approach within post-pandemic online education systems, beginning with foundational systems thinking concepts. It compares multiple systems thinking models, presenting the PPE model's straightforward way. PPE is a "back to the basics" prescriptive model that simplifies systems thinking and defines it as an optimization tool in online education. The study used the PPE approach and Quality Matters concepts as guidelines to analyze three case studies: a preK-12 online school, a degree college, and a corporate learning setting. Using open coding, the results identified commonalities across these reports, showing relationships between established associations to systems thinking principles and the need for more technological awareness, acceptance, and preparedness; improved readiness and adaptability of online systems; and increased learning management system utilization. The PPE model needs further grounded theory development to validate its application in diverse educational landscapes.

DOI: 10.4018/978-1-6684-7285-9.ch010

INTRODUCTION

In December 2019, the novel coronavirus (2019-nCoV) manifested as a zoonosis in the Chinese province of Hubei. By March 2020, the World Health Organization (WHO) declared the disease a global pandemic (WHO, 2020). As the condition continued to spread, a new and unforeseen concern arose, posing a grave threat to healthcare facilities around the globe. Consequently, while humanity fought the COVID-19 pandemic, it was also introduced to a new way of life. The epidemic significantly affected people's health, finances, and communities, and most of those affected were unprepared for most of its damage (Karakose, 2021). The global COVID-19 pandemic has catastrophically damaged healthcare, national economies, and schools.

As a result, many governments across the globe expended remarkable efforts to handle the numerous difficulties in fields such as healthcare, the economy, and education (Aduhene & Osei-Assibey, 2021; Buonsenso et al., 2021; Crawford et al., 2020; Desson et al., 2020; Fouad et al., 2020; Higginson et al., 2020; Hoekman et al., 2020; Nayak et al., 2021; Shmis et al., 2020; Tumwesige, J., 2020). The extensive spread of the COVID-19 virus has emphasized the urgent need for all educational institutions to make substantial modifications to their administration, curriculum, and pedagogical approaches. During the first weeks of the pandemic, many schools across the globe suspended in-person education, focusing instead on online and hybrid choices (Karakose, 2021; Singh et al., 2021). Most nations worldwide implemented distance education programs before, during, and after the COVID-19 pandemic to ensure that students' access to education would not be significantly hampered (Murphy, 2020; Tesar, 2020).

Thus, COVID-19 forced learning organizations to analyze their processes holistically for survival, and COVID-19 significantly affected these foundations. The traditional education systems before COVID-19 needed to prepare the people, processes, and environment (PPE) for such an urgent transition to online learning. In this chapter, we discuss the concept of systems thinking in online education, capitalize upon lessons learned during the COVID-19 pandemic, and provide recommendations for future learning organizations based on three case studies in different learning contexts: PreK-12, higher education, and in a corporate setting.

What Is Systems Thinking?

Systems thinking began as a theoretical framework that developed into an area of study, and subsequently, it became an organizational practice that helps learning organizations find solutions to systemic problems (Senge, 1990). Within this practice, a system consists of a collection of interdependent parts through feedback relationships and performs a specific task through their arrangements and connections with other parts (Meadows, 2008). In a system, the elements can be visible or invisible parts. However, no matter their visibility, the feedback relationships between them generate the functions or emergent properties that collectively create the system. Specifically, no single system component typically sets out to perform the system's functions (Meadows, 2008).

An Everyday Systems Thinking Example

A natural ecosystem is an example of a system. Therefore, a simplified representation of the forest ecosystem's components, connections, and roles follows, as shown in Table 1. Animals and plants are examples of biotic components, the living parts of an ecosystem, while soil, water, and air are abiotic components that together create this system. Notice how each relationship has different functions based on the connections between various aspects of the ecosystem. Additionally, the elements work together to create different outcomes. For example, animals eating plants or other animals sustain themselves (life maintenance), which could create a decline in other species' populations and create room for the proliferation of different species.

Table 1. A simplified representation of a system's components

System	Elements	Relationships	Functions
Forest ecosystem	Sun, air, soil, water, plants, animals	● Plants take in carbon dioxide from the air and water from the soil and make food in the presence of sunlight ● Some animals (herbivores) eat plants ● Some animals (carnivores) eat other animals ● Sun shines and causes transpiration, and forms clouds ● Clouds bring rain, and soil absorbs the water ● Plants and animals die and decompose to fertilize plants and trees	● Maintenance of life ● Emergence of new species ● Extinction of struggling species

Note. Illustrative example by Authors adapted from Meadows (2008)

This example demonstrates that systems thinking can be understood in simple terms. Next, the literature review explains more about systems thinking and provides some models to help readers understand how systems thinking could benefit different contexts.

LITERATURE REVIEW

Systems thinking is a framework for understanding the interconnectedness of systems in everyday situations (Hogan & Weathers, 2003). Additionally, systems thinking is dynamic and complex yet connects structural elements for reasoning, logic, and strategic arrangements, which transforms systems (Cabrera & Cabrera, 2019; Flood & Romm, 1997; Junpho & Kaewprapan Rosenkranz, 2018; Maani & Cavana, 2007; Senge, 1990). The least apparent system components, such as its purpose or function, are frequently the most significant factor in determining how the system will behave, and systems thinkers seek to understand this aspect of systems (Arnold & Wade, 2015).

For example, Chatterjee et al. (2022) discuss the course development process, which includes "an iterative, collaborative and an incremental process between the instructor and designer" in the planning phase, "feedback and guidance by the instructor" during the implementation phase, and constant feedback from students that informs design decisions (p. 101). Although Chatterjee et al. (2022) do not explicitly employ systems thinking, their research demonstrates how a course design's intangible components could impact a system's structure and function. Therefore, understanding hidden components like course design could lead to significant insights or improvements in education or online learning.

It is also possible to evaluate behavior by applying systems thinking models, which could contribute to understanding online education's complex, nonlinear relationships. For example, we find potential relationships with Cabrera and Cabrera (2019) in our preliminary model, People, Process, and Environment (PPE). The interrelated phenomena and dynamic relationships occur through continuous interactions within the PPE approach, where People interact with Processes and the Environment. The PPE approach also relates to Cabrera and Cabrera's (2019) emergence and network structure concepts. Online education's dynamic nature contributes to new teaching and learning methods that emerge from the system. Stakeholders could then share these techniques through network structures such as free learning websites and online learning communities.

Cabrera and Cabrera (2019) provide a foundational base in systems thinking, but other methods also consider and define it (examples: Richmond, 1994; Senge, 1990; Stave & Hopper, 2007). Although we highly value these diverse ideas, PPE relationships align with our research objective to understand complex OES. Therefore, focusing on PPE relationships is suitable for understanding systems thinking in education (Dhukaram et al., 2016). We have considered other perspectives while developing the model, which we discuss later in this chapter. Specifically, in this work, we evaluate our experiences during COVID-19, reflect on the lessons learned in the post-COVID educational domain, and link online teaching to PPE. We selected PPE because this approach draws attention to relationships between these aspects and is thus suited to educational systems' intricate, changing nature.

Nonetheless, other methods could provide helpful information. It may be beneficial to review those options in future research. There are several systems thinking models available in the body of knowledge available, such as System Dynamics (Forrester, 1961), Single-Loop and Double-Loop Models (Argyris, 1976), Spiral Model (Boehm, 1986), Ecosystems Model (Bronfenbrenner, 1994), Iceberg Model (Kim, 1999), Deming's The Plan-Do-Study-Act (PDSA) Cycle (The W. Edwards Deming Institute, 2021). Although several more exist in the literature, we briefly explain the six models listed above.

Forrester's (1961) system dynamics describe how a system works over time, allowing stakeholders to predict what may happen based on changes in variables potentially. This model is helpful because it can be applied to any system modeling various components and has a high level of complexity. Overall, system dynamics is a descriptive toolset that could predict future system behaviors based on current system structures (Forrester, 1961).

Single-Loop and Double-Loop Models are feedback models proposed by Argyris (1976). Single-loop learning helps people understand how specific actions lead to results. Double-loop learning helps people understand the assumptions they make when they examine how particular actions lead to results, which can lead to a deeper understanding of organizational behaviors. The W. Deming's PDSA Model uses single and double loops, emphasizing Process and feedback, thus helping people potentially evaluate their underlying assumptions (The W. Edwards Deming Institute, 2021). Both Argyris and Deming add value to systems thinking by providing these ideas. For instance, an interdependent process could include learners receiving feedback on their assignments from educators and giving feedback about their experiences.

The business community often calls Boehm's (1986) Spiral Model the software development lifecycle model that demonstrates the systems thinking principles like adaption and iteration. These principles are fundamental in online education because learning has a cyclical nature. Therefore, the Spiral Model can represent how online education needs constant adaptations and enhancement, mirroring iterative learning cycles.

Bronfenbrenner's (1994) Ecosystems Model, grounded in a system's ecological state, might illustrate all elements of the PPE Model, thus showing the interconnectedness between humans and the environment. In online education, the Ecosystem Model can evaluate the connectedness and interconnectedness of learners, instructors, educational materials, and technology platforms, showing disruption that impacts all or system parts.

Kim's (1999) Iceberg Model encourages stakeholders to understand more about mental models and underlying structures that cause observable occurrences. This model identifies the root causes of reoccurring problems that are difficult to solve. The Iceberg Model also takes a high-level approach that could apply to any system type.

In addition to these models, we acknowledge significant social and organizational pressures that could mold online education's delivery and design. A concept from an institutional theory called isomorphism indicates that organizations within a specific field may become more like each other over time due to those pressures, which include coercive, memetic, and normative forces (DiMaggio & Powell, 1983).

Concerning online education, educational institutions may produce similar education systems because they attempt to comply with industry norms and regulatory requirements. They may also mimic other successful programs (Mann, 2019). Therefore, due to isomorphism, which could increase homogeneity in online education (Mann, 2019), various educational institutions could adapt the PPE approach to fit their circumstances and streamline educational processes. Hence, students could have a standardized experience in an online environment, but it is also important to note that isomorphism is a potential limitation in innovation.

Although we mention some models here, we capitalize upon commonalities from several systems thinking models in this work, especially Dhukaram et al. (2016), who write about systems thinking in terms of "the interactions and relationships between the people, technology, processes and the organisations" through the Socio-Technical Systems (STS) approach for understanding education systems with case studies (p. 1). Our approach is more nuanced and focuses on interconnectedness and relationships; we are concerned explicitly with optimizing systems thinking for human activity in education systems. Additionally, we aim to simplify systems thinking through People, Process, and Environment, which may be easier to understand than other models. Finally, models like System Dynamics (Forrester, 1961) and the Iceberg Model (Kim, 1999) are descriptive, while the PPE approach is a prescriptive model, serving as a framework for creating and enhancing systems for improved outcomes.

When attempting to understand online education systems (OES), one must seek to understand their complex structures and components and how these systems function. However, it is essential first to understand online learning within those systems. Singh and Thurman's (2019) systematic review indicates multiple definitions of online learning. Still, most definitions have some common traits, specifically that learning occurs through a technology that delivers learning information and may occur synchronously or asynchronously (Singh & Thurman, 2019). We capitalize upon Singh and Thurman's (2019) definition with Meadows' (2008) system thinking perspective to operationally define an "online education system" as an intricate web consisting of multiple interconnected components that may include but are not limited to instructional materials and strategies, learning management systems, internet-based learning, and assessments. Together, these components could provide learning experiences delivered primarily through digital technology, either synchronous or asynchronous. Their feedback relationships and dynamic interactions contribute to these systems' effectiveness and learning outcomes. Consequently, outcomes within a digital education environment may hinge upon how these systems can adapt to learners' needs and changes in system processes. This comprehensive description provides the foundation to apply systems thinking in this field.

There is a total of six components that make up an OES (Hemphill et al., 2019). These include "online course delivery and support system, course development system, student academic support system, university-wide instructor support system, different program/departmental curriculum subsystems, and student personal support systems" (Hemphill et al., 2019, p. 18). These components comprise the interconnected parts of our proposed definition of an OES. For example, "online course delivery and support systems" could provide learning experiences through digital technology. Additionally, "student academic" and "student personal support systems" could serve as feedback loops, potentially allowing the system to adapt to the learner's needs. Ultimately, these components and their interactions could help stakeholders better understand how to improve OES learning outcomes.

Multi-faceted OES could benefit from stakeholders encompassing a deep insight into their complexities by viewing them through different lenses. To further examine these systems in more detail, Mao and Shearer (2019) outline a system-informed methodology for digital spaces that include design, infrastructure, and users. This depiction of OES and those found elsewhere in the literature provide evidence of this field's complexities.

Researchers have suggested various ways to describe these systems in acknowledging these nuances. For example, Sockman et al. (2019) provide one way to divide the OES into three subsystems: the macro, the meso, and the micro. The macrosystem comprises the community organizations encircling comprehensive institutional settings, the mesosystem consists of the school environment and community partnerships, and the microsystems are people classified as learners or community agency members (Sockman et al., 2019). In another example, Dzemydienė & Tankelevičienė (2010) show connections between structure and function through their multi-layered digital environment approach. Their "knowledge-based component architecture" divides into four learning management system subsystems: "development of distance courses," "supporting a study process," "organizing realization and control which identifies the registered users," and "logistics meant for planning the personnel, resources, finances, equipment, and information of the institution" (pp. 229-231). In our last example, Frick (2020) identified educational systems by their relationships: "teacher-student relationships" and "student-content relationships." "teacher-content relationships," "student-content relationships," "teacher-context relationships," and "context-context relationships" (pp. 695-698). Although Frick (2020) did not specifically write about OES, they included them in their comparison of past, present, and future systems, with the primary point being that these relationships are relevant in the technological age.

Each one of these classification systems could offer valuable perspectives based on their OES focus. What is even more important is that various authors have different ideas about how the whole relates to the parts, how to tell which elements are part of the system and which are not, and how "parts and wholes may have different meanings from different perspectives" (Cabrera & Cabrera, 2019, p. 13 as cited by Tamim, 2020). Within these frames, these classification systems do not meet our research objectives. Therefore, this study divides the online education system into three subsystems or levels: people, process, and environment, which we aim to optimize specifically for OES.

Systems Thinking in Online Learning: The PPE Model

In this chapter, we operationally define systems thinking as a theoretical framework and organizational practice that scrutinizes the relationships of people, processes, and environment (PPE) within systems. This approach establishes an effective model for determining solutions for a learning organization's severe systemic problems. Merriam-Webster (2022) defines people as "human beings in general or considered collectively." Consequently, People are a systems thinking domain that is strategic and essential for a complete system. The process domain is "a series of actions or steps to achieve a particular end" (Merriam-Webster, 2022). The environment is "the natural world, as a whole or in a particular geographical area, especially as affected by human activity" (Lexico.com, 2022).

Although people, processes, and environments (PPE) can be parts of other systems, we specifically evaluate PPE within complex online educational systems. Even though each PPE domain is significant, there is more importance in the relationships between each aspect than in each element individually, as is the nature of systems thinking. These domains serve as the principal foundation for forming this chapter. Therefore, we present the PPE model as a new perspective, a "back-to-the-basics" approach, simplifying existing complex models into an uncomplicated design.

The PPE model arose from the pandemic, and its design can attach to a series of models, thus streamlining them to create an innovative tool. The PPE model encompasses a message of simplicity, plus the method and action to improve online educational ecosystems potentially. Additionally, every connection in the PPE model constitutes an intersection, interconnection, and an intraconnection, reflecting the systems thinking perspective.

Earlier, we used a forest ecosystem as a metaphor because it may be easier to understand than systems thinking in social systems like education. Different parts of the forest, such as trees, animals, and weather, interact similarly to OES components, such as students, teachers, and technology. Like how the different parts of a forest ecosystem, such as organisms, landscapes, and climate, interact in a complex web of relationships, so do educational system components. These intricate interactions create an ever-changing system one cannot understand by analyzing each part separately. When stakeholders view education systems with this perspective, there are more opportunities to assess their strengths and weaknesses, thus providing an avenue to develop improvement plans.

Consequently, it is instrumental in establishing a framework for investigating educational systems that can precisely identify the elements, relationships, and resulting system functions. The "people" of an education system are the educators, students, and parents. The "process" refers to the activities of developing and delivering lessons, as well as assessing student progress. Relationships exist between these parts. Some examples could include parents sending their kids to school so that they can learn, teachers instructing students, and schools employing teachers. Like the economy, the functions become evident due to the interplay between its components. However, they may include, for example, imparting basic knowledge, ensuring a certain number of years of completed education, or acclimating people to a particular national identity.

Table 2. Online education system's components

System	Elements	Relationships	Functions
Online Education System	**People** (Parents, Teachers, Students) **Process** (Developing Material, Teaching, Evaluating), **Environment** (online Schools, Teaching Material, Online Technologies)	● Parents make admissions of their students in schools/ online schools ● Teachers of that school teach students through online technologies ● Teachers create/curate online learning material and share it with their students ● Teachers create online discussion forums for student interactivity ● Teachers evaluate their students using online technologies	● Dissemination of knowledge and training of skills ● Socialization ● Assurance of a certain number of years of education

Note. Table created by Authors.

PPE and Quality Online Learning

Systems must operate within a range of normalcy for learning organizations to meet their intended outcomes. This range usually consists of the conventional activities of an organization and its assumptions about normalcy within these activities (Uddin et al., 2019). In the event of disruption from the normal state, the "pre-event" is within the acceptable range, and the "post-event" following a significant disturbance, potentially could bring equilibrium to the system, albeit hopefully positive changes due to the reflection of learning experiences in the "post-event" phase (Uddin et al., 2019, p. 393).

Therefore, systems thinking in the context of the post-pandemic world considers the pre-pandemic time as the known acceptable range, thus providing a baseline for online systems. To demonstrate such a baseline, we reviewed several high-quality online standards applicable to various learning situations that were in use before the pandemic. Some of those standards include Quality Matters (QM) Standards (QM, 2018a), ISTE Standards (International Society for Technology in Education (2023), Quality Scorecard Suite (OLC, 2023), and the closely-related OSCQR developed by the State University of New York (SUNY) that evaluates individual courses instead of the quality of whole education systems that the OLC now manages in collaboration with SUNY (Online Learning Consortium, 2023; State University of New York, 2013). We selected the QM General Standards (GS) (QM, 2018a) because they are applicable in all three settings addressed in the case studies below. Additionally, we selected QM because these ideas may be easier to understand, with only eight GS (QM, 2018a), as compared to some standards like OSCQR, which describes 50 standards (Online Learning Consortium, 2023; State University of New York, 2013). There are multiple Specific Review Standards under each GS (QM, 2018a) that we did not mention because we are looking at high-level relationships between PPE and online learning (QM, 2018b; QM, 2018c; QM, 2018d). Therefore, listing every substandard would not match the chapter's focus. To be clear, we use QM GS (QM, 2018a) as an example to make it easier for readers who are familiar with and interested in improving online learning but may not be familiar with applying systems thinking principles in this context.

QM establishes a baseline for quality online and hybrid learning in the United States for PreK-12, higher education, and professional development settings (QM, 2018a). QM provides high-quality, research-based standards for assessing and developing online courses, curricula, and teaching (QM, 2018a). We propose that online standards like QM GS (QM, 2018a) may align with the systems thinking framework because they indicate high quality in online and blended learning at a high level. Additionally, QM's pursuit of quality online education relates to our systems thinking framework that seeks to optimize the interdependent components of educational systems. The People, Process, and Environment framework is a subframe of systems thinking, which like systems thinking, generally emphasizes the interdependence and interconnectedness of different system components. This method acknowledges that changes in some parts of the system can have an accumulation effect throughout the whole system. When stakeholders organize online education systems according to online standard lists like the QM GS (QM, 2018a), it is possible to draw parallels because of the inherent interconnectedness and interdependence of systems thinking components.

Different online standards exist, such as the QM K-12, higher education, and professional development rubrics (QM, 2018b; QM, 2018c; QM, 2018d). All have the following eight General Standards (except for K-12, which has "Instructor Support" added to the end of the seventh general standard as shown below) (QM, 2018a, Used under license. All rights reserved. Retrieved from MyQM):

1. Course Overview and Introduction
2. Learning Objectives (Competencies)
3. Assessment and Measurement
4. Instructional Materials
5. Learning Activities and Learner Interaction
6. Course Technology
7. Learner Support (and Instructor Support for K-12 only)
8. Accessibility and Usability*

Each standard relates to a specific PPE domain (Figure 1 and Table 3). Below we provide a visual chart and a text table to accommodate varying learner preferences. Both Figure 1 and Table 3 communicate the potential interconnectedness of the PPE domain with online learning standards, using QM GS (QM, 2018a) as an example.

For readers who prefer a text-based chart, we provide a brief explanation of how an online standard, like QM GS (QM, 2018a), interacts within each domain:

QM is one of several nationally recognized online standard sets. High-quality online standards provide a credible framework for online education, and thus, we use the QM GS (QM, 2018a) to support research about the PPE approach. This chapter presents a systems thinking analysis of fully online learning organizations, utilizing QM GS (QM, 2018a) as an illustrative example of such standards. We specifically focus on how COVID-19 significantly affected PPE in online education. The chapter highlights the central pieces within the PPE domains that connect to these standards in this online context, which intensified due to the pandemic. Additionally, combining the PPE approach and online standards like QM GS (QM, 2018a) could create a structure that supports challenging times in online education.

Online standards, like the QM GS (QM, 2018a), could connect to different parts of the PPE domains, providing a systems thinking perspective. For example, the *Course Overview and Introduction* (QM, 2018a) standard could define how educators and students interact (People), outlining the Processes for these interactions. Similarly, the *Learning Objectives* (QM, 2018a) might form a vital part of the Process because they are the output the system hopes to render. Considering *Assessment and Measurement* (QM, 2018a) could help online teachers measure students' progress (Process) through the Environment, which is the online educational platforms and tools. Next, the Environment domain houses instructional materials. When offered through an online platform, they could assist with forming students' experiences and thus facilitate the learning Process. *Instructional Materials* (QM, 2018a) could be part of *Learner Activities and Learner Interaction* (QM, 2018a), which could connect all PPE domains. Teachers and students interact (People) through learning activities (Process) within the online Environment. *Course Technology* (QM, 2018a) could be a significant part of the Environment domain because it builds the Environment and supports learning Processes and People using the technology. *Learner Support* (QM, 2018a) could link to the People domain because learners need accessible support services for their well-being and success. Finally, *Accessibility and Usability* (QM, 2018a) may represent another aspect of the Environment domain, emphasizing that the Environment must function in a way that facilitates (Process) learners (People) with varying abilities to meet their learning outcomes. This hypothetical interaction between the QM GS (QM, 2018a) and the PPE domains illustrates the potential interconnectedness of online education. However, these relationships can change depending on context and specific implementation.

Figure 1. An example of relationships between PPE approach and online standards: QM GS
This chart demonstrates the interconnectedness of online standards and People, Processes, and Environments.
Note. Chart created by Authors and adapted from QM (2018a). Used under license. All rights reserved. Retrieved from MyQM.

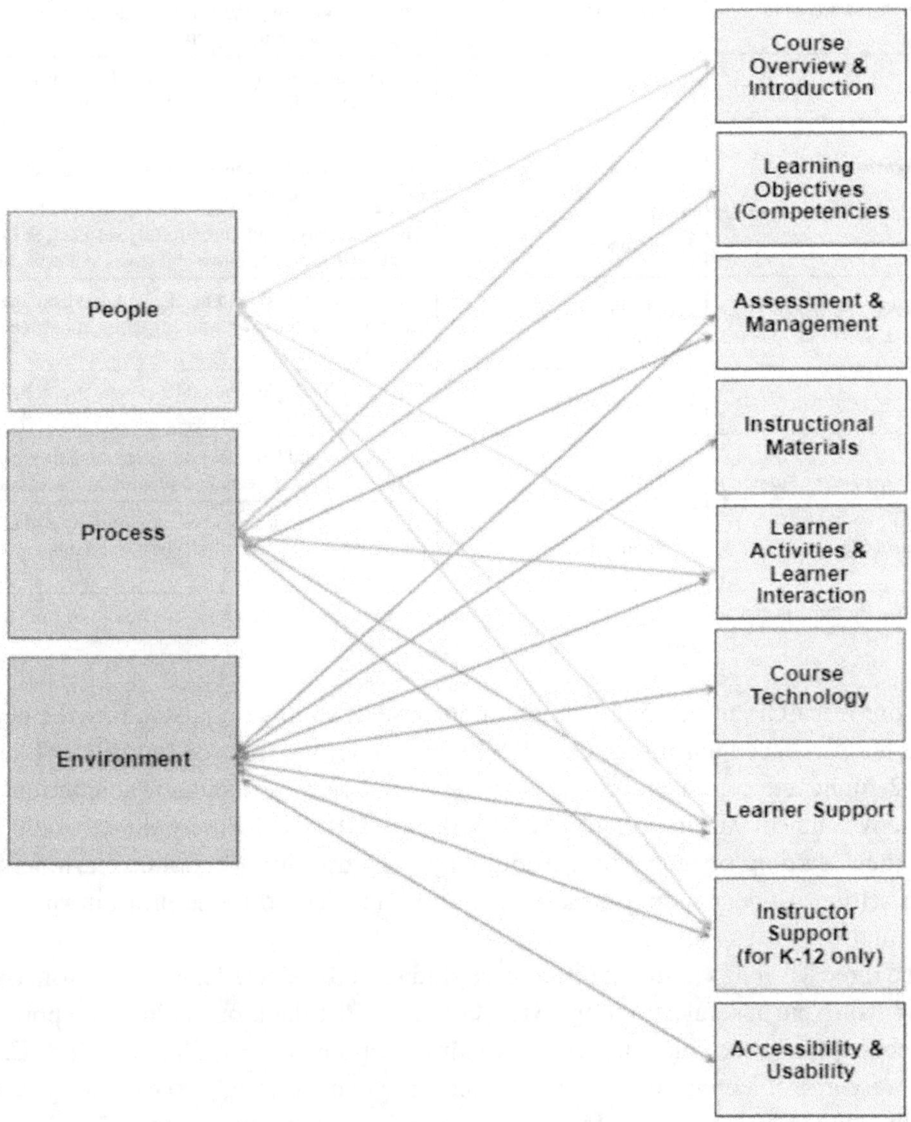

Methods

Narrative research describes the human experience expressed in text format. Additionally, narrative researchers use small sample sizes to create rich, open-ended discourse to investigate the meanings people give to their experiences (Josselson, 2010). Bruner (1991) states that narratives construct reality. Therefore, a suitable way to describe the systemic problems during the COVID-19 pandemic is to analyze the stories of those close to the experiences that crisply describe these systemic issues. Therefore, narrative research methods guided this study.

Table 3. Example of relationships between PPE approach and online standards: QM GS

QM General Standard	Related PPE Domain(s)	Description
Course Overview & Introduction	People, Process	This standard stresses establishing clear expectations for students and giving them a thorough introduction to the course, teacher and learning objectives.
Learning Objectives (Competencies)	Process	The learning objectives should be clearly articulated, measurable, and matched with course activities and evaluations.
Assessment & Measurement	Process, Environment	The goal of the evaluation should be to measure students' progress toward accomplishing the learning objectives and to offer relevant feedback.
Instructional Materials	Environment	Ting materials should be carefully selected, engaging, and connected with the course objectives and activities.
Learner Activities & Learner Interaction	People, Process, Environment	Course activities should be planned to encourage active learning, collaboration, and engagement between students and teachers.
Course Technology	Environment	Course technologies should be easy to use, reliable, and help students reach their learning goals.
Learner Support (& only for K-12: Instructor Support)	People	Students should have easy access to support services like academic support, technical support, and accessibility services.
Accessibility & Usability	Environment	The course design should be user-friendly and accessible to all students, according to the best standards for usability and accessibility.

Note. Chart created by Authors and adapted from QM (2018a). Used under license. Retrieved from MyQM.

We began our research journey with the initial purpose of simply understanding the challenges that educators and learners faced during the COVID-19 pandemic. We focused on distinct educational settings: PreK–12, higher education, and corporate education. Learning about any adaptations they used to mitigate their environment was important. We hypothesized that each environment would have distinct challenges in their learning organizations. Additionally, the transition to online learning might be different in each setting. Understanding these transition periods could be helpful information for future systemic planning.

We selected three learning environments as case studies and collected narratives from one individual in each setting using purposeful sampling. After collecting the data, our inductive approach led us to test the PPE concept in these contexts, which we aim to optimize for online learning. Using the PPE approach, our research objectives became more focused when we sought to expose systemic issues in each setting that may require strategies for improving online education in a post-COVID world. Therefore, the study's ultimate purpose was also to provide a theoretically grounded tool for improving OES.

We analyzed the narratives using an open coding process, a form of manual thematic coding, where we identified words and phrases according to specific themes. This process involved reading the narratives multiple times, underlining reoccurring words or themes. Afterward, we recorded our findings on sticky notes and grouped them. At that point, we could connect systems thinking concepts to the themes we found through this process. This hands-on approach allowed us to establish insight within each narrative and capture the shared experiences between the case studies. We display the results of our narrative analysis and the synthesized themes from all cases in the Discussion.

RESULTS: CASE STUDIES

The following three case studies report the disruptions of COVID-19 within different learning organizations and the solutions intended to mitigate their dislocation. Even though each case study represents a different kind of online learning environment, all case studies demonstrate how stakeholders' decisions in each setting significantly impacted learning.

Case Study 1: Traditional Students Shift to an Online PreK-12 School

When COVID-19 disrupted traditional learning organizations' systems, some were unprepared for the immediate, drastic shift in their instructional delivery systems, turning towards emergency remote teaching (Hodges et al., 2020). Some may perceive traditional learning as more effective than online learning (Kumari et al., 2021), but online learning can supplement blended learning in traditional K-12 classrooms (Zhang et al., 2019). They did not fully utilize the power of online education in their systems (Huck & Zhang, 2021).

One advantage of fully online Pre-K-12 schools over traditional schools was that online schools were prepared for online teaching and learning before the COVID-19 pandemic because online education is their learning model (Childs et al., 2022). Online schools often hire highly qualified teachers who are technologically savvy and well-versed in educational technology and tools (Nelson, 2022). Additionally, fully online Pre-K-12 students had more training to receive instruction in an online environment than students in traditional schools.

Opposite to what some might expect, the fully online PreK-12 school also experienced disarray during the pandemic but for different reasons (Participant A, personal narrative, February 17, 2022). In this case study, we report teachers' experiences at a fully online PreK-12th grade school in the midwestern United States to illustrate the breakdown in online education. Most of the problems stemmed from students accustomed to traditional learning environments withdrawing from their schools in record numbers to attend fully online schools.

Many local schools offered online classes during the COVID-19 pandemic and provided laptops and Wi-Fi hotspot boxes like the online school. However, some students and their families were unhappy with the online instruction received from their local brick-and-mortar schools. They thus sought alternative education options that met their needs. The families complained to the online schoolteachers that the brick-and-mortar teachers did not have adequate training to support their children and that some schools did not offer live teacher support. In some instances, the students could only contact their teachers by email and had to wait up to two days to receive an answer (Participant A, personal narrative, February 17, 2022). Students fell behind in their classes and lost motivation when they did not understand how to move forward without their teachers' help. In other instances, students had never used laptops and online curriculum, and no one helped them learn how to interact with the software.

The online school's leadership expected this might cause some disruption to their current systems but had not fully anticipated the effects of these decisions. Initially, the most troubling problem was the lack of teachers, principals, school enrollment counselors, and other staff required to support the new students offloading into the online school system (Participant A, personal narrative, February 17, 2022). In the 2019-2020 school year, 17,106 students enrolled by the first week of school (OSDE, 2019). However, in 2020-2021, a record-breaking 30,000 more students enrolled than in the 2019-2020 school year, making it the largest school system in the area in the 2020-2021 school year (Watson, 2020). The disproportionate increase in enrollment led to an immediate need to secure additional technology tools, such as laptops and wireless hotspots, for 30,000+ new traditional students. Many new students had limited experience using laptops or other technological devices to interact with their fully online curriculum and submit homework assignments. Additionally, some teachers' rosters had doubled in size compared to the previous year. These systemic changes equated to double the work and the time invested in teaching and testing students, including twice the benchmarks and standardized testing.

After the online school mitigated these initial hurdles, many traditional students and their families needed to adapt to online learning and needed extra structure and support. The teachers taught the students and families how to access their assignments, attend online synchronous meetings with their teachers and classmates, interact with the 24/7 tutoring service, complete projects in an online school, upload appropriate documentation demonstrating evidence of their learning, and more. They held monthly or bi-weekly training sessions in the evenings for all, followed by daily reminders about completing these tasks and succeeding in an online school.

Another problem is that some traditional students and families needed help learning to use the wireless hotspots that the school provided to them. Many students in poor, rural areas used the Internet for the first time and were unsure how to access it with the hotspots. Those who did know how to use them sometimes had trouble connecting to the Internet. The online school issued hotspots from two wireless carriers with different coverage areas, so not all hotspots worked in all areas. These technical problems caused the traditional students to fall behind in their studies, sometimes to the point of truancy.

Some students did not report when their wireless devices failed even though there was a sticker with a technical support phone number on them, and subsequently, did not complete any assignments due to these problems. They did not report technical difficulties to their school even though their teachers constantly reminded them how to connect to the Internet and where to obtain access if their hotspots failed. The teachers, unaware of the technical issues, contacted the students and their families and asked for screen time meetings, where students log in to a Zoom meeting and stay online with the teachers until they catch up on their work. When students did not come to the meetings, the teachers would often travel to the students' homes and attempt to collect work from them in person or observe the students working until they completed the required weekly assignments. After visiting students in person, the teachers discovered that the students needed help with their hotspots.

Some traditional students' families did not always offer adequate support to their students. They did not understand how to work with tech support and use technology in the school context. Furthermore, some parents depended solely on teachers in their previous traditional schools for their children's education. Therefore, when they enrolled their children in an online school, the families did not anticipate the need to provide additional learning support for their children's homework and technical difficulties, even though the online teachers gave frequent reminders. The teacher who provided the narrative often gave gentle reminders and wrote in an email, "Please know that online school is different than brick-and-mortar schools. You are now your child's primary teacher and are 70% responsible for their education. I will support you up to 30% of the time. Know that I am here for you if you have questions. I will help you." The teacher gently directed the parents to take responsibility for their children's technical needs: "I need you to contact tech support when your student has a technical problem and not wait for me to notice that your student is not working. Please let me know as soon as there is a problem, and I will deliver homework to you in person." Even still, the families sometimes did not understand their support roles when the teacher offered training sessions to new families. The teacher shared her screens to demonstrate the functions the parents needed to complete on behalf of their students.

The extra time invested in helping traditional students learn to attend school online impacted those who had participated in the online school in previous years. Some experienced online students received reduced teacher attention. The teacher needed to manage behavioral disruptions because some traditional students were unfamiliar with online class etiquette. For example, some students came to class without wearing appropriate attire. Some other traditional students were unsuccessful in creating a productive learning environment at home and would lie in bed, sometimes falling asleep during class. Even though there were behavioral and procedural disruptions, the influx of traditional students into the online school brought more diversity to the class.

The school usually focuses on students receiving dedicated individual attention each week from their teacher. However, this was impossible due to the enormous class sizes; teachers instead taught students in small groups to ensure that every student received equal attention. This shift in teaching strategy helped the teacher maximize her time. Additionally, some students new to attending school from home could make friends easier in the small group meetings, limiting their isolation.

Not only did the COVID-19 pandemic disrupt this online school, but all the People involved with it, including teachers, students, families, and administrative staff. However, as the traditional students learned how to engage with the online school, the teacher reported a richer experience for her students. The pandemic acted as a strength test for this OES. The lessons learned from the teacher's experiences, including teacher preparedness, adequate support, manageable class sizes, and classroom management techniques, shed light on the structural weaknesses. However, this case study also highlights the silver linings that arose from adaptation and creativity.

Case Study 2: Higher Education in an Online Setting in India

Online courses are common in developed countries; however, the COVID-19 pandemic compelled teachers in the developing world, including India, to use digital platforms for instruction and assessment to maintain student engagement. This drastic switch caused challenges the country had never experienced. The education sector responded to the crisis by innovating and developing new methods to ensure students' continued education. Consequently, UNESCO (2020) launched the Global Education Coalition program to help students take advantage of this opportunity. Additionally, the Ministry of Human Resources Development (MHRD), India, made available a variety of free, computer-based e-learning programs like SWAYAM Prabha, Study Webs of Active-Learning for Young Aspiring Minds (SWAYAM), and e-pathshala (Ministry of Education, Government of India, 2020).

Despite these relief efforts, the transition to online education presented unique challenges. Since using the Internet for school is still relatively new in India, many teachers have not had the opportunity to obtain the skills required for providing online instruction and assessment. However, when the pandemic triggered a nationwide lockdown, teachers were forced to use virtual classrooms due to a lack of alternatives. Therefore, they had to overcome a national need for domestic infrastructure and technological issues.

They needed foundational technology, such as laptops, smartphones, and Wi-Fi hotspots, to support efficient online education. However, buying these devices can be a significant investment in many developing countries, making them unattainable for people in certain parts of India. To put this in perspective, according to World Bank data, India's Gross National Income (GNI) per capita for 2021 was $2,230 (on Atlas method, current US dollars) (The World Bank Group, 2023). Comparatively, the average United States household income was approximately $70,784 (United States Census Bureau, 2022). Therefore, purchasing devices that cost hundreds of dollars could be difficult for some Indian families.

Considering these significant financial barriers, it is understandable that these unexpected changes presented challenges unique to this context. To highlight this point, we report the experiences of a tenured instructor and some of her colleagues at a degree college in Odisha, India, serving students in the Bachelor of Arts program with Education Honors. The instructor found systemic failure in remote learning in the areas that arose when the pandemic necessitated a fast shift to an online learning environment. These issues may be associated with a need for more expertise and training on using digital resources, as well as the senior colleagues' reluctance to change to online teaching. The decisions by senior staff exacerbated the problems of the chosen technological options.

Additionally, some of the colleagues and the students needed more technological literacy and experienced challenges due to their limited exposure to online learning tools, which we refer to as "low literacy levels." The limited digital literacy was not an indicator of their educational attainment or literacy rates but rather their understanding of the technologies and platforms used in OES. These circumstances emphasize the differences between traditional academic skills and modern digital competencies needed for online learning. Many instructors and students struggled with the foreign online environment, even though the instructors had high academic qualifications. Overall, the instructor experienced many difficulties during teaching and assessing student progress, including utilizing instructional methods that foster student engagement.

The Indian government mandated that educators participate in online learning opportunities. As a result, the college planned a meeting to decide how to use the video conferencing software and create lessons for online students. They ultimately selected Google Meet, and the instructor provided a tutorial on how to use Google Meet for virtual presentations.

Some colleagues had trouble in Google Meet, specifically managing an online class. For instance, one colleague reported, "After the meeting ended and we leave the meeting we found that the students did not leave the meeting and were involved in mischievous tasks." This circumstance illuminates the challenge of monitoring students' actions in this context. During the class, most students disabled their video and audio options, offering only limited responses to the instructor's prompts and online discussion posts. The instructor noticed that the students were not actively participating in the online discussions, so they thought it would benefit the students to create presentations. The instructor concluded that by implementing these teaching strategies, the students would receive help by enhancing their education and improving their ability to use and apply technological tools such as PowerPoint, screen sharing, and the whiteboard application.

The shared learning materials frequently disappeared from the WhatsApp group, leading to constant reposting. The instructor reported that it was common to see WhatsApp messages like, "Ma'am, could you please share the notes on Adjustment Mechanism in the group once again? Despite my best efforts, I have been unable to locate it." Consequently, the instructor created a Google Classroom for students and distributed its URL to circumvent these problems. Specifically, Google Drive served as a central repository for all the educational resources the instructor developed. Creating the central hub increased student engagement because the students started lively discussions about what they learned about the course materials and posted their questions.

The instructor also encountered issues with online classes. Students would often join and subsequently leave the session due to incoming calls on their mobile devices. The instructors and the students also experienced technical difficulties like students' limited access to smartphones, loss of network connection, and data pack problems. Therefore, the instructor found herself teaching the same topics numerous times, only covering some lessons planned for the semester. When there was only one phone in the students' houses, and one of the students' siblings or parents needed to use it, other students in the household could not log into their online classes. As a result, online class attendance was low. In addition, the government declared, "As the attendance of the students for online classes is very low, when the physical classes start, the teachers should re-teach their subjects." Accordingly, the instructor and her colleagues had more work to do.

There was also a need for more colleagues with adequate computer skills required to teach online courses. Therefore, both education and assessment became more challenging. The government did not provide professional development for the educators yet expected them to complete their duties without adequate computer literacy. Moreover, even senior colleagues struggled to take classes online.

To help teachers from various colleges gain the required skills, the instructor's college hosted a Google-based online workshop for educators from all over the country under the instructor's direction. The workshop featured nationally recognized professionals with experience and expertise. After the teachers started taking online classes, they realized quickly that their tasks would be complex and time-consuming. They would have to produce, gather, or download course materials, such as PDFs, images, and videos. Afterward, they needed to upload them to a WhatsApp group and create a Google Meet link for students. They also monitored online attendance and communicated with students about relevant group developments. Overall, the teachers made every effort to encourage students' active participation in their virtual lectures.

Since observing students in classes was impossible, evaluations could not take place. Therefore, when determining grades became complicated, the University would distribute notices to skip exams, such as in the second and fourth semesters. Instead, the instructor and her colleagues graded students based on their performance in earlier terms. Nonetheless, as COVID-19 persisted, the college still required students to provide email addresses and participate in online exams.

The college created a special email account just for the exam. The college administration decided to relocate a desktop computer and printer to the Undergraduate examination room to send questions to students via email, receive the answer scripts of students online, and print the answer scripts of students to evaluate the students. Students then received their exam questions via email from the college's address; they replied to the same address with their answers. Students took their exams in their homes, and the college did not employ an online proctoring system. Consequently, there was no way to verify the students' original work. Unlike the offline exams, the online exams occurred within the college which went to other colleges for review. As a result, teachers' bias impacted students' test scores due to various evaluation processes.

In conclusion, the shift to online education during the pandemic presented several challenges for instructors and students at the college in Odisha, India. Despite these difficulties, they demonstrated the potential of online education in this region. However, learning leaders must invest in the necessary infrastructure to successfully implement and utilize their OES. Additionally, the college could benefit from adding more professional development opportunities and increasing access to essential technological tools.

Case Study 3: Corporate Learning

COVID-19 disrupted corporate learning across every industry as governments worldwide shut down to combat the virus's rapid spread. These expeditious shutdowns forced some corporate learning organizations to rethink how they teach safety to their workforce. As a result, they expanded online learning to larger learner populations through online education. In this case study, we examine an organization that operates in the logistics and freight industry that attempted to shift its training delivery methods from primarily in-person training sessions to online learning. The company also aimed to ensure a quality online experience for its employees. Before the COVID-19 pandemic, the organization sustained a self-directed learning management system where employees could work independently on training needs. However, these online learners mostly worked remotely from home before the pandemic and did not represent this company's varied learning population's needs (Meador, 2021).

During COVID-19, learning leaders did not mandate that users self-enroll in training due to some workers needing more time to care for themselves and family members who became infected with COVID-19. Additionally, decision-makers relaxed the number of learning courses they required users to enroll in compared to before COVID-19 for these reasons. Ironically, due to the pressures of a strained workforce, a company specializing in logistics needed to manage the logistics of their workers completing safety training, potentially decreasing the overall safety of this organization.

Since some workers had customer-interfacing roles before the pandemic, they needed to continue operating in those roles. For example, customer-facing office workers could not work remotely. Additionally, people whose jobs required safety training, such as truck drivers handling and shipping hazardous substances, needed to demonstrate their proficiency before the company cleared them for driving. Therefore, despite remote training, instructional delivery for some job classifications requires permanent in-person instructor-led training (ILT), although modified for safety (Meador, 2021).

The organization incorporated safety measures for learning teams to attempt to curtail the spread of COVID and thus changed its ILT processes. For example, the organization previously held extensive training sessions to accommodate many people simultaneously with a class registry of 25 or more participants to warrant an instructor before the pandemic. However, it had to modify the learning environment, limiting classes to between five and ten participants, with highly regulated spacing between participants, per course per instructor. These circumstances caused an extreme backlog of learners waiting to enroll in face-to-face classes. Additionally, changes to learner-instructor interaction occurred. For example, instructors at one logistics organization provided paper-based support materials to learners before COVID-19, such as printed slide decks, paper handouts, and printed multiple-choice quizzes (Meador, 2021). These materials were digitized into online exercises and activities using automation embedded into courses and used more online resources. Due to rapid changes in organizational learning environments, learners not accustomed to taking company courses online were responsible for navigating exercises and activities. Adaptation and changes to the learning environment at one logistics organization satisfied new requirements for safety through limited face-to-face interaction between instructors and learners while decreasing the need to distribute paper materials (Meador, 2021).

COVID-19 allowed learning leaders to scan their landscapes for opportunities and evaluate departmental strengths and weaknesses. Even though learning leaders struggled to implement their online learning processes initially, they learned from their experiences during the pandemic. For example, learning leaders at a logistics company used dialogue and inquiry opportunities between leadership and learners to capture the voice of their customers (Meador, 2021). As a result, some valuable insights emerged, such as the usefulness of course automation from the perspective of training administration. Managers could efficiently run reports and verify which learners completed their required training, thus determining if they met the company's approved performance metrics while capturing basic information about learning opportunities that learners self-enrolled in based on their interests (Meador, 2021).

Additionally, instructors could use course data to evaluate their learning plans based on learner performance and company training needs. Another advantage was that more learners could complete their training quickly. As a result of these benefits, leadership in the organization looked to standardize and modernize current course material offered online to improve its quality of instructional material.

Organizational leadership committed to a training model incorporating broader organizational use of online course offerings besides required training. As part of this shift away from a primarily in-person training model, they evaluated ways to improve processes by strategically acquiring feedback from learners interacting with the online courseware via pre-post survey data. They also captured trainer or instructor feedback to determine the efficiency of the educational roadmap devised by organizational leaders.

The leadership and instructors also met online consistently to collaborate and determine plans and solutions for modernizing, process improving, or standardizing learning enablement opportunities for the organization with learner-provided feedback (Meador, 2021). Although the logistics company could not develop a fully online training model due to some training requirements, the benefit of a new hybrid learning model emerged for organizational leaders to transition training teams into learning hubs within organizations. They also incorporated more online courses besides mandatory training into their employee learning hierarchies, which could help mitigate future disruptions to their training systems.

Case Study Review

The three case studies presented in this chapter highlight the significance of systems thinking in online education. In Case Study #1, a sudden, drastic shift to online learning caused a ripple effect throughout the people, processes, and environment domains. Therefore, this study emphasizes an online system's interconnectedness, as depicted in Table 2 and Figure 1. Case Study #2 illuminated how adjustments to instructional practices impacted the People domain, specifically the instructors and the students, as shown in "Elements" under Table 2. Finally, in Case Study #3, the learning leaders adapted and expanded their use of course technology despite initial struggles, thus impacting the Environment domain, as shown in Table 3. Together, these case studies illuminate the interconnectedness of the People, Process, and Environment domains in online OES and demonstrate the significance of systems thinking in this field.

DISCUSSION

The case studies presented reveal lessons learned through adversity during COVID-19. However, linking PPE to specific online education standards could be helpful before we discuss the case studies. Therefore, we refer to the QM GS (QM, 2018a) as an example but recognize that other standards may also be suitable. The "People" domain connects to aspects such as teacher functions, learner engagement, and support personnel sustaining the teachers and students, which are related to QM GS (QM, 2018a) and other online standards. The "Process" includes procedures, methods, and systems involved in delivering online education, which relates to several QM GS (QM, 2018a), including course design, assessment, and learner support (QM, 2018a). It should be mentioned that QM GS (QM, 2018a) is a high-level list, and users may have to determine the fit of each categorical item based on organizational structure, referring to the Specific Review Standards related to each GS (QM, 2018b; QM, 2018c; QM, 2018d). Finally, the "Environment" focuses on internal and external conditions that could impact OES and could relate to QM GS (QM, 2018a) that address course technology and accessibility (QM, 2018a). Therefore, as we review the case studies, we encourage readers to consider aspects of PPE and online standards list QM GS (QM, 2018a), using elements of PPE potentially related to one or more of them.

Case Study 1

Even though the school was overloaded with traditional students from other school districts, the online K12 school district showed a relationship with QM GS (QM, 2018a) because their teachers and staff were proficient in online instruction during the COVID-19 pandemic, which addresses the People aspect of PPE. Since the online school was entirely online before the pandemic, they already knew how to engage students and maintain clear communication with technology. The online school district's courses related to QM GS (QM, 2018a) because their course designs, assessments, and learning objectives were already suited for an online format. The online teachers used multiple assessment methods to evaluate their students and provided individualized feedback to help students improve their skills. The environment was crucial in this online Pre-K-12 setting because the district needed to ensure all students could access their online work. Since the school provided resources and support services to assist students and their families and navigate the online learning environment, they showed a relationship with QM GS (QM, 2018a). However, despite the school's online expertise, it needed more technological infrastructure and resources. The online school's story is a lesson in disaster preparedness, keeping more supplies on hand than schools need today, which is challenging to do with limited funding in a public school.

Case Study 2

As the college switched to online learning, using online standards could have helped ensure their students had high-quality learning experiences. Since the college attempted to create online courses with learning goals and assessments, using standards that outlined learner support and accessibility to adapt the traditional teething methods to the online environment could have helped them be more successful in their endeavors. Only some college instructors embraced the people aspect of PPE by participating in professional development classes. Pointedly, some were reluctant to adapt to online learning. They could have created more engaging online content if they had been open to learning how to use technology for online courses. Others did their best to adapt their teaching style to online education and attempted to emphasize their learning objectives and provide feedback to students. Regarding processes, if the institution had implemented online standards for assessments, that would have ensured their assessments were relevant, reliable, and valid. They would have also offered various assessment methods to engage student learning and maintain academic integrity. These were valid concerns for the college because they did not know when students were paying attention or if they were the people submitting their exams. It would have also been helpful if instructors had provided timely feedback supported by clear rubrics, which could have improved the student's performance in online education. Concerning the environment, if the college had ensured that the learning platform was user-friendly and accessible to all students, their environment would have related more to QM GS (QM, 2018a) for accessibility and learner support. In addition, they would have offered online libraries, tutoring services, and technical support to assist students with succeeding in their courses.

Case Study 3

Considering online standards like QM GS (QM, 2018a) would have been helpful to the logistics company. Their online courses were only slightly positively aligned with the People aspect of PPE because the company provided training and resources for the employees to gain proficiency in online learning tools and platforms. However, since the company's online learners were primarily remote workers before the pandemic, their online courses expanded learning opportunities beyond traditional roles. For example, learners had opportunities to enroll in courses about soft skills and informational technology versus annual compliance course offerings. As for the Processes domain, the logistics company developed its online curriculum following agreed-upon and adopted standards with clearly stated learning objectives, relevant, reliable evaluations, and opportunities for workers to demonstrate their skills. The rapid shift from a series of online learning instances to having many online learning offerings for corporate learners could have caused inconsistency among design teams. One aspect that may have related to QM GS (QM, 2018a) was how learning leaders adopted standardized digital formatting of learning materials for learner support and standardized reports through data collection of employee feedback. Following this method, learning leaders continuously assessed and improved their training programs. Regarding the Environment domain, the organization used its online learning platform to varying degrees because some jobs required in-person training due to the extremely hazardous conditions of those jobs. Therefore, the organization could have improved learner engagement and interaction by considering QM GS (QM, 2018a) to support the learners accustomed to in-person training and interaction with their instructors.

Commonalities Between Case Studies: People, Processes, and Environment

Educators in K-12, higher education, and corporate learning can utilize these lessons in a post-COVID learning environment for quality online learning. These are as follows:

People

However small the changes, such as training parents to contact tech support or directing learners in a corporation to access learning remotely, these changes can have rippling effects on systemic change. In addition to altering individuals' interactions with the system, these dynamics also impact the system's general effectiveness and efficiency. The system slowly develops as more individuals adapt to these changes, reflecting new behavioral patterns. Therefore, the system's evolution could contribute to enhanced learning outcomes, increased technical literacy, and more substantial educational structures that could withstand unexpected discontinuities. Essentially, each change stimulates the educational system transforming it into highly versatile models suitable for 21st-century learning.

Leadership Actions

In all three scenarios, the leadership's decisions about online systems impacted the learning outcomes for learners of all ages. Furthermore, in some cases, the previous decisions of others in traditional learning environments also impacted the online systems. Despite each organization's challenges, all three case studies show some similarities in leadership. In each case study, the leaders swiftly adapted to the issues produced by the COVID-19 pandemic. They saw the need to shift to online learning rapidly and tried their best to ensure continuous education and training. All the organizations' leaders placed the safety of their students or employees as the utmost priority by switching to online learning and integrating precautions for in-person training (like in Case Study # 3). Unlikely leaders emerged to support online learning through technology integration. Additionally, the organizations realized their need for improvement based on their failed early pandemic efforts.

Provision of Encouragement and Rewards

Receiving rewards and encouragement is vital for skill development. These positive reinforcements boost learners' confidence and motivation to keep learning and applying their newly acquired knowledge. In each situation, the learners needed encouragement to persevere through uncertain times. The online PreK–12 school did its best to mitigate its circumstances but could not adequately support all students because of its overloaded system. Students and their families, especially those accustomed to brick-and-mortar schools, could have experienced frustration or feeling overwhelmed, possibly affecting the students' engagement in their new online environment. The second case study does not explicitly report on rewards or encouragement. However, since the college was already online, they may have missed opportunities to provide additional encouragement to students. Likewise, the third case study does not mention rewards or encouragement. However, corporate learners may have experienced reduced motivation or commitment to online learning if the logistics company did not provide incentives for adapting to new learning methods. Little encouragement could disrupt system equilibrium, impacting people's behavior and attitudes, fundamentally changing learning and engagement processes, and conceivably producing a less motivating learning environment.

Technological Awareness, Acceptance, and Preparedness

Effective leadership, awareness, and acceptance of new technologies and their usage in teaching and training are also essential. Technological awareness, acceptance, and preparedness are vital for successfully adopting and implementing technology. Specifically, people must be aware of technology before they can accept it to take action to implement and use technology in online education. We use Mishra and Koehler's (2006) TPACK framework to understand these circumstances.

The convergence of technological, pedagogical, and content knowledge can lead to effective education (Mishra & Koehler, 2006). To achieve maximum benefits from technology in online education, individuals must acknowledge how technology improves their educational roles and then accept it to gain full access to its benefits (Archambault & Barrett, 2010).

The online school aligned most with the TPACK of all three case studies. Before the pandemic, they had already acknowledged the value of technology, demonstrating their technological awareness and acceptance of technology. The teachers' online education credentials confirm their experience instructing in a virtual classroom. Furthermore, the school's curriculum was already accessible online, demonstrating their pedagogical and content knowledge within this context. Overall, the school responded to the pandemic at a systemic level.

The college initially needed TPACK framework alignment by their reliance on in-person instruction. It demonstrated limited technological knowledge for efficient online teaching rated limited technological knowledge for efficient online teaching. The dramatic change to online learning brought on by the pandemic further demonstrated the need for pedagogical expertise in the online setting. The procedure for transforming their coursework into an online environment also revealed knowledge gaps in their content knowledge. However, they showed systems thinking by immediately recognizing these problems and acting to resolve them. They organized the necessary faculty training, acquired the required technology, and built a support system for their students. Despite facing significant obstacles, they responded quickly and were able to adapt, which shows that they had some understanding of how the various technological, pedagogical, and content knowledge components interacted within the larger system.

TPACK was developed for K-12 but is applicable in corporate learning because instructors must know what to do as a part of their process routine (Archambault & Barnett, 2010). The logistics company had difficulties because some training required the learners' in-person attendance. However, by decisions were made to offer more training via online channels, they demonstrated some content knowledge. The company had trouble balancing the intricately interconnected parts of its system, especially considering the pandemic's demands for health and safety (Meador, 2021). However, they demonstrated adaptability and the ability to learn from their experiences. The pandemic tested each learning organization's readiness and adaptability. Each organization displayed varying levels of systems thinking and alignment with the TPACK framework, reflected in how it responded to the COVID-19 pandemic's challenges.

Process

Process modifications are not separate events but occur as system interactions. When organizations change their processes, such as transitioning from in-person to online courses, these modifications affect more than teaching and learning. The change specifically causes a ripple effect when these interconnected system changes occur. For example, modifying instructional methods affects instructor responsibilities. Learners interact with content and their peers and require administrative support. When processes adjust for online learning, this could enhance an organization's ability to recover from future system dysfunctions. Therefore, process changes could be opportunities to strengthen system nodes, which could make learning organizations better prepared to meet the dynamic needs of learners, instructors, and administration. Therefore, process changes illustrate system evolution, fostering a more flexible learner-centered environment.

E-Learning and Assessment

Online assessment is significant in education because it allows teachers to track students' progress, identify their strengths and shortcomings, and provide feedback to teachers and students in an online learning environment. Concerning the fully online PreK-12 school, the influx of traditional students into the online learning environment created the challenge of maintaining quality assessments with a significantly larger student population. At the Indian college, transitioning to online learning presented significant assessment challenges. Specifically, implementing online exams took much work due to limited resources and infrastructure. Additionally, they experienced academic integrity issues and problems with unfairness. Finally, in the third case study, the logistics company's experiences highlight the importance of assessment in professional learning. The assessment did more than just track individuals' progress but also ensured that safety training was practical and enhanced the company's operations. Even though the evaluation happened in three different contexts, all cases underline assessment's critical role in online education. All the learning organizations needed more flexibility and innovative assessment techniques to accommodate their unique demands. Additionally, the adaptations implemented by all three learning organizations demonstrate the potential of assessment and online education to enhance learning processes by providing timely feedback and fostering individualized learning.

Improve Readiness and Adaptability of Online Systems

The traditional systems needed to prepare educational environments for rapid course conversions in all three scenarios. Nevertheless, COVID-19 compelled people within systems to adapt to the online environment. For instance, the online PreK–12 school saw a spike in enrollment due to the pandemic, which overstretched the organization and put the technical infrastructure under pressure. Despite these obstacles, the school was able to adapt, showcasing its resilience and capacity to repair systemic flaws. In the second instance, the institution in Odisha, India, had severe difficulties converting to online instruction. A lack of infrastructure and resources caused system dysfunctions affecting teaching, learning, and administrative support. However, the college was able to adapt by leveraging current technologies, such as Google Meet and WhatsApp, to maintain and promote learning and communication. In the meantime, it created a more flexible environment focused on learners and improved its preparedness for upcoming challenges. Similarly, despite the early challenges, the logistics company adjusted to changes and strengthened its system, increasing its readiness for any upcoming interruptions. Despite difficulties, all three learning organizations successfully modified and enhanced their systems, demonstrating adaptability and resilience. These experiences emphasize the importance of readiness and adaptability in online learning systems and the opportunity for process modifications to strengthen system nodes and better meet the changing demands of learners, instructors, and administration.

Proper Training for All

There was also an acute need for proper training of teachers and students in PreK-12, higher education systems, trainers, and employees in corporate learning organizations. In the first case study, students and parents needed training to navigate online education for different reasons effectively. The teachers required help adapting their instructional methods to suit larger class sizes in online learning, which required additional training and support. The teachers attempted to train the parents to help their children navigate the online system to interact with their teachers and other students effectively because they were accustomed to relying on teachers for their children's educational outcomes. The students, especially the traditional ones, needed the training to interact with their coursework and behave during online meetings. In the second case study, the need for proper training was even more pronounced because many instructors needed technical knowledge to teach online effectively. They had to learn how to use online learning systems for instruction and assessment quickly and learn techniques that kept their students engaged in their learning. In the third case study, the logistics company needed to make changes and how employees interacted with their learning content which required extensive training and support. Overall, the case studies cases demonstrate the need for process modification to improve training for all. In the future, we recommend adding training in all these systems and several follow-up training sessions to ensure the correct implementation of online standards in learning organizations.

Environment

The online environment includes many internal and external factors impacting learning organizations. These factors could consist of internal factors, such as teachers' abilities to teach online students effectively, students' understanding of navigating online learning sites, and administration and support staff's ability to uphold online learning. Furthermore, external factors such as technological advancements, societal changes, economic conditions, and unanticipated global events could impact online learning environments.

Technology-Enabled Environments

Learning organizations should upscale their technology integration and be supported by technology to improve already high-quality learning settings. For example, digital tools for creating and sharing digital content should be available in the workplace. In addition to digital devices, learning organizations should ensure a robust technological infrastructure that includes high-speed internet access and sufficient hardware and software. These aspects are the backbone of online learning. Professional development training is also crucial to upscaling technology integration, including integrating it into learning processes, using online tools effectively, and troubleshooting common technical problems. In the PreK-12 school, teachers and students relied heavily on online learning platforms and tools for learning and assessment, which was vital to online learning. Similarly, the college instructor had to adapt to online learning systems for teaching and assessment quickly. They needed access to online content and sharing tools to improve their online course delivery. Technological availability and digital literacy were crucial factors that impacted their transition to online learning. Finally, in the case of the logistics company, their employee training needed additional technology integration. Several learners were waiting to enroll in face-to-face safety classes, and integrating more technology into their learning processes could have alleviated this bottleneck. In each case, upscaling technology integration could have improved the quality and effectiveness of the online learning environment and eased the stress of transitioning to online learning. When learning organizations provide access to enhanced infrastructure and digital tools for creating and sharing digital content, learner experience an enhanced learning environment.

Learning Management System Utilization

A learning management system (LMS), a software application that manages educational courses, training programs, materials, or learning and development programs through proper documentation, administration, tracking, reporting, automation, and delivery, is desirable in an online learning environment. Many LMSs also provide options for communication between students and teachers, collaboration, and give access to audio-visual tools that enhance learning. A common theme across all the case studies is using Learning Management Systems to Foster online education delivery. In the fully online PreK-12 school, the teachers and students depended on the LMS structure for their courses, and teachers used it for content delivery and assessing student progress. Similarly, the logistics company used an LMS to provide standardized training to its employees, demonstrating the value of an LMS in academic and corporate learning environments. The college instructors' experiences further illustrate the need for LMS. Instead of using a highly integrated LMS, they had to piece together their online learning using tools such as Google Meet, Google Classroom, and WhatsApp to simulate fully integrated LMS functions. However, this provisional system experienced many challenges. For example, they lost several vital resources and had difficulty tracking students' progress. These concerns further emphasize how streaming online education through an LMS provides benefits such as improved educational processes and resource management.

Recommendation Summary

Initially, the research hypothesis was that distinct educational settings (PreK-12, higher education, and corporate education) would each have unique challenges in transitioning to online learning during the COVID-19 pandemic. Our investigation found common themes that emerged despite each learning organization's distinctive characteristics. During our analysis of the case studies, we used systems thinking and online standards for K12, higher education, and continuing and professional education standards to evaluate how COVID-19 impacted online learning environments. We aim to develop recommendations for future online learning strategies that mitigate the challenges faced during the pandemic, aligning with modern education's evolving needs. Thus, learning organizations can create high-quality online education by adhering to specific online standards, such as QM GS (QM, 2018a). We propose that online standards combined with a systems thinking mindset could assist learning organizations in improving their online learning platforms, creating resilient, innovative systems that could better meet the demands of students in the modern digital world. Our findings demonstrate how important access, flexibility, and engagement in online learning environments can be. Ultimately, the four themes that emerged from our research include strengthening technological infrastructure, enhancing digital literacy, implementing comprehensive learning management systems (LMS), and building a culture that can adapt well to changing systems. Although the last theme, particularly the need for a resilient, adaptable culture, bolsters the other three themes. Learning organizations could build an effective online learning environment with this foundation.

Below, we explain the strategies that could improve online learning based on these themes. Together, these themes could be a part of a systems thinking plan that strengthens OES in a dynamic world.

Strengthen Technological Infrastructure

Online learning relies heavily on technologically solid infrastructures in all environments, including high-speed internet access and adequate hardware and software. Organizations could enhance their technology integration for better learning environments. This includes opening and facilitating access to digital content creation and sharing tools and their seamless integration into the educational process. By using dependable technology, instructors can more effectively complete their course objectives, conduct appropriate assessments, and give students a reliable and consistent learning environment.

Enhance Digital Literacy

Educators should receive professional development training to advance technology integration. Training should concentrate on troubleshooting common technical issues, using online tools effectively, and integrating technology into learning processes. It could make the switch to online learning easier and raise the general effectiveness and standard of the educational process. Educators can better articulate learning objectives and develop strategies to meet them by improving their digital literacy. Additionally, as educators can better facilitate online discussions and collaborative activities, increased digital literacy can support active learning and learner interaction.

Implement Comprehensive Learning Management Systems (LMS)

A comprehensive LMS plays a pivotal role in an online learning environment. It manages courses, training programs, and materials. LMS also facilitate communication and collaboration and provide access to audio-visual tools, enhancing learning. The experiences from the case studies underscore the importance of an integrated LMS that can streamline online education and provide improved resource management. Institutions should consider investing in an LMS that offers broad functionality, ease of use, and the ability to support their specific needs. The case study emphasizes the importance of accessibility and usability in course technologies. Moreover, students can access and engage with course material effectively with the help of a properly implemented LMS and adequate training about how to use LMS.

Adapt to Changing Environments

Institutions must embrace agility in adapting to internal and external factors that threaten the People, Process, and Environment domains. This includes adjusting teaching methods, administrative support, and learner interaction with online content. It also involves being responsive to societal changes, economic conditions, technological advancements, and unexpected global events, which could significantly impact online learning environments. By being adaptable, organizations can provide consistent learner support, adjust assessments as needed, and ensure that educators measure learning outcomes with accuracy and fairness.

CONCLUSION

During this chapter, we examined systems thinking principles and the PPE approach within the context of online education systems. We then connected PPE to online standards, using QM GS (QM, 2018a) as an example to analyze three case studies, considering the interactions of traditional and online systems. As a result, we provide recommendations for improving online education in a post-pandemic world.

We suggest that leaders empower people by allowing them to make decisions and implement changes that support dynamic learning systems. Additionally, leaders should offer opportunities for those in ancillary roles, such as parents, to learn how to provide more support in educational systems. Providing adequate technological resources and literacy during the transition to online environments is essential. Finally, educational leaders should have contingency plans to build resiliency into online systems that do not meet learning outcomes. These steps will help foster and enhance online learning environments.

Although we have strived to provide a comprehensive analysis, we recognize our study's limitations. All findings are based on three case studies within unique contexts. Additionally, these learning organizations' experiences may only represent some learning organizations transitioning to online education. Furthermore, the unforeseen COVID-19 pandemic led to a rapid shift in online learning required to continue online education, which introduced unprecedented challenges that may not always be present during normal times. Finally, we are not QM-certified instructors and thus did not attempt to certify any online courses during this study but instead used QM GS (QM, 2018a) as a guideline for our work. Additionally, the simplistic and potentially effective PPE model requires grounded theory development to validate its appropriateness in diverse online educational landscapes.

Therefore, we recommend continued testing of the PPE approach combined with online standards in future systems thinking studies as stakeholders continue to reimagine systems thinking in a post-pandemic world. This approach allowed us to comprehensively review each case study to understand online educational systems' interconnectedness deeply. Future researchers could gain more insight into this field's complexities and challenges by concentrating on the interactions between People, Process, and the Environment. We hope our research will lead to more recommendations for adapting to systemic changes and a more resilient, practical approach to online education. With a systems thinking mindset, there are opportunities to develop holistic, innovative research to ensure greater educational outcomes for all learners in a post-COVID world.

Note: We are not connected to QM, and our views are not necessarily theirs. Since we used QM concepts as part of this project, QM requires us to publish this statement: "The QM Rubrics have been developed and are regularly updated through a rigorous process that examines relevant research, data, and practitioner perspectives. They consist of Standards supported by detailed Annotations explaining the application of the Standards. They are intended to support the continuous improvement of courses with constructive feedback provided by trained and certified Peer Reviewers using a specific review protocol. Additionally, the QM Rubric and QM Course Review process are focused solely on course design. QM recognizes that design is only one influential factor in online learning quality and that this quality can be impacted by other aspects of online learning, such as online teaching, student readiness, institutional support, and technology. This research includes an adapted form of the Standards from the Quality Matters Rubric" (QM, 2018a).

REFERENCES

Aduhene, D. T., & Osei-Assibey, E. (2021). Socio-economic impact of COVID-19 on Ghana's economy: Challenges and prospects. *International Journal of Social Economics, 48*(4), 543–556. doi:10.1108/IJSE-08-2020-0582

Archambault, L. M., & Barnett, J. H. (2010). Revisiting technological pedagogical content knowledge: Exploring the TPACK framework. *Computers & Education, 55*(4), 1656–1662. doi:10.1016/j.compedu.2010.07.009

Argyris, C. (1976). Single-loop and double-loop models in research on decision making. *Administrative Science Quarterly, 21*(3), 363–375. doi:10.2307/2391848

Arnold, R. D., & Wade, J. (2015). Notably, the least obvious part of the system, its function or purpose, is often the most crucial determinant of the system's behavior. *Procedia Computer Science, 00*, 2–10.

Bronfenbrenner, U. (1994). Ecological models of human development. In International Encyclopedia of Education (2nd ed., Vol. 3). Elsevier.

Bruner, J. (1991). The narrative construction of reality. *Critical Inquiry, 18*(1), 1–21. doi:10.1086/448619

Buonsenso, D., Roland, D., De Rose, C., Vásquez-Hoyos, P., Ramly, B., Chakakala-Chaziya, J. N., Munro, A., & González-Dambrauskas, S. (2021). Schools closures during the COVID-19 pandemic: A catastrophic global situation. *The Pediatric Infectious Disease Journal, 40*(4), e146–e150. doi:10.1097/INF.0000000000003052

Cabrera, D., & Cabrera, L. (2019). What is systems thinking? In J. M. Spector, B. B. Lockee, & M. D. Childress (Eds.), *Learning, Design, and Technology: An International Compendium of Theory, Research, Practice, and Policy*. SpringerLink. doi:10.1007/978-3-319-17727-4_100-1

Chatterjee, R., Juvale, D., He, L., & Lundy Evans, L. (2022). Chapter 10 back to design basics: Reflections, challenges, and essentials of a designer's survival kit during a pandemic. In V. Dennen, C. Dickson-Deane, X. Ge, D. Ifenthaler, S. Murthy, & J. C. Richardson (Eds.), Global Perspectives on Educational Innovations for Emergency Situations (pp. 99–108). Association for Educational Communications and Technology. doi:10.1007/978-3-030-99634-5

Childs, J., Grooms, A., & Mozley, M. P. (2022). Hidden in (virtual) plain sight: A charter district's focus on attendance during COVID-19. *Education and Urban Society*. Advance online publication. doi:10.1177/00131245211065414

Crawford, J., Butler-Henderson, K., Rudolph, J., Malkawi, B., Glowatz, M., Burton, R., Magni, P. A., & Lam, S. (2020). COVID-19: 20 countries' higher education intra-period digital pedagogy responses. *Journal of Applied Learning & Teaching, 3*(1), 1–20. doi:10.37074/jalt.2020.3.1.7

Desson, Z., Lambertz, L., Peters, J. W., Falkenbach, M., & Kauer, L. (2020). Europe's COVID-19 outliers: German, Austrian and Swiss policy responses during the early stages of the 2020 pandemic. *Health Policy and Technology, 9*(4), 405–418. doi:10.1016/j.hlpt.2020.09.003

Dhukaram, A. V., Sgouropoulou, C., Feldman, G., & Amini, A. (2016). Higher education provision using systems thinking approach – case studies. *European Journal of Engineering Education*, *43*(1), 3–25. doi:10.1080/03043797.2016.1210569

DiMaggio, P. J., & Powell, W. W. (1983). The iron cage revisited: Institutional isomorphism and collective rationality in organizational fields. *American Sociological Review, 48*(2), 147–160.

Dzemydienė, D., & Tankelevičienė, L. (2010). Multi-layered architecture of the adaptable architectural distance learning system. *Technological and Economic Development of Economy, 15*(2), 229–244. doi:10.3846/1392-8619.2009.15.229-244

Environment. (2022). In *Oxford Lexico*. Lexico.com.

Flood, R. L., & Romm, N. R. A. (Eds.). (1997). *Critical systems thinking: Current research and practice*. Plenum Press.

Forrester, J. W. (1961). *Industrial dynamics*. MIT Press.

Fouda, A., Mahmoudi, N., Moy, N., & Paolucci, F. (2020). The COVID-19 pandemic in Greece, Iceland, New Zealand, and Singapore: Health policies and lessons learned. *Health Policy and Technology, 9*(4), 510–524. doi:10.1016/j.hlpt.2020.08.015

Frick, T. W. (2020). Education systems and technology in 1990, 2020, and beyond. *TechTrends, 64*(5), 693–703. doi:10.100711528-020-00527-y PMID:32838405

Hemphill, H., Hemphill, L., & Runquist, R. (2019). Improving completion in online education systems: An application of systems thinking. In M. J. Spector, B. B. Lockee, & M. D. Childress (Eds.), Learning, Design, and Technology (pp. 1–36). Springer International Publishing. doi:10.1007/978-3-319-17727-4_104-1 doi:10.1007/978-3-319-17727-4_104-1

Higginson, S., Milovanovic, K., Gillespie, J., Matthews, A., Williams, C., Wall, L., Moy, N., Hinwood, M., Melia, A., & Paolucci, F. (2020). COVID-19: The need for an Australian economic pandemic response plan. *Health Policy and Technology, 9*(4), 488–502. doi:10.1016/j.hlpt.2020.08.017

Hodges, C., Moore, S., Lockee, B., Trust, T., & Bond, A. (2020). *The difference between emergency remote teaching and online learning*. Educause Review.

Hoekman, L. M., Smits, M. M. V., & Koolman, X. (2020). The Dutch COVID-19 approach: Regional differences in a small country. *Health Policy and Technology, 9*(4), 613–622. doi:10.1016/j.hlpt.2020.08.008

Hogan, K., & Weathers, K. C. (2003). Psychological and ecological perspectives on the development of systems thinking. In *Understanding Urban Ecosystems* (pp. 233–260). Springer. doi:10.1007/0-387-22615-X_15

Huang, C., Wang, Y., Li, X., Ren, L., Zhao, J., Hu, Y., Zhang, L., Fan, G., Xu, J., Gu, X., Cheng, Z., Yu, T., Xia, J., Wei, Y., Wu, W., Xie, X., Yin, W., Li, H., Liu, M., ... Cao, B. (2020). Clinical features of patients infected with 2019 novel coronavirus in Wuhan, China. *Lancet, 395*(10223), 497–506. doi:10.1016/S0140-6736(20)30183-5 PMID:31986264

International Society for Technology in Education. (2023). *ISTE Standards*. ISTE.

Josselson, R. (2010). *Narrative research* (N. Salkind, Ed.).

Junpho, M., & Kaewprapan Rosenkranz, W. (2018). A study of systems thinking process: a review of the literature. *Proceedings of International Conference on Technology and Social Science (ICTSS 2018).*

Karakose, T. (2021). The impact of the COVID-19 epidemic on higher education: Opportunities and implications for policy and practice. *Educational Process: International Journal, 10*(1), 7–12. doi:10.22521/edupij.2021.101.1

Kim, D. (1999). Introduction to systems thinking. Pegasus Communications (now Leverage Networks Inc.).

Kumari, S., Gautam, H., Nityadarshini, N., Das, B. K., & Chaudhry, R. (2021). Online classes versus traditional classes? Comparison during COVID-19. *Journal of Education and Health Promotion, 10*(457). Advance online publication. doi:10.4103/jehp.jehp_317_21 PMID:35233404

Maani, K. E., & Cavana, R. Y. (2007). *Systems thinking, system dynamics: Managing change and complexity* (2nd ed.). Pearson Education Canada.

Mann, B. (2019). Compete, conform, or both? School district responses to statewide cyber charter schools. *Journal of School Choice, 14*(1), 49–74. doi:10.1080/15582159.2019.1566996

Mao, J., & Shearer, R. L. (2019). Technology affordance in online learning: A systems thinking and system dynamics theoretical framework. In M. Spector, B. Lockee, & M. Childress (Eds.), *Learning, Design, and Technology*. Springer. doi:10.1007/978-3-319-17727-4_106-1

Meador, M. (2021). *Learning organization in a time of crisis: A case study of logistics and freight, llc and covid-19* [Dissertation, The University of Memphis]. ProQuest Dissertations Publishing.

Meadows, D. H. (2008). *Thinking in systems: A primer.* Chelsea Green Publishing.

Ministry of Education, Government of India. (2020). *Students to continue their learning by making full use of the available digital e-Learning platforms – Shri Ramesh Pokhriyal "Nishank" Union HRD Minister appeals to students, teachers, and parents to follow Janata curfew on 22nd March 2020.* Press Information Bureau, Government of India.

Mishra, P., & Koehler, M. J. (2006). Technological pedagogical content knowledge: A framework for teacher knowledge. *Teachers College Record, 108*(6), 1017–1054. doi:10.1111/j.1467-9620.2006.00684.x

Murphy, M. P. (2020). COVID-19 and emergency eLearning: Consequences of the securitization of higher education for post-pandemic pedagogy. *Contemporary Security Policy, 41*(3), 492–505. doi:10.1080/13523260.2020.1761749

Nayak, J., Mishra, M., Naik, B., Swapnarekha, H., Cengiz, K., & Shanmuganathan, V. (2021). An impact study of COVID-19 on six different industries: Automobile, energy and power, agriculture, education, travel and tourism and consumer electronics. *Expert Systems, 39*(3), 1–32. *Wiley Online Library Database Model, 2023.* Advance online publication. doi:10.1111/exsy.12677 PMID:33821074

Nelson, C. E. (2022). *9 skills that make great online school teachers.* Connections Academy; Pearson.

Oklahoma State Department of Education (OSDE). (2019). *Student enrollment in Oklahoma public schools tops 700,000.* Oklahoma State Department of Education.

Online Learning Consortium (OLC). (2023). *OLC Quality Scorecard Suite.* Online Learning Consortium.

People. (2022). In *Merriam-Webster.*

Process. (2022). In *Merriam-Webster.*

Quality Matters (QM). (2018a). *Home.*

Quality Matters.

Quality Matters (QM). (2018b). *Standards from the Quality Matters Continuing Education and Professional Development, Sixth Edition.* Quality Matters.

Quality Matters (QM). (2018c). *Standards from the Quality Matters Higher Education Rubric, Sixth Edition.* Quality Matters.

Quality Matters (QM). (2018d). *Standards from the Quality Matters K-12 Rubric, Sixth Edition.*

Richmond, B. (1994). Systems dynamics/systems thinking: Let's just get on with it. In *International Systems Dynamics Conference*, Sterling, UK. 10.1002dr.4260100204

Senge, P. (1990). *The fifth discipline, the art and practice of the learning organization.* Doubleday/Currency.

Shmis, T., Sava, A., Nunes Teixeira, J. E., & Patrinos, H. A. (2020). *Response to COVID-19 in Europe and Central Asia.* World Bank Group Education.

Singh, J., Steele, K., & Singh, L. (2021). Combining the best of online and face-to-face learning: Hybrid and blended learning approach for COVID-19, post vaccine, & post-pandemic world. *Journal of Educational Technology Systems, 50*(2), 140–171. Sage Journals. doi:10.1177/00472395211047865

Singh, V., & Thurman, A. (2019). How many ways can we define online learning? A systematic literature review of definitions of online learning (1988-2018). *American Journal of Distance Education, 33*(4), 289–306. doi:10.1080/08923647.2019.1663082

Sockman, B. R., Clossey, L., Carducci, O. M., Batson-Magnuson, L., Muzure, D., White, G., Wehmeyer, A., Green, B. A., & Wells, H. (2019). Systems thinking as a heuristic for the implementation of service learning in a university. In M. Spector, B. B. Lockee, & M. Childress (Eds.), *Learning, Design, and Technology* (pp. 1–26). Springer., doi:10.1007/978-3-319-17727-4_91-1

State University of New York. (2013). *OSCQR.* Online Learning Consortium; State University of New York.

Stave, K. A., & Hopper, M. (2007). What constitutes systems thinking? A proposed taxonomy. *25th International Conference of the System Dynamics Society.*

Tamim, S. R. (2020). Analyzing the complexities of online education systems: A systems thinking perspective. *TechTrends, 64*(5), 740–750. doi:10.100711528-020-00538-9

Tesar, M. (2020). Towards a post-COVID-19 "new normality?": Physical and social distancing, the move to online and higher education. *Policy Futures in Education, 18*(5), 556–559. doi:10.1177/1478210320935671

The W. Edwards Deming Institute. (2021). *Enriching society through the Deming philosophy| PDSA cycle.*

The Deming Institute. (n.d.). https://deming.org/explore/pdsa/

The World Bank Group. (2023). *GNI per capita, Atlas method (current US$) - India.*

Tumwesige, J. (2020). *COVID-19 educational disruption and response: Rethinking e-learning in Uganda.* University of Cambridge.

Uddin, M. S., Routray, J. K., & Warnitchai, P. (2019). Systems thinking approach for resilient critical infrastructures in urban disaster management and sustainable development. In E. N. Farsangi, I. Takewaki, T. Y. Yang, A. Astaneh-Asl, & P. Gardoni (Eds.), *Resilient Structures and Infrastructure* (pp. 379–415). Springer. doi:10.1007/978-981-13-7446-3_15

UNESCO. (2020). *COVID-19 response toolkit - UNESCO. Globaleducationcoalition.unesco.org.* UNESCO.

UNESCO. (2020). *COVID-19 response toolkit - UNESCO. Globaleducationcoalition.unesco.org.* UNESCO.

United States Census Bureau (USCB). (2022). *Income in the United States: 2021.* U.S. Department of Commerce.

Virtual Learning Leadership Alliance, & the Digital Learning Collaborative. (2023). *The Standards.* QM Quality Matters.

Watson, S. (2020). *EPIC releases county-by-county enrollment numbers.* Epic Charter Schools.

World Health Organization (WHO). (2020). *Naming the coronavirus disease (COVID-19) and the virus that causes it.*

Zhang, Y., Li, X., Zhu, L., Dong, X., & Hao, Q. (2019). What is a smart classroom? A literature review. *Perspectives on Rethinking and Reforming Education*, 25–40. doi:10.1007/978-981-13-9439-3_2

KEY TERMS AND DEFINITIONS

Environment: External conditions and factors that could influence or interact with the system contribute to the formation of a system's development and behavior. These could include physical space, human constructs such as culture, or larger systems that impact smaller systems.

Interconnectedness: Refers to the different system components' relationships and dependencies, emphasizing how system components could significantly impact other system parts.

Intersection: Refers to the space where various system components meet and interact, which could reveal emergent system properties.

Online Education System: Facilitates remote teaching and learning in a structured virtual environment and usually consists of learning management systems, instructional content, assessments, communication tools, and student support tools such as help functions.

People: Individuals interacting and contributing within a system, which could include students, teachers, school administrators, student support staff, and other stakeholders interacting with the system.

Process: The actions or methods used to obtain a specific system outcome or goal, which can involve coordinating system resources and people to achieve that result.

Systems Thinking: A comprehensive approach to understanding interactions and relationships through a system's components. People can resolve complex issues by acknowledging a system's dynamics and the possible impact of its actions.

Chapter 11
Redefining Success:
Humanizing Teachers and Students

Sarah Beach
University of Virginia, USA

Helen H. Min
University of Virginia, USA

ABSTRACT

The ramifications of the global coronavirus pandemic challenged the barriers between professional and personal lives, revealing that even in professional settings, individuals involved in systems are human, with increasing mental and physical health needs. Systems thinking encourages holistic rather than outcomes-oriented thinking, and the U.S. education system would benefit from reorienting understandings of success under this framework. This study employs Critical Systems Heuristics to explore interview data collected from teachers during the pandemic. Findings indicate that teachers and students yearn to function within a system where they can bring their whole selves. This chapter explores how success is currently defined for students and teachers and what a reimagined system could look like. This new system would move away from standardized methods of measuring all students and teachers against the same outcomes and consider expanding agency and collaboration with a focus on growth. Barriers and facilitators to sustaining such a system are considered.

INTRODUCTION

For too long, systems of education have relied on metrics such as standardized tests as ways to evaluate both students and teachers within the education system. Reynolds and Howell (2010) indicated that one of the "traps" in conventional thinking, as opposed to systems thinking, is the "focus on outcomes (and thus only on what can be measured)" (pp. 5-6). Systems thinking encourages holistic rather than outcomes-oriented thinking, and thus the United States (U.S.) education system would benefit from reorienting the current understanding of success.

DOI: 10.4018/978-1-6684-7285-9.ch011

The unprecedented outcomes of the global coronavirus pandemic challenged the barriers between professional and personal lives, as work from home became the norm. With the crossing of this boundary, an underlying truth surfaced: even in professional settings, individuals involved in systems are human. While the statement seems obvious, the ramifications of humanizing workers has been and remains significant in its implications across industries, particularly with increasing rates of mental and physical health needs since the onset of the pandemic. This chapter examines how the pandemic challenged existing boundaries in the education system and how lessons learned from 2020-2021 may help us reimagine the success and wellbeing of teachers and students. Utilizing data collected from teachers during the pandemic, the findings of this research indicate that when considering success from a systems-based perspective, it is apparent that teachers and students yearn to be seen as whole people, and that redefining success holds the potential to transform the education system for increased sustainability in a post-pandemic world.

The context of the pandemic has shifted roles and responsibilities within the education system, adding additional stress and constraints. Teachers are switching careers, schools are struggling to support teachers, and students are left in the flux of navigating uncertainties.

Teacher Mass Exodus

Teachers have long faced some of the highest rates of occupational stress (Gallup, 2014) with a quarter of schoolteachers rating their experience of stress at work as *very or extremely stressful* (Kyriacou, 2000). By 2021, in the height of the pandemic, this number increased to nearly one in four teachers not only feeling stress, but wanting to leave the classroom (Zamarro et al., 2021). Stressors have been exacerbated by staffing shortages which have contributed to increased teacher workloads with decreased support (Lieberman, 2021), multiple changes to instructional demands (Zamarro et al., 2021), and challenges to the personal lives of teachers and students due to the pandemic (Steiner & Woo, 2021). Without meeting the complex needs of both teachers and students, the educational system suffers from increased burnout and turnover of teachers. In the long term, this may contribute to a lack of consistent and qualified teachers, poor teacher performance (Greenberg et al., 2016), and negative impacts in students' academic outcomes and social adjustment (Oberle & Schonert-Reichl, 2016).

Professional Development

Professional development (PD) is one avenue to provide teachers support. Teacher PD has been generally conceptualized as incorporating six critical features for effectiveness: sustained duration, collaboration, teacher buy-in, subject specific content, expert trainers, and opportunities for practice (Desimone, 2009; Walter & Briggs, 2012). Much of this research has focused on identifying characteristics of teacher PD with aims to improve student achievement based on test scores, rather than teacher focused changes. Not only have these recommendations for effective teacher professional development faced criticism (Sims & Fletcher-Wood, 2021; Kennedy, 2016), they are not sufficient to meet the needs of teachers, students, and schools in an education system transformed by the global pandemic.

Since the pandemic, the demand for PD training, particularly applying trauma-informed approaches and for social and emotional development of students, has increased. This is likely due to the disruptions to student social and emotional development during the pandemic. Local, state, and federal policies have increasingly categorized psychological trauma as a public health epidemic and responded by endorsing trauma-informed approaches (Children's Law Center, 2021). Often referred to as "trauma-informed care" or "trauma sensitive," trauma-informed approaches use core principles of safety, trust, collaboration, and empowerment to encourage professionals to consider the trauma histories of those they work with and respond to emotional needs rather than punishing behaviors (Levenson, 2017). Trauma-informed approaches in schools have led to dramatic reductions in student behavior challenges (Dorado et al., 2016) and increases in knowledge about the consequence of trauma on learning among school personnel (Anderson et al., 2015; Perry & Daniels, 2016).

There is an increasing need, however, for trauma-sensitive approaches that consider teachers' social and emotional competencies to cope with the challenges associated with their work, especially regarding the added stressors that they have faced since the onset of the pandemic. Existing research on school-based PD that focuses on the psychoeducation of professionals (e.g. wellbeing, mindfulness, trauma-sensitivity) is often conducted with aims to bolster teachers' ability to support students and focus on student-related outcomes (e.g. Thomas et al., 2019), despite evidence that staff who provide support to youth with trauma experiences may have an increased risk of experiencing adverse consequences stemming from their empathic engagement (Anderson et al., 2022; Hatcher et al., 2011). For example, Borntrager et al.'s (2012) study of 6 public schools in the Northwestern U.S. revealed that 229 school staff members reported high levels of secondary traumatic stress akin to those found among mental health workers. While the preliminary findings of this study are limited because of convenience sampling of school staff members who came to a conference addressing secondary traumatic stress, results suggest there is a need among school personnel to mitigate the challenging impacts of caring for students who have experienced extreme adversity. Addressing this stress is essential for both teacher and student success.

Student Success

Since the beginning of No Child Left Behind, educators and academics have investigated the competencies students need for the 21st century (e.g., Partnership for 21st Century Skills) and on accountability metrics measuring success (Kay & Boss, 2021). These have focused primarily on standardized test scores, with some movement toward including additional "soft skills" such as communication, creativity, collaboration, and critical thinking developed by Partnership for 21st Century Skills, "Seven Survival Skills" (Wagner, 2010), global competencies (Reimers & Chung, 2019; Zhao, 2012), and digital citizenship competencies (Tamayo, 2016).

Unfortunately, within the education system, students are often quantified and dehumanized through test scores. Students' varied funds of knowledge (Moll et al., 1992) are not tested or valued when success is limited to the results of one end-of-year standardized assessment. This was challenged during the pandemic with the permission from the U.S. Department of Education to waive standardized testing requirements (Devos, 2020). While standardized test scores following the re-opening of schools indicate that there was a "learning loss" for many students (e.g., Locke et al., 2021), other metrics of skills gained throughout the pandemic were not measured, such as problem-solving, increased collaboration, and adapting to new circumstances such as online learning. This introduces a tension between the concepts of success explored more in the results below.

Conceptual Framework

Historical Perspective and Systems Thinking

While the context of the pandemic is imperative in understanding the current challenges facing the U.S. education system, considering the historical perspective of how the system came to be is also beneficial. While modern technology, thought, and expression have continued to grow and diversify at exponential rates, the education system is still largely rooted in ideas from the advent of industrial times. As machines were being developed, reductionistic thought thrived (Ackoff, 1974). Society sought ways to simplify and make life easier. After the Industrial Revolution, society began expanding again, with considerations of multiple ways of knowing; however, the education system has continued to prioritize the values of efficiency and competition, making little room to incorporate expansionist thinking (Jennings, 2020).

Systems thinking considers not just individual parts that make up a whole system, but rather the system itself as a unit and the roles that each of its sub-parts play within that whole (Ackoff, 1974). Considering the components of a system addresses problem solving from a macro perspective, seeing not just single, isolated problems, but what Ackoff (1974) coins as a *mess* of problems that are interrelated and complex, creating other problems as they interact with the system (p. 21).

Historically, much of the burden of the flaws of the system have been placed upon teachers (e.g., Ackoff, 1974; Guggenheim et al., 2011; National Commission on Excellence in Education, 1983). Despite acknowledgement of the intertangled mess of problems, teachers are often burdened with additional responsibilities or villainized for the shortcomings of schools without recognition that they are often operating within an antiquated system. This chapter seeks to investigate the benefits of a critical systems heuristics to address some of the concerns that leave teachers feeling disempowered.

Critical Systems Heuristics

Within the web of systems thinking, the researchers of this study approach this work through the lens of Critical Systems Heuristics (CSH; Ulrich, 1996; Ulrich & Reynolds, 2010). Because of power dynamics within systems of education, this framework of systems thinking, which analyzes data through the lens of the teachers involved in the system, is particularly helpful.

Ulrich and Reynolds (2010) suggest that there are three major reasons for using CSH: to better understand the big picture, to promote better understanding from different perspectives, and to aid in reflection to promote change. Through CSH, people can understand that the way they perceive and navigate the world likely differs from others involved within a system. The CSH framework allows these differences to be uncovered and interrogated so that sources of contention and misunderstandings rooted in perception can be alleviated. Table 1 below is an adaptation of CSH to the system of education adapted from Ulrich's (1996) framework. The focus in this study is particularly on the problems/stakeholding issues, but these findings will be discussed in the context of the entire education system.

This study's application of the conceptual framework allows for the comparison of the system *as it is* versus *how it ought to be.* A "normative" educational system observes the system in the present status quo, while the "reimagined" educational system envisions what the system should be. The juxtaposition between the normative and reimagined systems allow a consideration of the complicated *mess* of the current education system to be explored as means "of coping with the circumstances as best one can" (Reynolds & Howell, 2010, p. 5).

Table 1. Critical Systems Heuristics framework for a system of education

	Roles/Stakeholders	Role Specific Concerns/Stakes	Problems/Stakeholding Issues	
Values & Motivations for Improvement	*Beneficiary* Who is/ought to be the **beneficiary** of the education system?	*Purpose* What is/ought to be the **purpose** of the education system?	*Measure of Improvement* What are/ought to be the **measures of success** for the education system?	
Power/ Control	*Decision Makers* Who is/ ought to be in **control of the conditions of success** in the education system?	*Resources and Attributions* What conditions of success are/ ought to be **under control** of the education system?	*Environment* What conditions of success are/ ought to be **outside the control** of the decision makers?	Teachers (The Involved)
Knowledge	*Planner/Expert* Who is/ought to be **providing relevant knowledge and skills** for the education system?	*Expertise* What are/ought to be the relevant **new knowledge and skills** for the education system?	*Guarantee* What are /ought to be the assurances of **successful implementation**?	
Consequence/ Legitimacy	*Witness* Who is/ought to be representing the interests **of students** in the education system?	*Emancipation* What are/ought to be opportunities for students to have expression and freedom from the constraints of the educational system?	*Worldview* What space is/ought to be available for reconciling different worldviews regarding the education system among teachers and students?	Students (The Affected)

METHODS

Through CSH, empirical data was analyzed to help better understand how teachers are thinking about success within the system for themselves and their students. The following research questions guided the investigation:

1. How do teachers make meaning of success for themselves?
2. How do teachers make meaning of success for their students?
3. How do teachers want to define success?

The data for this study were collected in two different settings. One set of data included 24 lead teachers in K-12th grade classrooms across the United States collected in April 2021. Teachers who completed a demographic survey and interview consent form through a flier link posted on social media (i.e., Facebook, Instagram, and Twitter) were contacted by email. To maximize variation in the sample, nine interviews were selected from the larger dataset. The interviews chosen represented five teachers who identified as male, four as female, five as White, two as Hispanic/Latinx, two as Asian, three at the elementary level, three at the middle school level, and three at the high school level.

The other was collected in March through May 2020 and included six teachers selected through convenience sampling within a single school division: one at the elementary level, four at the middle school level, and one at the high school level. Of these teachers, five were White and female, and one was Black and male. While neither of these samples are representative of the U.S. teaching population, together they allow for a deeper understanding of teachers' meaning making by seeing where the definitions of success converge and diverge among teachers in different settings with different experience levels.

To analyze the data, the qualitative coding processes by Kalpokaite and Radivojevic (2019) were applied. In the first *inspection cycle* (Kalpokaite & Radivojevic, 2019, p. 50), the researchers read and coded the data using content analysis. All instances where teachers referred to students in their interviews or discussed student measures of success were coded as student data. Interview content that specifically referred to teachers or measures of teachers' success was coded as teacher data. This differentiated analysis about meanings of success for teachers and students. Next, in the *coding cycle* (Kalpokaite & Radivojevic, 2019, p. 51), a priori codes from the CSH framework, "success is" and "success ought to be," were applied to student and teacher data. "Success is" was applied when teachers discussed how the system operates currently, and "success ought to be" was used when teachers noted how they wished for the system to be. To establish a consistent coding procedure, the researchers collaborated on coding the first interview and subsequently conducted individual coding for the remaining interviews. The researchers then employed elaborative coding, identifying several recurring themes, including "professional development," "burn out," and "barriers," and added corresponding codes to distinguish the data. To ensure the accuracy of the coding, the researchers reviewed all the coded data and resolved any ambiguities or uncertainties that arose. This elaborative coding approach enabled the researchers to thoroughly analyze the data and identify key themes and patterns.

Next, in the *categorization cycle* (Kalpokaite & Radivojevic, 2019, p. 52), the data were grouped into codes to form preliminary categories. Focused coding was then employed to assign relevant data to these categories. Subsequently, major categories were developed by combining related codes. The data were reviewed for confirming and disconfirming evidence. Finally, in the *modelling cycle* (Kalpokaite & Radivojevic, 2019, p. 52), the researchers examined the relationship between the themes and the CSH framework, which culminated in the identification of the final themes presented in the subsequent sections.

RESULTS

The following section reviews the findings of how teachers define success as is and as they wish it to be, both for themselves and their students. See Table 2 for a summary of results.

What Teacher Success Is

Within the system as is, teachers defined their own success in three major ways: completing all parts of their job, helping students achieve high test scores, and managing their classrooms. These three components were essential not because teachers necessarily saw them as most important but because they believed these were the indicators of success that were incentivized by the system.

Doing All Parts of the Job

Teachers responded that teacher success encompassed taking on myriad administrative and paperwork requirements to document student participation and success. This included clerical work such as submitting attendance, grading, reporting grades, completing Individualized Education Programs (IEPs) for special education students, and developing personal professional goals. One teacher explained that a stressful component was remembering all the things that you are expected to do:

Did you turn your attendance in in the first 10 minutes or did you forget and have they called you six times? Or did you forget your lunch duty? Or have you done your grades weekly and turned in? And have you called the 15 children who are failing your class because they're not doing their work?[1]

Table 2. Summary of results

Themes	Illustrative Quote
Teacher success is defined by	
Doing all parts of the job	"Did you turn your attendance in in the first 10 minutes or did you forget and have they called you six times? Or did you forget your lunch duty? Or have you done your grades weekly and turned in? And have you called the 15 children who are failing your class because they're not doing their work?"
Test scores	"That's what we've been told over and over, like basically our jobs are tied to our scores. Our accolades are tied to our scores. We have been told whether we are a good teacher or not is tied to our scores."
Classroom Management	"[Students] can't be too happy because if they're too happy and having fun if they're laughing. And God forbid if they're out of their seats and jumping around, something is wrong. They cannot be having too much fun."
Teacher success ought to be defined by	
Student success	"Part of that responsibility of a teacher is to provide…educational guidance, moral guidance, to shape students' lives in a more meaningful, positive manner."
Authentic relationships	"We see our students every day. We work with our students; we get to know our students. We have to build relationships with our students on a level that no one else will. We have to form a bond with them for them to be willing to learn from us"
Student success is defined by	
Doing well on standard, rigidly defined, isolated skills	"We are judging children and teachers on a one snapshot kind of deal… if on that one particular day, a student is upset, or they aren't good test takers, or they just aren't interested, then that shows that they're failures."
Student success ought to be defined by	
Feelings of success	"We want them to take responsibility for their learning to become adults of responsibility. A part of that is learning how to look honestly at your own work, at your own performance, your own yardstick. Where are you; how are you growing?"
Agency and exploration	"You'd have to get to know your students first to see what they're interested in… there'll be more of a process as opposed to just, well, here's a nonfiction article about why birds fly to the south. Well, who cares? You know, we need more interest."
Growth and strengths-focused	"There's very little time for them to explore and find out what they're really good at. And a lot of them have arrived at the conclusion at this young age that they're not good at anything and that really breaks my heart because they're all good at something and they don't see it because it's not valued by the educational system. "
Authentic problem solving	"Let's give them real life problems. Let's give them authentic consequences and authentic audiences. And let's see. Let's open up opportunities for them to show us things that maybe we didn't even think to assess in the first place."
Barriers to Success	
Holding on to 'tried and true" methods out of fear	"The question is, does my kid still love reading enough to come back and want to do it next year? And that's not the question I'm going to ask as a parent. I'm going to ask, 'Did my kid pass?'"
"Out of touch" with the ideal	"You know, I'm so out of touch with what ideal might look like."
Teacher time is valuable and limited	"If teachers have too much of a workload and not enough time, I feel like that could be traumatic as well, where it turns into teacher burnout. They feel like they have so much to do, but then they won't have enough time and energy left to invest in the kids and teach the kids. So, I feel like the administration at the school is very important. And for them to listen to teachers and help the teachers with either more time, less meetings, or more support is really important."

Administrative work expected of teachers included calling and communicating with parents to update them on student's academic progress and behavior, overseeing students during lunch or recess, and communicating with other colleagues to ensure that students receive necessary support. Schools and administrators adjusted for the pandemic in different ways. While some teachers were relieved of their duty to oversee students during lunch, other teachers were required to stay on a video call with all of their students during lunch and recess times.

In addition, teachers explained that their success encompassed providing resources for their school community such as giving snacks to students who came to school hungry, sharing identified external resources for students and their families, and working to facilitate social and emotional development of students. Some teachers even explained that they took on the role of parents, therapists, social workers, nurses, caregivers, mental health, and first aid responders, exemplified by one teacher's remarks that part of a teacher's role is "making sure that [students'] basic needs are met." The demand on teachers' social and emotional support increased during the pandemic, as teachers were often the closest point of contact with students' and their families. Thus, some teachers took on the responsibility to drive to the homes of students to drop off lunches, laptops, books, and printed materials. Further, teachers were often first to know about the challenges faced by the students and their families stemming from pandemic related deaths and hardship (e.g., parents laid off from work, etc.)

Test Scores

Teachers reported that their success is measured by the performance of their students on standardized assessments and teacher created exams. Standardized assessments include exams created by state departments of education or national testing agencies that measure whether students are performing at their grade level based on standards detailed by the state or national agency. The teacher created tests measure students' performance on a subject at the beginning of the school year through diagnostic tests, assess students' progress throughout each lesson through summative tests, mark students' mastery of a learning objective or standard through formative assessments, and measure students' overall understanding of a subject through summative or benchmark exams.

This emphasis on the outcome of student test scores, often synonymous with student achievement, became a central priority in measuring teachers' success because of what the administration and school districts communicated directly, nonverbally, and through teachers' pay systems. One teacher reflecting on their first few years in the profession remarked that they were seen as successful if they followed the curriculum given to teachers by the school districts and followed the progression of the assigned textbooks. Teachers reported that there is a clear gap between what teachers think will be successful for students and how the school administration or district officials measure success. One teacher explained:

Well, it just feels cruel. What we do to core teachers. In terms of setting up evidence that they've taught, and in the end, no matter what other stuff is done inside the classrooms, we kind of come down to the end and go, "Okay, what are those [state standardized test] scores?"

Here, the teacher deplores the standardized testing system on which "core teachers," teachers who teach tested subjects like English and math, are gauged for their success. The teacher explained that other actions that the teacher conducts for their students and their learning are irrelevant if the students are not tested in that content area.

On one hand, while teachers knew that their value as teachers relied more heavily on their relationships with students and other intangible aspects (e.g., classroom culture, students' love of learning, students' practice of social and emotional skills), they were told that their jobs were dependent on the state standardized test scores of students in their class. This inevitably became a big stressor for teachers:

That's what we've been told over and over, like basically our jobs are tied to our scores. Our accolades are tied to our scores. We have been told whether we are a good teacher or not is tied to our scores.

Similar to other teachers, this teacher highlighted the repetitive emphasis placed on students' academic achievement as measured by the standardized exams. Through this messaging, administration and district leaders communicated a connection between teachers' occupational worth, professional efficacy, and financial stability with students' scores on the end-of-year exams. Details relevant to the exam, such as the teachers' ability to prepare students for the exam or students' growth on the exam based on their baseline scores, were usually not accounted for and did not factor into the measure of teachers' success as it related to their job security, professional achievement, or financial reward.

The pressure to prepare students to score highly on the standardized exams encouraged teachers to make curricular decisions that would result in higher test scores for students. For example, teachers found it important to teach test taking strategies, made teaching the standards on which the exams were based a priority in lesson planning, and taught to the average student. Teachers used words like "teaching to the top of that bell curve" and "cluster management" to describe their effort to move students through tested material efficiently before the date of the exam. Further, since the students are compared to other students nationwide regardless of their socioeconomic, English as a second language, or immigrant status, teachers' approach to teaching had to account for producing outcomes irrespective of their backgrounds.

Some teachers reported feeling the pressure to ignore what students need and/or ignore their personal teaching approach to teach what is needed to score highly on tests. The urgency of covering test content and the stress of the exam outcomes impacting their professional performance led to teachers' feelings a lack of ability to change their lesson plan if the material was beyond students' zone of proximal development, not keeping the students' interests, not matching students' emotional availability, and/or not enhancing their test taking ability. Teachers also felt they often had to forgo their personal teaching approach to cover material even if it meant students would be overwhelmed by the amount of content or lack understanding about the connection between concepts and skills.

The emphasis on the standardized exam also resulted in teachers' shift in understanding of their own measure of student success. Only the skills that were tested by the standardized exams were perceived as foundational. If the subject had already been tested, teachers deemed learning in that area as complete and not necessary for further teaching. Despite the lack of consensus among teachers on whether the tests actually measure students' understanding of the standards, teachers mentioned using the test as a measure for their own professional success. For example, one teacher explained that when scores are low teachers would "go in and go 'okay, I must be doing something wrong.' And then they started developing these goals that eat away at their own skills and capabilities as opposed to solving some problem." Teachers viewed their own success based on the scores of students, changed their teaching strategies in response to these scores, and blamed themselves for the outcomes of these tests.

Teachers argued that qualifying and quantifying teaching and learning as measures for success was misguided. One teacher explained:

[State Standardized Assessments] represent to me a lot of things that I think are a misguided attempt to quantify and qualify teaching. I'm not sure that I feel that they are accurate and capable of doing what the powers that be want them to do.

Here, the teacher explained that the standardized exams represented one of many misguided attempts to measure success of teachers and teaching. Among these measurements of success included normative perceptions of successful student behavior. This reliance on testing was complicated during the pandemic as students were no longer sitting in the classroom during assessments and end-of-year state tests were canceled. The pandemic helped illuminate that the testing culture in schools is pervasive. Even when tests were canceled, many teachers still felt very tied to teaching as if teaching to the test, as discussed more in the *Barriers* section below.

Classroom Management

While understanding the shortcomings of qualitatively and quantitatively measuring student success, teachers simultaneously found themselves buying into the exact measures they wanted to reject. One way teachers identified buying into normative perceptions of success was by holding classrooms to the normative standards of successful student behaviors. One teacher described in a sarcastic tone:

They can't be too happy because if they're too happy and having fun, if they're laughing, and God forbid if they're out of their seats and jumping around, something is wrong. They cannot be having too much fun.

Although she was saying this statement facetiously, this teacher later noted that she did often internalize this judgment and apply it to her own and other classrooms. Successful student behaviors involved sitting in their seats, silent listening to the delivery of content, and seriousness about learning. Teachers used phrases like, "shackled to those chairs", "formal instruction", and "not out of their seats" to describe the behavior of students in a classroom perceived as successful by administrators or outside observers. During school closings, new methods of interacting and engaging with students were necessary to connect with students via online platforms, which some teachers navigated more successfully than others.

What Student Success Is

Doing Well on Standard, Rigidly Defined, Isolated Skills

It follows that if teacher success is measured on student outcome data, that student success is defined similarly, usually by grades and standardized state assessments. Students' test scores often determined whether the student was seen as successful. Teachers recognized that this system was flawed:

We are judging children and teachers on a one snapshot kind of deal... if on that one particular day, a student is upset, or they aren't good test takers, or they just aren't interested, then that shows that they're failures.

Typically, these standardized exams that are the major measure of success do not account for students' test scores over the school year, their ability to express understanding through another medium (e.g., essay, project), or their physical and/or mental wellbeing on the day of the exam. This teacher notes the limitations of this systematic measurement, indicating that this exam only looked at "one snapshot" of a student and that through this one measure, students were deemed as "failures" or successes.

The inclusion of material on these exams is usually not set by classroom teachers, but by those outside of the local school. One teacher noted:

It's pretty cut and dry because it's set by the state. Our benchmarks are released [standardized state] tests. And so in terms of feedback, what students get is a score, right? And they know which questions they missed. And theoretically, they know which skills they struggled with.

By limiting the students to measures of success imposed by the standardized exams, the feedback students received in response to their learning was also limited. The binary of getting the question correct/incorrect or passing/failing gave the impression that students no longer needed to work on the skills they had correctly answered and that learning could be reached in a binary of achieved/unachieved.

Because the normative view of student success limited students to their performance on standardized exams, learning within the classroom focused on tested content. Across elementary and secondary levels, students learned tested skills rather than skills that may be helpful in other life contexts. For instance, an elementary teacher shared:

In language arts, we're still teaching guide words. When was the last time I looked up a word in the dictionary and had to use the guide words? Should people know a basic ABC order? Yes, but do they really need to know which three words fit on the dictionary page? And I just think we're a little far away from reality.

Here the teacher exemplifies the tension they feel between setting a student up for success within the school system and setting them up for success in life. Since the students' success will be measured from their ability to perform on an exam, however, the content that they learn in school is also constrained to the material that will be covered on the exam. This conception of success was particularly lacking in capturing necessary skills that students learned during the pandemic, such as adapting to online school, problem solving, and additional modes of communication. For a more in depth discussion of the constraints of the system, see the *Facilitators* section below.

Reimagining Teacher Success

Interestingly, the confines of the system and the lack of power felt by teachers meant that many of them had a difficult time reimagining what success might look like in the ideal. The concept is discussed more in *Barriers* below. However, when given time for reflection, teachers brought up the idea that they feel success when their students are successful, though not always in traditionally defined ways. Additionally, developing authentic relationships with students was a marker of success that most teachers shared.

Successful Teachers Have Successful Students

Teachers consider themselves successful when their students are successful. In the following section, reimagined pathways to student success will be explored. When teachers can facilitate student success in those areas, they themselves will be successful.

While teacher and student success are currently often defined in terms of rigid data within their subject areas, the following quotation is illustrative of what teachers across elementary and secondary levels indicated, that teachers are concerned with student development outside of their acquisition of knowledge about their specific subject:

Part of that responsibility of a teacher is to provide…educational guidance, moral guidance, to shape students' lives in a more meaningful, positive manner.

Teachers want to define their success in a multitude of ways other than those currently constrained by the system. In order to reimagine what success could look like, facilitators discussed below need to be in place so that teachers would have the freedom to be defined as successful when acting in ways aligned with their ideals.

Authentic Relationships

An additional clear indicator of when teachers felt successful was the characteristic of establishing authentic student-teacher relationships. A majority of teachers in this study discussed how these relationships not only facilitated student academic success but were a success in and of themselves.

We see our students every day. We work with our students; we get to know our students. We have to build relationships with our students on a level that no one else will. We have to form a bond with them for them to be willing to learn from us, and so, in doing that, we form a connection with them, and we sort of take them under our wing. You know I joke all the time that the kids I'm going to have are my students. And that's because it's true.

In this quote, it is evident that the teacher considers not only this relationship important for his students to learn, but also significant for himself. His joking about his students being his children is indicative of the personal fulfillment and success he feels from forging those strong relationships with students.

This relationship is not without benefits in terms of student learning. As one teacher noted:

If the kid knows that you care about them, and the kid knows that you're there for them, and that you're not treating them like a number, and that you're not treating them like a grade on a sheet. You know their name, you know what they like, you know what they dislike, you know what they're doing outside of school. In my experience, they will, like, move heaven and earth to try and do whatever you ask them to do.

While the relationship itself is important, this also indicates that these relationships are a facilitator for student success and therefore teacher success.

Reimagining Student Success

When teachers discussed how they wished to define success for students, they talked about how they wanted students to have agency to learn about things that they were interested in and to focus on growth and their academic and/or vocational areas of strength. In doing this, students could foster their own feelings of success, which tied to feelings of self-worth.

Feelings of Success

One piece of accountability that is often missing from school definitions of student success is how successful students feel. In order for this idea to be incorporated, the system needs to not only depend on external assessments but allow space for internal inventories for students to reflect upon and measure their own successes. As one teacher explains:

We want them to take responsibility for their learning to become adults of responsibility. A part of that is learning how to look honestly at your own work, at your own performance, your own yardstick. Where are you; how are you growing? ...Whether or not you're giving 100%. You know where you have areas to improve and you also know where you do well. So, I think that we need that type of assessment as well...not just an extrinsic evaluator and extrinsic assessment, but an internal one too.

This is not without measurement or accountability, but the accountability looks different than the rigid system in place today. She elaborates:

Measuring anybody's knowledge and gifts and skills on a single day, in a single way, at a single moment, against a single standard, with a single rubric— that's [she makes a tsk noise] that's incredibly reductionary... It diminishes us tremendously.

Many teachers confirmed that they believe in continued measures of success through tests and projects, but that one summative test does not allow for students to show the ways in which they are successful. When tests can be a piece of gathering information about success, but not seen as the sole indicator, it will set students up for greater success.

Agency and Exploration

Teachers repeatedly discussed that student success will not look the same for every student. This was often discussed in opposition to the rigid system teachers are educating within. In order for students to be able to explore what's interesting to them, they first have to be exposed to those ideas. As one teacher explained this by saying, "One of the ultimate goals [for students] is college readiness and college, but I don't think that college is for everyone, so I think that exposing students to different ideas for future paths is super important." If teachers are to provide students with a variety of ways to be successful, they also need to provide them with the agency to explore and learn about ideas that they are interested in. To illustrate this point, teachers said things like:

You'd have to get to know your students first to see what they're interested in. Let's say, for example, this year we have a student who's really interested in lawn care and excavating and that sort of thing. Or if kids are interested in, say, video games. Okay, you could find some articles about video games, but it could be more about how they're designed or how they're created. So there'll be more of a process as opposed to just, well, here's a nonfiction article about why birds fly to the south. Well, who cares? We need more interest.

The contradiction to the normative system is evidenced here. While student interest can be incorporated to some degree in the normative system, the freedom to deeply explore those interests is limited.

Even more important than maintaining interest, though, was how presenting information that is not promoting student success is impacting students' self-conception. Teachers spoke about how the lack of student agency along with strict standards can harm self-esteem because constant failing and need for improvement is detrimental to developing self-worth. In an ideal situation, one teacher shared she would prefer:

a much more student directed sort of a layout. I think that there will be a lot more room for kids to explore different types of vocations, occupations, hobbies, interests in order to find their niche because there's a lot of students that I know don't have that opportunity. They don't have it in school right now because we're so busy force feeding them what they have to know, and they don't have it at home because they don't have that support from whoever they live with.

And so, I think that that's probably something that they would have to start when they were really young, is empowering them. To say this is what I'm good at, and this is what I'd like to do, and this is what I'd like to do with that, and this is what I'd like to try, and teaching them that trying things is okay and deciding that you don't like them is okay, too.

This agency to explore would lead to students who were able to grow specifically in their strengths, which is related to the next theme related to reimagining student success.

Growth and Strengths-Focused

In a reimagined system, students may have the freedom to not spend the majority of their school lives constantly remediating their weaknesses, but instead realizing their strengths and allowing for continued growth in those areas. Here, assessment could facilitate decision making and growth not just for the teachers but students, too, as opposed to just a summary of learning. One teacher spoke of the contradiction of this ideal to the normative system, "I like to see something that builds a student's understanding of what they're going to do or what they could do as opposed to something that says this is what you learn. Period."

This concentration on achievement rather than growth is in large part due to the systems in place. This short-sighted focus is harmful to students because it encourages a pattern where they see themselves as failures due to the system of assessment. As one teacher explained that when focused on state standardized assessments, *there's very little time for them to explore and find out what they're really good at. And a lot of them have arrived at the conclusion at this young age that they're not good at anything, and that really breaks my heart because they're all good at something, and they don't see it because it's not valued by the educational system. My kids that can draw beautifully and are so creative, that doesn't really do them any good on their [state standardized assessments]. My kids that can take apart an engine and put it back together, that doesn't matter to anybody giving a test anywhere. And so to them, that's not important that they can do that. And it breaks my heart because to me it's stripping them of their sense of self-worth.*

Here, students' feelings of success are an important component of their actual success, and lack of success is seen to attribute to low conceptions of self-esteem.

Authentic Problem Solving

Several teachers noted that the education system would benefit from a shift to focus on fixing authentic problems, saying things like:

Let's give them real life problems. Let's give them authentic consequences and authentic audiences. And let's see. Let's open up opportunities for them to show us things that maybe we didn't even think to assess in the first place.

This idea further confirms the first theme, because when students can ask authentic questions and problem solve to find answers, it will assist them in feeling more successful. In turn, this will support the reimagined version of teacher success, which is defined by helping students to find success in these ways. Here, we see an opportunity for schooling in a post-pandemic world, where problem-solving issues that students face can become part of their schooling experience.

DISCUSSION

Barriers to Success

There are a number of ways that the current system and all of the actors within it perpetuate the current definitions of success that are held. These barriers will need to be addressed via supportive staffing and professional development models detailed below in order to move towards the reimagined definitions of success.

Holding on to "Tried and True" Methods out of Fear

One of the biggest barriers to success is the fear of failure. Teachers are often uncomfortable letting go of ways of teaching that have "worked" in the past, even if a new method might allow for higher levels of success for their students.

As an example, one teacher shared:

I think I'm just as guilty of it as any lawmaker or parent or the community that I complain about not supporting. I'm part of that same thing. I want to know if my kid has done well. "Is my kid an A student?"

And the teacher going, 'Man! Your kid really got so much better at writing!'

'Well, did they pass?!' And that doesn't necessarily need to be the question. The question is, does my kid still love reading enough to come back and want to do it next year? And that's not the question I'm going to ask as a parent. I'm going to ask, 'Did my kid pass?'

The system itself constrains what is prioritized when considering success because of the measurements of assessments and the accountability tied to them. Teachers shared that even when they believed certain skills (e.g., writing) were more important than others (e.g., guide words), they prioritized teaching the skills they knew would be on the test even if they weren't particularly helpful to the students' overall knowledge and development. In the pandemic, opportunities were given for teachers to move away from these ways of teaching when state standardized testing requirements were waived. Many teachers, however, continued to teach in the ways most familiar with them. This is related to the next barrier about ideal systems.

"Out of Touch" With the Ideal

When asked about what an ideal education system might look like, one teacher laughed, exclaiming, "You know, I'm so out of touch with what ideal might look like!" Several other teachers had similar remarks when asked about reimaging success. This is evidence that the system is so powerful, that many within it are not able to think or dream about what could be better. Rather, they just put their heads down and do what is required. During the pandemic, there were some freedoms from the system given, but not capitalized upon. This indicates the need for this systems thinking perspective on the system of education.

Being entrenched in the system as-is often leads to viewing students through a deficit lens. Even though, through interview data, it is obvious that teachers want what is best for their students, they would then report things such as "they're never going to pass the fifth grade [standardized state] test."

Further, teacher success is confined to the limitations of the existing system. While teachers have agency in their classrooms, to build relationships with their students and to make decisions on what and how to implement lesson plans and classroom systems, they are unable to engage meaningfully with the systems that drive problems that are beyond teachers' control. For example, in communities where rates of community violence is high, students are at increased risk of exposure to trauma and its related symptoms (e.g., avoidance, numbness, irritability). Students who express these symptoms in the classroom or are triggered within the school day, are more likely to struggle to meet the normative standards for success as defined by standardized assessments. Teachers are also likely to struggle to support the students with adverse experiences in supporting standardized measures of success, as the source of their trauma stems from outside of the classroom. Teachers are thus asked to figure out ways to survive, exist, and/or succeed within the confines of these parameters. Furthermore, schools are also limited to measures of success that are within the confines of the existing system. Schools are not designed to support students who have needs that are non-academic, because success is not measured with non-academic standards.

Teacher Time Is Valuable and Limited

This reimagined way of defining student success, where there is not one clear pathway, requires more planning time and a different kind of teacher stress and burden. Even in the school system as it is, all the requirements of teachers outside of teaching can be a barrier to success (e.g., meeting students' non-academic needs, staff shortages requiring additional coverage time, etc.). As one teacher explained:

If teachers have too much of a workload and not enough time, that could be traumatic as well, where it turns into teacher burnout. They feel like they have so much to do, but then they won't have enough time and energy left to invest in the kids and teach the kids. So, I feel like the administration at the school is very important. And for them to listen to teachers and help the teachers with either more time, less meetings, or more support.

Stress and pressure to be successful impacts actual success, and administration will need to develop clear communication systems in order to not continue to burden teachers with more expectations without additional measures of support, as discussed more below.

Critical Systems Heuristics (CSH)

The results of the study suggest that a holistic perspective of the educational system is necessary. Assessment has become synonymous with a measure of accountability, distancing itself from its primary purpose of identifying areas of student growth and learning. Assessment and accountability themselves have become the ends rather than a means to the ends. With the cancellation of state-mandated testing in 2020 and for many in 2021, an opportunity for thinking of schooling outside of testing opened. The principles of CSH recommend a critical evaluation of the system's current state and the desired outcome. This approach enables a comprehensive analysis of the system, both means and ends, and identifying discrepancies from original intentions. A summary of this discussion is found in Table 3.

The values and motivations driving improvements in the education system often prioritize the system itself over the needs of the students. The high-pressure of top-down control means that actors within the system are compelled to make considerations different from those they would prioritize in an ideal world. Thus, the beneficiaries of the system become schools themselves, because they stand to gain more than the students. The purpose of schooling remains contested (e.g., Labaree, 1997) and this influences how policies are operationalized. However, the purpose of school *should* be learning how to learn, which will benefit students in their roles as human beings first, as well as citizens and employees. The results of this study reveal that test scores are a primary measure of success. Nonetheless, teachers yearn for a system in which they can leverage their expertise to help students recognize their own success and guide students on that pathway to success.

In terms of power dynamics and decision-making in education, state accountability systems hold the majority of the decision making power, yet local school leaders and teachers are often better positioned to understand and respond to the unique needs of their students. Despite having some control over how they deliver material, teachers are still constrained by the measurements of success. In the ideal system, teachers would have greater control over various facets of the system, allowing them greater pedagogical freedom. While outside the scope of this study, it is important to note the financial climate of schools also presents challenges. As funding is often distributed based on local taxes, schools in the poorest neighborhoods often receive the least funding. Although funding mechanisms such as Title I provide some additional money, it is not sufficient to address the lack of resources external to the school. An ideal system would prioritize the redistribution of resources to provide more support to schools serving students in need of more academic and material support.

Table 3. CSH of the public education system

	Roles/Stakeholders	Role Specific Concerns/Stakes	Problems/Stakeholding Issues
Values & Motivations for Improvement **Is** **Ought to be**	*Beneficiary* The **beneficiary** of the education system? Schools Students	*Purpose* The **purpose** of the education system? Contested Student learning	*Measure of Improvement* The **measures of success** for the education system? High test scores Learning as acknowledged by the student & teacher
Power/ Control **Is** **Ought to be**	*Decision Makers* **Control of the conditions of success** in the education system? State accountability systems Teachers & local school leaders	*Resources and Attributions* Conditions of success **under control** of the education system? Classroom pedagogy Classroom pedagogy	*Environment* Conditions of success **outside the control** of the decision makers? School resources based on tax/ neighborhood wealth Distributions of wealth so that under-resourced neighborhoods have the most resourced schools
Knowledge **Is** **Ought to Be**	*Planner/Expert* **Providing relevant knowledge and skills** for the education system? State or nationally developed curriculum Educator and student co-constructed curriculum that cover overarching state standards	*Expertise* Relevant **new knowledge and skills** for the education system? State or nationally developed curriculum / experts brought in for PD Teachers with local knowledge	*Guarantee* Assurances of **successful implementation**? Top-down management through observations and rankings Time and space for collaborative setting of goals, development of, and implementing knowledge and skills
Consequence/ Legitimacy **Is** **Ought to Be**	*Witness* Representing the interests **of students** in the education system? Varies based on school All stakeholders	*Emancipation* Opportunities for students to have expression and freedom from the constraints of the educational system? Outside of schooling Built within the system with opportunities for agency and exploration	*Worldview* Space available for reconciling different worldviews regarding the education system among teachers and students? Normatively suppressed Built into the system as a lab for working out conflict outside of school.

Source: Adapted from Ulrich and Reynolds (2010)

In addition, this power and control is related to the control of knowledge. While it is often quipped "knowledge is power," the adage "power is knowledge" is more appropriate. States and federal agencies, far removed from classrooms, are often responsible for curriculum development. Educational decisions are often made by business and policy experts with limited input from teachers. While teachers appreciate guidelines about what to teach, as discussed in the results, their involvement in the co-construction of the curriculum with their students would enhance student engagement and retention of the learned content. Similarly, teachers' professional development should be co-constructed based on relevant research, providing them with the opportunity to apply the acquired knowledge and skills. Rather than relying on top-down monitoring and evaluation, teachers should be responsible for their own implementation, as discussed further in the *Facilitators* section below. In this model, teachers in Networked Improvement Communities (NICs) would serve as the primary means of implementation, replacing additional initiatives that add to teachers' workloads.

Furthermore, the educational system needs to consider the effects on students. Although some schools adequately represent students' interests, this varies widely across schools. All stakeholders should prioritize students' needs and interests, which should be the primary driver of learning in this reimagined school. Currently, students' opportunities to express their opinions on the constraints of schooling are limited, often to outside-of-school spaces. Schools should strive to provide these opportunities within the school by ensuring that the diversity of the school body is represented in the school staff. Lastly, teachers are regarded as experts while students are often seen as the passive recipients of knowledge. In this reimagined system, students should frequently be given the opportunity to discuss and reconcile differing worldviews with their teachers and fellow students. The increased polarization in politics underscores the importance of this skill, and schools should serve as a laboratory to practice how to communicate with those who hold different beliefs, with the aim of negotiating next steps, rather than relying on top-down commands from teachers. The situating of these findings within the CSH framework shows the ways that the system as-is has deviated from the ideal.

Implications and Future Directions

When considering the transition from what is to what ought to be, the pandemic can be leveraged as an example of emancipatory space. The skills and knowledge learned during the pandemic are not easily measured on a standardized test, and so "learning loss" was identified as a major issue in school systems around the country. However, what are the skills that students and teachers did learn during that time that will serve them well in the future? How can we learn from the shift away from the standardized system, in which students sit in seats and receive information from teachers, toward this reimagined system. Below, facilitators for this new measurement of success are considered.

Facilitators for Success

Teachers and students alone cannot change the system to redefine success. There are components of the system that need to be amended in order to support these innovative perspectives. CSH encourages an investigation of who holds power and who is advocating for those who are without. These facilitators would support the reimagined versions of success by moving toward a common vision across system stakeholders.

Supportive School Staffing. One of the most important staffing decisions in order to facilitate this new reimagined success for teachers and students is the administration. This is especially true at the building level, but administration at the district, state, and national level also impacts this reimagined state. Administrators who support teachers and allow them the freedom to try new things will create an environment with less fear and more willingness to change. Alternatively, administrators who do not take this approach become a barrier to implementation. Many teachers noted that unsupportive administration was a major part of teachers experiencing burnout during the pandemic. Key characteristics of a supportive administration include allowing for flexibility, encouraging an environment where it is safe to try and explore new ways of teaching, and helping teachers define success for themselves.

An interesting staffing role mentioned in interviews included adding support staff not only for students but that teachers could use as well. Teachers, who were interviewed during the pandemic, expressed the role of burn-out in teachers' choices to leave the classroom. A way of helping support teachers would be to allow access to mental health support during the school day where teachers would be able to discuss issues related to their classrooms with staff who were able to understand their context and offer support in ways that are feasible within their school environment.

Humanizing Teachers and Treating Them as Professionals. Teachers are more likely to find success when they are treated as professionals who know what students' needs are and how to meet them. Again, this requires teachers to be given the flexibility and freedom to experiment. Changing ideas and exploring new ways of teaching will not always be successful in the first iteration.

Professional Development. Within this new model, allowing teachers to work collaboratively and build community is of importance. One evidence-based way of doing this is through Professional Networked Improvement Communities (NICs; Feygin et al, 2020). In these communities, teachers have agency to collaborate and create their own ideas of improvement while implementing and collecting data around those ideas through a Plan, Do, Study, Act model. This allows teachers to act as the professionals they are and create space and protocol to allow them to problem solve together in community rather than always relying on outsider expertise.

PD starts in pre-service settings and primes teachers to think within the system. Once teachers enter the school system, PD has the potential to reinforce how things are or challenge the status quo. As discussed in the literature review, there is currently a lack of PD that addresses non-academic needs of teachers, the necessity of which was exacerbated by the pandemic. Remedying this could facilitate both support and success for teachers. Allowing teachers to gather information and try new approaches to trauma-informed care for themselves and their students has the potential to alleviate some of the burdens being placed on teachers. However, it is important to note that this is not a silver bullet solution. There are key systemic issues that are impacting teachers as those involved in the system, and changes at the system level must occur to fully create higher levels of success for teachers and students alike.

Success for Who? An additional consideration needed in moving away from standardized, rigid definitions of success is that it opens space for disagreement. When the belief that every individual is unique is central, it may be difficult to have consensus on what success looks like by every teacher or every student. Parents and teachers are hesitant to break away from a system especially when concerns around evaluation are still tied to the "old" system, as illustrated by the results of this study. With new flexibility, administration needs to be prepared for disagreement over what success is and have pathways in place for helping to assuage tensions around differing definitions of success. There are systems that are successfully navigating these tensions, such as networks of Project-Based Learning schools that utilize rubrics and professional development communities. Additionally, allowing space for these conflicts in schools and processes for mitigating them could allow for the development of a skillset that is much needed today.

Future Research

Limitations of this work include the use of a relatively small number of teachers within the sample. Future research in which a system such as this one was allowed to be implemented without constraint and studied is an important next step. Focusing on outcomes as understood by teachers and students, rather than external quantitative measurements, will be important to see the success of this way of being. This is a major shift, which is sure to have bumps and flaws along the way. Longitudinal research will be needed to track the process and so that others in similar contexts can learn from the successes and pitfalls of those who try it out.

CONCLUSION

While the normative perceptions of teacher and student success are largely rooted in rigid and standardized measures (i.e., standardized testing, managing behaviors, and completing job related paperwork), teachers and students are yearning to function within a system where they can bring their whole selves. In utilizing CSH, and considering not only how the system is, but how it ought to be, success can be reimagined. In this reimagined system, student success is measured both externally and internally, providing students the agency to explore their interests and focus on growth, especially in areas where they already have innate abilities. This approach respects students as humans with the capacity for their own decision-making. When teachers can promote these kinds of student success through authentic relationships, this can be a means of their own success. In order to facilitate this, supportive school staffing, especially administration, is key. Additionally, teachers need to be seen as whole people who bring their personal and professional selves into the teaching profession and can be trusted to make decisions that propel student success. Models of professional development that allow for teachers to respond to their own humanity and the humanity of their students through trauma-informed approaches and communities have potential to sustain this reimagined system. As one teacher shared about teaching in the pandemic:

School can, learning can take a lot of different forms. It could go a lot of different directions; it can use a lot of different forms. And so if it can be those things, why on earth wouldn't we let it?....We've broken the mold. The question is now what?

The "now what" supported by this research is to humanize teachers and students and redefine success, so that individuals are not just a cog in a machine but a moving individual within a beautiful, diverse landscape. CSH helps this reframe with considerations about how to realign the values and motivations toward student learning, movement of power, control, and knowledge creation to a more local level, and an opening of space to allow students to confront the consequences of the system, where they can practice agreement and disagreement that they are sure to face in the "real world." This post-pandemic system may not be easily measured or tracked, but when we allow individuals room for agency and growth, the collaboration and new creation of this reimagined system is one in which creative solutions to the *mess* of global problems, inside and outside of education, can better be addressed.

REFERENCES

Ackoff, R. L. (1974). *Redesigning the future: A systems approach to societal problems*. John Wiley & Sons.

Anderson, E. M., Blitz, L. V., & Saastamoinen, M. (2015). Exploring a school-university model for professional development with classroom staff: Teaching trauma-informed approaches. *School Community Journal*, *25*(2), 113–134.

Anderson, K. M., Haynes, J. D., Ilesanmi, I., & Conner, N. E. (2022). Teacher professional development on trauma-informed care: Tapping into students' inner emotional worlds. *Journal of Education for Students Placed at Risk, 27*(1), 59–79. doi:10.1080/10824669.2021.1977132

Borntrager, C., Caringi, J. C., van den Pol, R., Crosby, L., O'Connell, K., Trautman, A., & McDonald, M. (2012). Secondary traumatic stress in school personnel. *Advances in School Mental Health Promotion, 5*(1), 38–50. doi:10.1080/1754730X.2012.664862

Children's Law Center. (2021). *A path forward transforming the public behavioral health system for children, youth, and their families in the District of Columbia.* Author.

Desimone, L. M. (2009). Improving impact studies of teachers' professional development: Toward better conceptualizations and measures. *Educational Researcher, 38*(3), 181–199. doi:10.3102/0013189X08331140

Devos, B. (2020, March 20). *Key policy letters signed by the Education Secretary or Deputy Secretary.* U.S. Department of Education. https://www2.ed.gov/policy/gen/guid/secletter/200320.html

Dorado, J. S., Martinez, M., McArthur, L. E., & Leibovitz, T. (2016). Healthy Environments and Response to Trauma in Schools (HEARTS): A whole-school, multi-level, prevention and intervention program for creating trauma-informed, safe and supportive schools. *School Mental Health, 8*(1), 163–176. doi:10.100712310-016-9177-0

Feygin, A., Nolan, L., Hickling, A., & Friedman, L. (2020). *Evidence for networked improvement communities: A systematic review of the literature.* American Institute for Research. https://www.air.org/sites/default/files/NIC-Systematic-Review-Report-123019-Jan-2020.pdf

Gallup. (2014). *State of American Schools.* Retrieved from http://www.gallup.com/ services/178709/state-america-schools-report.aspx

Greenberg, M. T., Brown, J. L., & Abenavoli, R. M. (2016). Teacher stress and health effects on teachers, students, and schools. Edna Bennett Pierce Prevention Research Center, Pennsylvania State University.

Guggenheim, D., Kimball, B., Chilcott, L., Strickland, B., Canada, G., Rhee, M., Weingarten, R., Roland, E., Richman, B., Finton, G., Cassidy, J., Roberts, K., Beck, C., & Legend, J. (2011). *Waiting for "Superman.* Paramount Home Entertainment.

Hatcher, S. S., Bride, B. E., Oh, H., King, D. M., & Catrett, J. J. F. (2011). An assessment of secondary traumatic stress in juvenile justice education workers. *Journal of Correctional Health Care, 17*(3), 208–217. doi:10.1177/1078345811401509 PMID:21571748

Jennings, P. A. (2020). *Teacher burnout turnaround: Strategies for empowered educators.* WW Norton & Company.

Kay, K., & Boss, S. (2021). *Redefining student success: Building a new vision to transform leading, teaching, and learning*. Corwin Press.

Kennedy, M. M. (2016). How does professional development improve teaching? *Review of Educational Research, 86*(4), 945–980. doi:10.3102/0034654315626800

Kyriacou, C. (2000). *Stress-busting for Teachers*. Nelson Thornes.

Levenson, J. (2017). Trauma-informed social work practice. *Social Work*, *62*(2), 105–113. PMID:28339563

Lieberman, M. (2021). How staff shortages are crushing schools. *Education Week*. Retrieved from https://www.edweek.org/leadership/how-staff-shortages-are-crushing-schools/2021/10

Locke, V. N., Patarapichayatham, C., & Lewis, S. (2021). *Learning loss in reading and math in US schools due to the COVID-19 Pandemic*. Istation.

Moll, L. C., Amanti, C., Neff, D., & Gonzalez, N. (1992). Funds of knowledge for teaching: Using a qualitative approach to connect homes and classrooms. *Theory into Practice*, *31*(2), 132–141. doi:10.1080/00405849209543534

National Commission on Excellence in Education. (1983). A nation at risk: The imperative for educational reform. *The Elementary School Journal*, *84*(2), 113–130. doi:10.1086/461348

Oberle, E., & Schonert-Reichl, K. A. (2016). Stress contagion in the classroom? The link between classroom teacher burnout and morning cortisol in elementary school students. *Social Science & Medicine*, *159*, 30–37. doi:10.1016/j.socscimed.2016.04.031 PMID:27156042

Perry, D. L., & Daniels, M. L. (2016). Implementing trauma—informed practices in the school setting: A pilot study. *School Mental Health*, *8*(1), 177–188. doi:10.100712310-016-9182-3

Reimers, F. M., & Chung, C. K. (Eds.). (2019). *Teaching and learning for the twenty-first century: Educational goals, policies, and curricula from six nations*. Harvard Education Press.

Reynolds, M., & Holwell, S. (2010). *Introducing systems approaches. Systems approaches to managing change: A practical guide*. Academic Press.

Sims, S., & Fletcher-Wood, H. (2021). Identifying the characteristics of effective teacher professional development: A critical review. *School Effectiveness and School Improvement*, *32*(1), 47–63. doi:10.1080/09243453.2020.1772841

Steiner, E. D., & Woo, A. (2021). *Job-related stress threatens the teacher supply: Key findings from the 2021 State of the US Teacher Survey*. RAND.

Tamayo, P. D. (2016). *Report to the legislature: Digital citizenship recommendations*. Office of Superintendent of Public Instruction.

Thomas, M. S., Crosby, S., & Vanderhaar, J. (2019). Trauma-Informed Practices in Schools Across Two Decades: An Interdisciplinary Review of Research. *Review of Research in Education*, *43*(1), 422–452. doi:10.3102/0091732X18821123

Ulrich, W. (1996). *A primer to Critical Systems Heuristics for action researchers. Independent Research Monograph*. Centre for Systems Studies, University of Hull.

Ulrich, W., & Reynolds, M. (2010). Critical systems heuristics. In *Systems approaches to managing change: A practical guide* (pp. 243–292). Springer. doi:10.1007/978-1-84882-809-4_6

Wagner, T. (2010). *The global achievement gap: Why even our best schools don't teach the new survival skills our children need-and what we can do about it*. Academic Press.

Walter, C., & Briggs, J. (2012). *What professional development makes the most difference to teachers.* Oxford University Press. https://clie.org.uk/wp-content/uploads/2011/10/Walter_Briggs_2012.pdf

Zamarro, G., Camp, A., Fuchsman, D., & McGee, J. B. (2021). *How the pandemic has changed teachers' commitment to remaining in the classroom.* Brookings. Retrieved from https://www.brookings.edu/blog/brown-center-chalkboard/2021/09/08/how-the-pandemic-has-changed-teachers-commitment-to-remaining-in-the-classroom/

Zhao, S. (2012). Global competency leadership and an empirical study of Chinese managers. In China's reality and global vision: Management research and development in China (pp. 104-124). Academic Press.

ENDNOTE

1 Quotes have been edited slightly for readability, editing out fillers such as repeated words, "um," and "you know."

Chapter 12
Schoolhouse as Learning Organism:
Making Education Research Part of a K12 Girls' School's Systems Thinking

Sarah Margaret Odell
The Hewitt School, USA

Maureen Burgess
The Hewitt School, USA

ABSTRACT

Before the COVID-19 pandemic, K12 schools largely remained rooted in hierarchical modes of operation, in-person and high-touch learning models, and cultures of learning that largely met those in the middle and mainstream, keeping systems oriented to the status quo. However, the signs of a burgeoning crisis in American schools were evident, as Sarah found through her graduate studies at U-Wisconsin, where the school of education long knew that a teacher shortage was beginning to create problems for staffing in rural parts of the state. Indeed, teacher education programs with once highly competitive admissions were struggling to find applicants. Informed by work of the Carnegie Foundation for the Advancement of Teaching, Hewitt is committed to the plan/do/study/act model of improvement science to ensure that Hewitt evolves in both practice and culture from the mindset of educational institution to that of a learning organism.

INTRODUCTION

Before the Covid 19 pandemic, K12 schools largely remained rooted in hierarchical modes of operation, in-person and high-touch learning models, and cultures of learning that largely met those in the middle and mainstream, keeping systems oriented to the status quo. However, the signs of a burgeoning crisis in American schools were evident for their reliance on a profession that is historically feminized and underpaid, subject to political crosswinds, and frequently over-burdened and under-resourced. Indeed,

DOI: 10.4018/978-1-6684-7285-9.ch012

teacher education programs with once highly competitive admissions were struggling to find applicants. The Covid-19 pandemic, the black@school racial reckoning, and the polarizing political discourse around teachers and curricula laid bare and exacerbated the cracks in the fragile ecosystems of schools (May, 2021; Mitchell, 2022; Roegman, et al, 2022). Teachers quit in record numbers, and according to a June 2022 Gallup poll, teachers are the most burned out group of professionals in the United States (even compared to healthcare professionals). A national conversation about "teacher burnout," gaps in student knowledge, and, indeed, the very purpose of K-12 education in a post-information age is driving the latest version of a crisis narrative for American schooling. Through the course of this chapter, we will describe how we thought about leaning more heavily into our mission of educating girls to be gamechangers and ethical leaders who forge an equitable, joyous future by using data and research to inform a feminist approach to leading our school and educating our students.

In order to navigate these competing priorities and contested histories to best meet the needs of students and faculty today, we argue that schools must shift to more flexible organizational models with dynamic feedback loops. Through a framework of systems thinking, a commitment to the improvement science method, and an approach rooted in feminist praxis, we argue that schools can move toward more inclusive models for students, faculty, and families that allows for diverse voices and lived experiences to shape the curriculum and pedagogy in the context of an ever-changing world and the dynamic challenges we face as local communities and global interdependencies.

At the start of the pandemic, our K-12 school for girls and young women embarked on a path to redesign the traditional "schoolhouse" as a learning organism. If a schoolhouse is a learning organization, structured hierarchically, a traditional model might place students at the base of the pyramid, with the principal, superintendent, or head of school at the top. Institutional hierarchies determine who learns what, from whom, and when. The larger the system is, the more complex the hierarchy. School systems, large and small, range from districts to single schools, from accrediting organizations to standardized tests and the college application process. If these kinds of systems tend to be largely static and regulatory, relying on the hierarchy to manage the daily operations but also to minimize disruptive change, we should ask ourselves: Why? Because those living and working within the system utilize mental models that shape not only how we see the world but how we decide to take action (or not take action) (Senge, 1990). How we see the world drives how we act, and most schools are fundamentally organizations for children that are shaped by the ways that adults see the world. The adult-centered and hierarchical model for schooling closes off essential pathways for feedback necessary to continuously evolve in ways that ensure voice and belonging for everyone in the community, especially children and faculty (Apple, 1979).

Children, especially as they move from early childhood to elementary and beyond, have little voice in directing the content and shape of learning experiences, despite decades of research on the importance of genuine inquiry and curiosity, teacher-student relationships, and effective (and ineffective) models of feedback and assessment. Similarly, many teachers feel they have insufficient time and support to reflect on their practice and learn ways to enhance and deepen their work as professionals, and others struggle with the onus of mandated curricula or outdated training rooted in tired "classroom management" practices. With the very real fear of infection and death looming over classrooms during the Covid-19 pandemic, the practices of the traditional schoolhouse were reinforced, challenged, and disrupted. We argue that the wellbeing and sustainability of a school community relies on the ability of the organization to embrace and learn from what continues to be revealed from this time of challenge and disruption.

As our school shifted to emergency remote and hybrid learning models, and then into a world forever altered by the Covid-19 pandemic, we continued to develop, implement, and revise our theory of a systems-thinking model for school leadership and community. We have reorganized the school's academic structure to break down disciplinary silos–eliminating department chairs, adding instructional coaches, and creating a K-12 curriculum advisory council with department representatives serving two to three year terms. The school's learning and innovation team (our instructional coaching model) and curriculum advisory council partner with the academic team (division and assistant heads of school) to ensure greater faculty voice and interdisciplinary planning of curriculum. Finally, our school established the Center for Gender and Ethical Leadership in Society in the fall of 2022, which oversees the work of all of these teams as a means of transforming our school into a learning organism rooted in feminist praxis and the improvement science method. Finally, our school established the Center for Gender and Ethical Leadership in Society in the fall of 2022, which oversees the work of all of these teams as a means of transforming our school into a learning organism rooted in feminist praxis and the improvement science method.

THEORETICAL FRAMEWORK

Organizational Theory as Status Quo: Structural Functionalism in Educational Leadership

Bolman and Deal (2021) suggest that there are four frames that organizations operate within: factories, families, jungles and temples and carnivals. These are not suggested frames–not ways that the authors recommend that institutions run, but rather an observation on how organizations exist. These frames pull on work in sociology, economics, management science, psychology, political science, and anthropology. As applied to schools, these frameworks may explain the particular behaviors of leaders within an institution, but they don't necessarily provide a path on how to move an organization forward in making progressive and effective change.

Much of the organizational theory that the field of education has drawn from, i.e. texts that are taught in educational leadership programs, and in schools of education, come more broadly from organizational theory not tied to education (Bolman & Deal, 2017; Morgan, 2006; Perrow, 1986; Pfeffer, 1997). This theoretical work, while important, has not done much to shift the paradigm away from thinking about organizations as neutral entities. The organization simply is, and this theory is we are measuring it. As Capper (2019) points out, many of these texts are grounded in structural functional and interpretivist epistemologies. Capper writes:

Educators expect staff to routinely challenge students with a rigorous, intellectually rich curriculum. Yet, educational leaders in the midst of their day-to-day work may intellectually drift, their leadership informed by a bland intellectual diet consisting of practitioner publications that most often align with status quo thinking. Because of their epistemological unconsciousness..., practitioners often fail to realize that what is touted as "new" or "innovative" in education often emanates from status quo (i.e. structural functional) epistemologies. (p. 16)

It is easy to see how much organizational theory can reify the status quo, just by looking at the studies that are frequently cited or seen as foundational: social systems theory (Getzels & Guba, 1955); needs hierarchy theory (Maslow, 1955); needs satisfaction questionnaire (Porter, 1964); total quality management

(Deming, 1988); Iowa studies (Lewin et al, 1939); Ohio state studies (Halpin & Winer, 1957; Hemphill & Coons, 1957), theory x&y (McGregor, 1957), Michigan studies (Likert, 1961), contingency theory (Fielder, 1967), leadership grid studies (Blake and Moulton, 1968), situational leadership (Hersey & Blanchard, 1969), path-goal theory (Evans, 1970), leadership style continuum (Tannenbaum & Schmidt, 1973). If a scholar in the field of educational leadership turned in a literature review with no studies featured after 1973, nobody would look at it twice. And yet, we continue to teach leader practitioners from studies that are now more than fifty-years old.

The argument pervades in both educational leadership programs and in practice that more complex and current organizational theories are too complex to be made useful. What practicality might a school leader find in queer theory? The truth is that school leaders need critical theories to meet the demands of today's complex school organizations and the broader world in which they operate. Schools face increasing pressure to resist becoming more inclusive places that examine how power hierarchies exist and function to keep some students at the margins (Lewis & Diamond, 2015). School segregation by race and class persists across the country, especially in cities and suburbs, with students of color and students living with poverty persistently and perniciously underserved. To educate all students for the future they deserve, schools and their leaders must be guided by a praxis that serves all, one that necessarily must include multiple epistemologies aimed at liberatory learning.

Feminist Epistemologies

As where organizational theory clings to rationality and an order within institutions (Bolman and Deal, 2021), feminist poststructural epistemologies encourage us to see the world through more relational and caring frames (Gilligan, 1982; Lather, 1991; Luke & Gore, 1992). Scholars like Charol Shakeshaft (1987, 2011), Margaret Grogan (1996, 2000; Grogan & Shakeshaft, 2011; Grogan and Smith, 1998), Catherine Marshall (1997; Marshall & Young, 2013) and Whitney Sherman Newcomb (2005; 2014; Sherman Newcomb & Niemeyer, 2015; Sherman & Beatty, 2010; Sherman & Wrushen, 2009; Newcomb & Mansfield, 2014) have all written extensively on feminist approach to leadership as an alternative to the White patriarchal norms of school leadership.

Jennie Weiner (Weiner & Burton, 2016; Weiner, Cyr & Burton, 2019; Burton, Cyr and Weiner, 2020) has integrated organizational theory with a feminist epistemology that looks at how women, and Black women in particular, are excluded and marginalized in school leadership structures. April Peters (Nash & Peters Hawkins, 2020), Judy Alston (2005), and Terri Watson (2020) have all pushed the field of educational leadership to recognize the important and unique contributions that Black women's perspectives and experiences have made to the schoolhouse. These scholars recognize and elevate the voices of *othermothers* (Hill Collins, 1990), and situate their specific positionality within the schoolhouse. Black feminist thought, initially written about by Patricia Hill Collins (1990), is the point north for these scholars writing about Black women leading schools. Hill Collins writes:

This distinction between knowledge and wisdom, and the use of experience as the cutting edge dividing them, has been key to Black women's survival. In the context of race, gender, and class oppression, the distinction is essential. Knowledge without wisdom is adequate for the powerful, but wisdom is essential to the survival of the subordinate. (p. 208)

There are no mentions of gender, race, and class in the work of Bolman and Deal (2021). Courses on educational leadership often make little mention of leader identity (Nicolaides and Gaynor, 1992). In fact, Nicolaides and Gaynor write:

[T]he course content [of these syllabi] subscribes to a perspective that socializes graduates intellectually and theoretically to mainstream interpretations of educational administration...[T]eaching in these courses is limited to topics and themes shaped by traditional perspectives...Alternative perspectives... were neither systematically or consistently incorporated into the courses examined. Issues such as those dealing with race, gender, ethnicity, and social class were underrepresented. (pp. 262-263)

Thus, when one comes to understand leading schools as systems thinking, rarely are they given a critical identity lens through which to think about the organization. We believe that this is malpractice always, and because we work in a girls school, we believe it is our responsibility to think about systems with a feminist leadership epistemology.

Systems Thinking and Schools That Learn

Systems thinking, in essence, is an understanding of the whole in relation to its parts, and an understanding of parts as interdependent in comprising the whole. As an approach, systems thinking looks for patterns and interrelationships among all elements of a system in order to make sense of the whole. In other words, the whole is never just the sum of its parts and a change in a relationship among two elements of the whole may have multiple causes and impacts. Systems thinking resists linearity for a networked and reciprocal understanding of how systems work. The origins of contemporary systems thinking lie in various fields, and different histories point to figures in biology, computer science, engineering, anthropology, and philosophy, among many others, to tell the story of its theories, methods, and approaches (Stacy, Griffin, Shaw, 2000; Monat & Gannon, 2015; Shaked & Schecter, 2017; Cabrera et al, 2021; McGlackan-Byrne et al, 2023). Peter Senge (1990) argues that systems thinking is the "fifth discipline" for his theory of a "learning organization," the umbrella under which the other four disciplines operate: personal mastery, mental models, shared vision, and team learning.

Like much of systems thinking and organizational theory, Senge's hugely popular text operates in a gender and race neutral context, relying on a highly structured model of how an organization learns, one which replicates the values of white patriarchy. This, according to Haim and Shaked (2017), reflects the modernist epistemology that underlies much of the literature on systems thinking approaches, which rely on a closed set of models, frameworks, and activities or tools intended to be applicable across contexts. They argue for non-structured systems thinking as part of their argument for their Holistic School Leadership approach, allowing for contextual interpretation and engagement with real world dynamics that are characterized by little structure, imperfect information, and unforeseen exceptions" (p. 49). They offer a framework not a theory, and their qualitative study does not engage with the epistemologies described by Capper (2019).

Only in recent decades have educational leadership models rooted explicitly in systems thinking and feminist postmodern epistemologies begun to emerge (Irby, Brown, Duffy, Trautman, 2001; Brown & Irby, 2003). Complexity theory, one mode of systems thinking, provides that bridge between systems thinking theory and a feminist praxis that embraces indeterminacy, fluidity, and interdependency, as it focuses on a complex adaptive system that can adapt in response to changes in its environment. Haim & Shaked (2017) write:

Complex adaptive systems are also characterized by a tension between chaos and order. The systems are chaotic in that for every pathway more than one course of action can be taken–leading to unpredictability and uncertainty at any given moment–and ordered in that patterns, consistency, and equilibrium will arise. (pp. 29-30)

By allowing for the uncovering of the voices of those marginalized by dominant theories of school change and leadership, a complex adaptive system informed by critical epistemologies will liberate possibilities for genuine change and innovation rather than impose tactical strategies that reinforce the status quo. And this is how we are using improvement science at Hewitt to ensure that we remain committed to listening for the voices that may not be the loudest, but are integral to our understanding the patterns of a system and our ability to accept uncertainty and perhaps a moment of chaos, to identify what steps are needed to thrive as a learning organism not just an institution for learning. A core principle of improvement science is systems thinking to make visible complexities that are not yet seen or seen but ignored (Bryk, Gomez, Grunow, LeMahieu, 2015). Without such openness and reflection, and without a deep commitment to critical epistemologies, improvement science can devolve into tactical improvements and rituals of accountability, such as conducting surveys but doing little with the data (Bryk, 2020). The pandemic was indeed a maelstrom that changed our school's environment in the space of a few days. As so often happens, we began a journey of school innovation rooted in feminist praxis before March of 2020, and that journey has been and will contine to be unexpected, collaborative, uncertain, and, perhaps most importantly, adaptive.

METHODS AND DATA

The Hewitt School as a Learning Organism: What We Did

To begin this move toward becoming a learning organism, our school applied for and was awarded an Edward E. Ford educational leadership grant in the fall of 2020 in support of our proposal to hire a learning and innovation researcher who would work closely with faculty and academic leaders on building a model of ongoing research into our school's practices following the improvement science methodology. Dr. Sarah Odell joined the community in July 2021, spending the first academic year working with a design team on an innovative high school program rooted in youth action research and committed to educating young women in the context of real world challenges and questions related to our school's mission. She also began to build a team of internal practitioner researchers and scholars from higher education to partner with us in conducting studies of our program and pedagogy through a feminist critical lens.

The school has provided release time for several faculty across our three divisions, and they serve as learning design coaches who work closely with faculty on implementation of research-informed practices in their classrooms. This learning and innovation team is core to Hewitt's commitment to improvement science, as they provide coaching in the design of learning experiences that effectively engage our students and provide continual feedback loops aimed at growth and progress. They are on the ground in our classrooms, listening to faculty and students as we move through cycles of change and reflection. Some changes are tweaks, like establishing a student-centered way of beginning class so that learners can transition, settle, and focus. Others are significant, like Hewitt's transition to competency based learning in order to provide our students with feedback on knowing, doing, understanding, and valuing. All interventions are informed by our team's research into our practices and our curiosity about the ways we are listening to our institution–its many voices and complexities, both human and systemic.

A key element in our process of listening includes an annual survey, which we began administering in December 2020 (in the midst of the pandemic). Now, every Spring Hewitt students and faculty and, starting in 2023, parents/guardians, take a comprehensive perception survey administered by YouthTruth. The survey covers topics on engagement, academic challenge, culture, belonging and peer collaboration, relationships and college and career readiness. In May, Sarah receives the results of the survey and spends a month digesting them and writing a short white paper for Hewitt's senior leadership team to help identify a few key areas of focus as well as areas of progress. Sarah provides both strengths and places of improvement to the team, so that the senior leadership team can incorporate that data in setting institutional priorities for the upcoming academic year. Sarah also presents the findings of the YouthTruth Survey to our faculty before the start of the academic year, and has our faculty grapple with the findings of the survey. We are in the process of developing a more robust process for faculty to work with the YouthTruth survey results, similar to the ways students have learned to use the data.

Using the YouthTruth survey results, Sarah, now the Director of Learning and Innovation Research, identified a few key projects to address points of growth for the 2022-2023 academic year. In both of the years that the YouthTruth survey was administered, Black girls at Hewitt identified that discipline at the school was administered unfairly. Hewitt is working with Assistant Professor of Educational Leadership at the University of Delaware, Lauren Bailes, who has done extensive professional development with public school districts on how to create more restorative and equitable discipline practices. Dr. Bailes is also a mixed methods researcher, and is using the raw YouthTruth data to see if other intersecting identities (sexuality, socioeconomic status, academic ability) are creating a more acute disparity in certain groups of Black students. The Klingenstein Center for Independent and International School Leadership at Teachers College, Columbia University has provided a seed grant to support Dr. Bailes' work and includes two full days of professional development for Hewitt faculty to address these disparities.

The YouthTruth survey results revealed two findings that we believe are intertwined: students expressed a desire for greater academic challenge, and faculty expressed a greater desire to be seen as professionals. To address both of these findings simultaneously, we are working with Dr. Frances Rust, who previously ran teacher education at the University of Pennsylvania Graduate School of Education and New York University Steinhardt School of Culture, Education and Human Development. We created a course release for 12 instructional coaches across K-12, where they teach 50% of the normal course load and have 50% of their time allotted to getting into their peers' classrooms. Dr. Rust works closely with our coaches to help them develop questions about their own teaching practice, which then serves as a teaching tool for our instructional coaches to work with their peers. Coaches are working toward creating a portfolio that documents either the process of working through their own critical reflections in their teaching or what it was like to work through a critical reflection with a colleague. At this end of this year, we hope to have a living library of projects that our coaches have worked through that will then serve as a resource to current and future Hewitt teachers on how to continue to refine their teaching. We are hoping to see results in both students feeling challenged in their classes, and teachers feeling recognized as professionals.

As part of our commitment to a feminist praxis, and keeping gender at the center of our teaching inquiry, we are partnering with psychologist Carol Gilligan to serve as a site for her to revisit her groundbreaking work on girls, which is now thirty years old. Dr. Gilligan served on Sarah's dissertation committee and trained her as a qualitative researcher. In the Spring of 2022, Dr. Gilligan expressed interest in returning to her work on girls, chronicled in *Meeting at the Crossroads*. Dr. Gilligan has been to campus three times as preparation for the pilot phase of her study, which will begin in January, 2023. She has met with students, faculty, and parents to discuss her work as well as provide professional development around her work. Dr. Gilligan taught a two-day workshop with one of her former graduate students on the Listening Guide method of research to faculty at schools participating in the study, as well as faculty who are facilitating youth action research at their schools. Even though Dr. Gilligan is coming to learn from our students, all of our community members are learning from her. We are learning to challenge our assumptions around how girls are supposed to behave in school: quiet, obeying orders, and what Dr. Gilligan constantly reminds us, "girls are taught to sacrifice their voices to have relationships." That is the idea we now hear most often repeated at Hewitt, how do we teach our girls that they do not have to make that choice, between having a voice and having relationships. Carol calls this developing a "healthy resistance." And even though her study has not yet begun at Hewitt, her research and her *voice* reverberate in our teaching and school governance practices.

Dr. Terri Watson at City University of New York has extensive experience studying Black girls in schools and Black women leaders in schools. Dr. Watson is meeting regularly with our Black Student Union as part of a project to better understand what is working for Black girls in school. By asking the questions of how Black girls are mistreated in schools, researchers often frame these young women's experiences with deficit language, sometimes even before beginning the research. Dr. Watson is framing her study around bell hooks' theory of love, and asking our students what teachers are doing to make them feel seen, supported, and loved. This will have impact on belonging for Black girls at Hewitt, and will give teachers actionable feedback on what practices are working well. This feedback will also make those faculty who engage in those practices feel seen and heard.

Finally, we have made the decision to put the YouthTruth survey results in the hands of our students. We know, intuitively, that they will see points in the data that the adults do not. Their priorities in the findings will also be different for the students. Three faculty at Hewitt, David Heller, Liz Brennan and Sarah Odell worked together to create a year long social science course, where students learn about education research and how to conduct it. David teaches history at Hewitt and Liz teaches math. This opportunity has also served as professional development for the three faculty, as they have each brought their disciplinary expertise to the role, and different ways of knowing regarding data. They also will help shape future opportunities for faculty, particularly members of the Curriculum Advisory Council, to engage with data from the professional community survey.

During a four day deep dive, where students in action research were released from their regular courses, they examined the YouthTruth data and came up with a research question. They will spend the remainder of the 2022-2023 school year investigating that question through interviews, observations and analysis. The question that they came up with, "what does success mean at Hewitt?" will shape their interviews and observations which will culminate in a presentation of the students' findings to teachers and school leaders in June. The students have had the opportunity to work with several of our research partners, including Terri Watson and Carol Gilligan, to expand their thinking about research, methods and doing research in community.

In addition to these five projects, which are now housed under a Center for Gender and Ethical Leadership in Society, we made the decision to create a post-doc position. Part of the thinking in creating this position, and becoming the only independent school in the United States with a post-doc, was that it would enable us to be a responsive research hub and keep our work from becoming static. The idea is that we will have a post-doc who works with us for two years, before continuing on to a permanent position. This means that as part of our ethos as a learning organism, our base of expertise will shift every two years. If there is a need, whether it is a finding that surfaces in our YouthTruth data or elsewhere, we have the ability to bring in someone with research expertise in that arena. Our current post-doc, Daron Cyr, has expertise in elementary education and in parent engagement. These are two areas of research expertise that those of us in learning and innovation leadership did not have.

Schools most often get caught being reactive. Some of that cannot be helped–as the day unfolds, needs will arise based on the interactions between students and teachers, teachers and parents, parents and school leadership. But how many schools feel like they have created the infrastructure to become responsive? If the Covid-19 pandemic showed us anything, it's that we can no longer survive on being reactive. It's hard to imagine right now what it would have looked like to become responsive in the moment when the country shut down, but their were glimmers of responsiveness in the moments when teachers brought food to students who were food insecure, or schools figured out how to get devices to kids who didn't have computers. Needs were met, sometimes, before a problem had arisen. If we are able to make this shift to being a learning organism instead of a learning institution, and being willing to shift to needs identified by our community, we will ultimately be a healthier school by taking stock of our strengths and having a plan for how we meet our challenges.

When I think about the power of this work, I think about the members of our Black Student Union waiting in the lobby for Terri Watson. She had only visited Hewitt a few times at that point, but the girls eagerly awaited her arrival sitting on a bench in our lobby. When Dr. Watson came in, they all stood up, ran over to her and hugged her. She embraced them just as eagerly. Our researchers have become part of the organism, and our students and faculty are seeing them as colleagues, mentors and collaborators. They have relationships at the school and have become part of the fabric of our day-to-day.

DISCUSSION

At this point, we have well over twenty-five (out of approximately 85) faculty members across three divisions serving on our Curriculum Advisory Council and Learning and Innovation team, and they provide faculty input on curricular and pedagogical decisions as part of our cultivation of the school as a learning organism. CAC representatives meet regularly with departments to share information, collaborate on curriculum planning, and gather input to bring back to the full council. Learning and Innovation team members work individually and in small groups with faculty on their pedagogy and learning experience design, often in interdisciplinary and cross-divisional ways. As our school and our city continue to build a future beyond the pandemic, we are working toward a balance of hierarchy to run daily operations and steward our school's mission and vision with a systems thinking approach to school improvement, an approach that must take into account the experiences and perspectives of all members of the community. Complex systems require both a hierarchy and a network in order to sustain and adapt continuously to change and innovation (Kotter, 2014).

Similarly, the projects that Hewitt's GELS center has underway will inform decision-making in multiple areas of the school, from pedagogy to classroom dynamics to equity work, and our learning will feed the next phase of a plan-do-study-act cycle (Bryk, et al 2015). Students, faculty, and GELS researchers will partner with the school's curriculum, innovation, and academic teams to assess what is working well and where improvement is needed. Our design for both continuity and flexibility in the staffing of roles like the post-doc position and CAC team allow for bringing different expertise and perspectives into the conversation. Stability and adaptability are mutually informing goals in our model of feminist educational leadership, both in times of dramatic contextual change, such as a pandemic closing down schools for months at a time, and in times of reflection and improvement to foster and sustain an equitable and joyful community for all learners, adults and children, faculty and administration, families and school research partners.

Coming out of the Covid-19 pandemic, schools must become more responsive. They also need to think of themselves more as learning organisms and less as learning organizations. Many of the hierarchies we depended on to run the day-to-day operations of school evaporated overnight. Teachers and school leaders are burned out from the last two years and leaving the profession at record numbers (Balingit, 2022). One of the implications of our work in becoming a learning organism at Hewitt, will be the impact on faculty retention. One of our initial observations is that faculty have seemed initially resistant to doing anything additional in their duties. After two years of wiping down tables, having to monitor school lunch and be present with students (a time when faculty used to be able to decompress and be with one another), little additional tasks have eroded a willingness to do more than what faculty perceive as their day to day job.

But one of the data points that we are thinking about was after Dr. Gilligan's faculty meeting, several faculty reached out to us to ask how they could get ahold of *Meeting at the Crossroads*. There was so much enthusiasm, that Sarah sent out a school wide email asking if faculty would like to do a book study. Within minutes, 25 faculty responded saying that they would like to discuss *Meeting at the Crossroads*. What this has shown us is that mission driven, responsive and intellectual work is still engaging. Faculty have a desire to be treated like thinking professionals who are seen as having expertise. Could this type of responsive work, which is deeply intellectual, hold the key to keeping teachers engaged during this inflection point? We think so.

Just published research (Hansen, Sabia & Schaller, 2022) shows that teen suicide rates have *increased* since schools have reopened. Hansen, Sabia & Schaller (2022) point to the pandemic as disrupting school bullying, which has now resurged since schools have reopened. What this points to is that status-quo in our schools was never good enough for most of our children, and a "return to normal" shouldn't be the goal. Hewitt had already put in motion, many of the pieces that led us to becoming a learning organism before the pandemic happened, but the pandemic accelerated our work's urgency. We must become a more responsive school, one that continually looks at how we are doing. Are our faculty fulfilled? Do our students feel a sense of belonging? Are we living our mission?

CONCLUSION

The truth is that we will always be working toward getting better at all of these questions. We will always be working toward realizing our mission, and making sure that the people who populate our school feel seen and fully realized in their identities. But the work is never done. Being responsive is our best bet at meeting our mission, and it means that we will always be learning how to teach better, how to serve better and how to educate girls better. Education isn't static, and an intervention that works today may not meet the needs of tomorrow. The sooner we recognize this, the sooner we can focus on becoming agile and meeting the needs of the people in our schoolhouse. This will be a shift for our students and for our educators, as we often think once a lesson or a way of approaching a topic works, we don't need to revisit it. This is a new way of doing school: collecting data, responding to it, and thinking about feminist theories to inform how we plan our work. But it is our best hope to meet the challenges in the field of education right now.

REFERENCES

Alston, J. A. (2005). Tempered radicals and servant leaders: Black females preserving in the superintendency. *Educational Administration Quarterly*, *41*(4), 675–688. doi:10.1177/0013161X04274275

Apple, M. (1979). *Ideology and curriculum*. Routledge. doi:10.4324/9780203241219

Balingit, M. (2022). Shortages of staff and equipment continue to plague schools, new data shows. *Washington Post*. https://www.washingtonpost.com/education/2022/12/06/teacher-supply-shortage-free-lunch/

Blake, R., & Moulton, J. (1968). *Corporate excellence through grid organization development*. Gulf Publishing Company.

Bolman, L. G., & Deal, T. E. (2021). *Reframing organizations:Artistry, choice, and leadership*. Jossey-Bass.

Brown, G., & Irby, B. (2003). The synergistic leadership theory: contextualizing multiple realities of female leaders. *Journal of Women in Educational Leadership*, 75. Accessed online: https://digitalcommons.unl.edu/jwel/75

Bryk, A. (2020). *Improvement in action: advancing quality in america's schools*. Harvard University Press.

Bryk, A., Gomez, L., Grunow, A., & LeMahieu, P. (2015). *Learning to improve:how America's schools can get better at getting better*. Harvard University Press.

Burton, L., Cyr, D., & Weiner, J. (2020). "Unbroken, but bent": Gendered racism in school leadership. *Frontiers in Education*, *5*, 52. doi:10.3389/feduc.2020.00052

Cabrera, D., Cabrera, L., & Midgley, G. (2023) .The Four Waves of Systems Thinking. In Routledge Handbook of Systems Thinking. London: Routledge.

Capper, C. A. (2019). *Organizational theory for equity and diversity: Leading integrated, socially just education*. Routledge.

Deming, W. E. (1988). *Out of the crisis*. Massachusetts Institute of Technology.

Evans, M. G. (1970). The effects of supervisory behavior on the path-goal relationship. *Organizational Behavior and Human Performance*, *5*(3), 277–298. doi:10.1016/0030-5073(70)90021-8

Fielder, F. E. (1967). *A theory of leadership effectiveness*. McGraw-Hill.

Fox, D. (2009). An Overarching Framework for Sustainability. *Built Environment, 35*(3), 302–307. doi:10.2148/benv.35.3.302

Getzels, J., & Guba, E. (1955). Role conflict and personality. *Journal of Personality, 24*(1), 74–85. doi:10.1111/j.1467-6494.1955.tb01175.x PMID:13296016

Gilligan, C. (1982). *In a different voice*. Polity Press.

Grogan, M. (1996). *Voices of women aspiring to the superintendency*. State University Press of New York.

Grogan, M. (2000). Laying the groundwork for reconception of the superintendency from feminist postmodern perspectives. *Educational Administration Quarterly, 36*(1), 117–142. doi:10.1177/0013161X00361005

Grogan, M., & Shakeshaft, C. (2011). *Women and educational leadership*. Jossey-Bass.

Grogan, M., & Smith, F. (1998). A feminist perspective of women superintendents' approaches to moral dilemmas. *Journal for a Just and Caring Education, 4*(2), 176–192.

Halpin, A. W., & Winder, B. J. (1957). A factorial study of leader behavior in descriptions. In R. M. Stodgill & A. E. Coons (Eds.), *Leader behavior: Its description and measurement* (pp. 39–51). The Ohio State University Bureau of Business Research.

Hansen, B., Sabia, J., & Schaller, J. (2022). *In-person schooling and youth suicide: Evidence from school calendars and pandemic school closures*. National Bureau of Economic Research. Accessed online: https://www.nber.org/papers/w30795

Hemphill, J. K., & Coons, A. E. (1957). Development of the leader behavior description questionnaire. In R.M. STodgill & A.E. Coons (Eds.), Leader behavior: its description and measurement (pp. 6-38). Columbus, OH: The Ohio State University Bureau of Business Research.

Hersey, P., & Blanchard, K. (1969). Life cycle theory of leadership. *Training and Development Journal, 23*(5), 26–34.

Hill Collins, P. (1990). *Black feminist thought: Knowledge, consciousness and the politics of empowerment*. Routledge.

Irby, B., Brown, G., Duffy, J., & Trautman, D. (2001). The synergistic leadership theory. *Journal of Educational Administration, 40*(4), 304–322. doi:10.1108/09578230210433409

Kotter, J. (2014). *Accelerate: building strategic agility for a faster moving world*. Harvard Business Review Press.

Lather, P. (1991). *Getting smart: Feminist research and pedagogy with/in the postmodern*. Routledge. doi:10.4324/9780203451311

Lewin, K., Lippitt, R., & White, R. K. (1939). Patterns of aggressive behavior in experimentally created social climates. *The Journal of Social Psychology, 10*(2), 271–299. doi:10.1080/00224545.1939.9713366

Lewis, A., & Diamond, J. (2015). *Despite the best intentions: How racial inequality thrives in good schools*. Oxford University Press.

Likert, R. (1967). *New patterns of management*. McGraw-Hill.

Luke, C., & Gore, J. (1992). *Feminist and critical pedagogy*. Routledge.

Marshall, C. (1997). *Feminist critical policy analysis I: Perspectives from K-12*. Falmer Press.

Marshall, C., & Young, M. (2013). Policy inroads undermining women in education. *International Journal of Leadership in Education*, *16*(2), 205–219. doi:10.1080/13603124.2012.754056

Maslow, A. H. (1970). *Motivation and personality*. Harper & Row.

McGlacken, B., Larkan, F., Mannan, H., & Kodate, N. (2023). Introduction to systems thinking. In F. Larkan, H. Mannan, N. Kodate, & F. Vallières (Eds.), *Systems Thinking for Global Health* (pp. 1–17). Oxford University Press.

McGregor, D. (1957). *The human side of enterprise*. McGraw-Hill.

Monat, J., & Gannon, T. (2015). What is systems thinking: A review of selected literature plus recommendations. *American Journal of Systems Science*, *4*(1), 11–26.

Morgan, G. (2006). *Images of organizations*. Sage.

Nash, A. M., & Peters-Hawkins, A. L. (2020). For us: Towards an intersectional leadership conceptualization by Black women for Black girls. *Journal of Educational Administration and History*. Advance online publication. doi:10.1080/00220620.1785403

Nicolaides, N., & Gaynor, A. K. (1992). The knowledge base informing the teaching administration and organizational theory in UCEA universities: A descriptive and interpretive survey. *Educational Administration Quarterly*, *28*(2), 237–265. doi:10.1177/0013161X92028002005

Perrow, C. (1986). *Complex organizations a critical essay* (3rd ed.). Scott Foresman.

Pfeiffer, J. (1997). *New directions for organization theory: Problems and prospects*. Oxford University Press.

Porter, L. (1964). Attitudes in management: Perceived deficiencies in need fulfillment as a function of job level. *The Journal of Applied Psychology*, *46*(6), 375–384. doi:10.1037/h0047808

Senge, P. (1990). *The Fifth Discipline: The art and practice of the learning organization*. Crown.

Shaked, H., & Schechter, C. (2017). *Systems Thinking for School Leaders: Holistic Leadership for Excellence in Education*. Springer International Publishing. doi:10.1007/978-3-319-53571-5

Shakeshaft, C. (1987). *Women in educational administration*. Sage.

Shakeshaft, C. (2011). Wild patience: The glass ceiling for women in school administration. In *Sage handbook of educational leadership*. Sage. doi:10.4135/9781412980036.n15

Sherman, W., & Wrushen, B. (2009). Intersecting leadership knowledge from the field: Diverse women secondary principals. *Journal of School Leadership*, *19*(2), 172–199. doi:10.1177/105268460901900203

Sherman, W. H. (2005). Preserving the status quo in renegotiating leadership: Women's experiences with a district-based aspiring leaders program. *Educational Administration Quarterly*, *41*(5), 707–740. doi:10.1177/0013161X05279548

Sherman Newcomb, W. (2014). *Continuing to disrupt the status quo? New and young women professors of educational leadership*. Information Age Publishing.

Sherman Newcomb, W., & Niemeyer, A. (2015). African American women principals: Heeding the call to serve as conduits for transforming urban school communities. *International Journal of Qualitative Studies in Education : QSE*, *28*(7), 786–799. doi:10.1080/09518398.2015.1036948

Stacey, R. D., Griffin, D., & Shaw, P. (2000). *Complexity and management: Fad or radical challenge to systems thinking?* Routledge.

Tannenbaum, R., & Schmidt, W. (1973). How to choose a leadership pattern. *Harvard Business Review*, *51*, 162–180.

Watson, T. N. (Ed.). (2020). A Seat at the Table: Examining the impact, ingenuity, and leadership practices of Black women and girls in PK – 20 Contexts. Journal of Educational Administration and History.

Weiner, J., & Burton, L. (2016). The double bind for women: Exploring the gendered nature of turnaround leadership. *Harvard Educational Review*, *86*(3), 339–365. doi:10.17763/1943-5045-86.3.339

Weiner, J., Cyr, D., & Burton, L. (2019). *How "identity-blindness" and other microaggressions shape the experiences of black women in principal preparation programs. Journal of Research on Leadership Education.*

Chapter 13
Reframing Systems to Thrive:
Teacher Shortages in Post-Pandemic Schools

Jennifer Bailey
The University of Texas at Tyler, USA

Gary J. Miller
The University of Texas at Tyler, USA

Forrest Kaiser
The University of Texas at Tyler, USA

ABSTRACT

This chapter explores systems thinking through the lens of open systems within the national teacher crisis in post-pandemic schools. The COVID-19 pandemic exacerbated a long-standing, significant teacher shortage, and there have been several federal, state, and local initiatives to address the challenges educators are facing. Through an exploration of the literature, this chapter provides insight to the historical context of teacher shortages and provides recent measures across the nation that have been introduced to encourage teacher recruitment and retention. Additionally, this chapter presents practitioners—school leaders and teacher leaders—as voices from the field to contextualize experiences of the five federal recommendations in mitigating the teacher crisis, providing insight to the daily challenges in meeting the needs of their schools. As a concluding reflection, the authors propose a potential solution through the open systems lens to build collective capacity in addressing the post-pandemic teacher shortage crisis.

INTRODUCTION

The COVID-19 pandemic provoked crises within all social and economic sectors, casting a harsh light on the vulnerabilities and challenges across our health care system. The ripple effect following the initial pandemic crisis exposed the inequalities ingrained in our nation's schools, including the national teacher shortage. As school leaders navigated new policies and procedures to keep students and employees safe

DOI: 10.4018/978-1-6684-7285-9.ch013

upon returning to in-person learning, they simultaneously struggled with staffing demands between substitute teacher availability and filling teacher vacancies. The national media highlighted various methods across the country that state legislation and local education agencies engaged in to combat these challenges, ranging from increased salaries and stipends for new hires to national guardsmen/women serving as substitute teachers while on duty. During the 2022-23 school year, staffing gaps were projected to be higher, prompting even more decisive action from some states, such as waiving licensure or degree requirements to fill teaching positions. Other decisive actions included a number of rural school districts in Texas switching to a four-day school week in response to shortages. Research indicated a historical decrease in teacher candidates (Garcia & Weiss, 2019; Sutcher et al., 2016). The decrease in teacher candidates may be due to declining enrollment in university-based teacher preparation programs, low-wages, hard-to-fill positions in specialized content areas, and legislation restricting instructional content. However, the residual impact of the pandemic exacerbated the shortages and presented a new reality for schools across the nation. The teacher shortage prompted reflective and responsive action as school leaders entered new territory into not only how they can recruit and retain teachers, but also how to build systems of support for the entire school community with new initiatives designed to mitigate the staffing challenges.

In considering this from a systems-thinking perspective, general systems theory (von Bertalanffy, 1968) suggests that a system is a sum of its parts—self-correcting and self-regulating through dynamic and continuous interactions. Metaphorically, this is likened to semi-permeable membranes that are bombarded with inputs or demands, in which the system selectively absorbs and processes inputs and then returns an output into the environment. By further examining through an open systems perspective, the input is filtered through a social frame of feedback mechanisms that gauge the environment and then subsequently respond to the environment's reaction to the output (Marion & Gonzales, 2014). Therefore, it can be argued that a system disconnected from its environment or not in sync with its environment will fail to thrive.

Through an open systems lens, the state and local responses to the national teacher crisis, like policy changes for waiving licensure requirements to military serving as substitutes, may be aligned to inputs. Accordingly, many school leaders are pivoting in post-pandemic schools, departing from historical hiring practices to fill vacancies with the new input opportunities. However, considering that student performance outcome expectations, or outputs, remain the same, there are considerations in the potential challenges that school leaders may face within the feedback mechanisms to support the needs of their new faculty while meeting the demands of the school environment. Here, the new inputs to mitigate the national teacher shortage prompt an examination into the experiences of school leaders as they reframe their systems to promote a thriving environment.

The aim of this chapter is to provide a comprehensive view of the national teacher shortage, the subsequent responsive efforts, and the implications for school leaders within post-pandemic schools. Here, this chapter is divided into four parts, first situating the reader within the open systems construct of the school system. From there, the authors then frame the inputs of teacher shortage concerns through a synthesis of the literature around historical and pre-pandemic teacher recruitment and retention efforts. Third, the authors explore state legislative, university-based, and local agency measures across the nation designed to support 2021-2023 shortages through newly available input measures. The final section brings forth the voices of current school leaders navigating the reframing of their systems and feedback mechanisms to meet the needs of their schools. Here, the authors situate the school leaders' voices to 1) contextualize localized environmental responses of the national recommendations to mitigate the teacher shortage crisis and 2) present implications and recommendations for improved practice.

OPEN SYSTEMS: THE ENVIRONMENTAL INFLUENCE

Open systems theory gained popularity in the 1960s, during an era in which society began shifting from individualized responsibility to collective accountability. As a result of this shift, the open systems perspective maintains that the organization is responsible for successes and failures, not the individual. Accordingly, Marion & Gonzales (2014) ascertain that "systems do not act in a vacuum; they do not control their own destinies. They are products of their environments" (p. 74). Further, open system theorists believe that the organization has greater control over the environment through collective capacity, as it comprises sub-systems that perform specific tasks that interact with one another. As a result, if a sub-system is damaged by its environment, then there is a domino effect, requiring adjustment to the peer sub-systems. Here, the peer sub-systems push back on the damaged counterpart to regain homeostasis (Marion & Gonzales, 2014).

Open Systems Construct Within Schools

Open system theorists maintain feedback is a crucial component to responding to the environmental demands that influence sub-system regularity. Similarly, Gurr, Drysdale, and Goode (2022) explain open systems within schools as a cycle with continuous feedback that inform each stage: inputs, which lead to transformation that subsequently creates various school and student outcomes, or outputs. Here, the driving force behind feedback within open systems comes forward in the frame of "mapping," which is similarly situated to the concept of schemata. Marion and Gonzales (2014) explain:

Organizations sense their environments with the eyes, ears, nose, touch and taste of their employees. The open system assigns meaning to a given stimulus by relating the stimulus to a preexisting concept map. If it has no preexisting map for defining a given bit of information, that information may be either figuratively (if not actually) invisible to the system, or it may be mapped into an improper category. (p. 76)

During a crisis, multiple inputs are introduced into the environment, representing significant influences from the crisis and intentional efforts to mitigate challenges. Historically, schools, and more specifically, educators, have demonstrated the ability to construct new maps to promote the feedback cycle and lead transformation efforts during times of change. However, it can be argued that there are multiple crises at hand within post-pandemic schools, garnering significant stimuli without preexisting maps. In these unprecedented times, examining the organizational actors and influences involved in the teacher shortage crisis through an open systems lens may help to understand, evaluate and improve the outcomes for schools more strategically and systematically. To this point, we cannot create new maps nor shift practices to reframe systems without first examining historical trends in teacher shortages. The following sections provide a historical overview of the environmental components (input/feedback/output) influencing teacher shortages and subsequent efforts to address the growing crisis among the nation's schools.

REVIEW OF THE LITERATURE: HISTORICAL INPUT OF TEACHER SHORTAGES

Although greatly accelerated during the pandemic, teacher shortages are not new. American schools have come a long way from the small single-teacher schoolhouse where the curriculum was loosely assembled to meet local needs and provide foundational academic skills. Throughout the 19th and early 20th century, state and federal government entities slowly gained interest in providing a universal public education that emphasized a common American culture, provided a more standardized curriculum, and reduced poverty among a growing nation (Kober & Rentner, 2020). Primary authority for public education started with the states who then shared responsibility with local school boards, who would then be accountable for developing, administering, and supervising curriculum and instruction (Toch, 2011).

With significant local control of education standards, disparities in access and quality has led to a call for a more significant federal role in public education. The passage of the Elementary and Secondary Education Act of 1965 (ESEA) prompted a significant shift in federal responsibility that focused on assisting students in poverty through federal funds targeted toward specific programs and services. Signed into law by President Lyndon B. Johnson, the act was part of his "War of Poverty" and set the stage for future, greater federal authority to come (Fritzberg, 2012). In 1979, the national agenda on education once again increased in influence as Congress formed the Department of Education to improve quality, ensure equal access to education, support state and local efforts, and increase accountability (U.S. Department of Education, 2010).

Changing Expectations: Modern School Reform

Efforts toward meaningful school reform have played an essential part in increasing educational equity and improving student outcomes in U.S. schools. A side effect of rigorous accountability policies and sanctions for low performance in U.S. schools has been a decrease in recruitment and retention rates for teachers willing to take on the greater individualized responsibility for the requirements outlined in the policy (Ingersoll et al., 2016). Because of the potential stress tied to meeting accountability targets, teacher attrition often impacts high-poverty schools the most (Carver-Thomas & Darling-Hammond, 2017; Holme et al., 2018).

The modern national school reform movement is marked with significant federal control and increased accountability. Rising international competition following World War II led national leaders to question the quality and organization of public education. Released in 1983 by a commission of 18 members from education, government, and the private sector, the report "A Nation at Risk" outlined a substantial and systematic failure in American schools. The eroding national standing of American schools was considered to be intentionally mediocre, self-sustaining, unorganized, and mismanaged (National Commission on Excellence in Education, 1983). Due to the economic and social concerns delineated in "A Nation at Risk," reforms of this period were reflective of cold war ideologies and focused on employable skills, social stability, meritocracy, and equal educational opportunity (Tozer et al., 2015).

When national values shifted toward racial equality, it encouraged reformers to seek educational improvements and improve access to high-quality programs for all students. A push for national standards followed commission recommendations, as reformers believed that program quality would improve by creating rigorous academic standards for achievement that were tied to progress measured standardized tests (Futrell, 1993; Kessinger, 2007). Gains were made in student performance despite critical reports that stated the increased accountability forced schools to meet minimum standards rather than promote higher-level learning. One example of the gains achieved was in math performance on the National Assessment of Educational Progress (NAEP) which increased significantly during this time (Tirozzi & Uro, 1997).

In 1994, Congress passed the Goals 2000: Educate America Act which outlined a series of voluntary model standards and strategies. Goals 2000 listed eight specific outcomes to be achieved by the start of the 21st century including measurable academic competency in core areas, increased graduation rates, adult literacy, drug-free environments, and advancement of U.S. standing in international education (Schwartz et al., 2000). Realizing that the high academic targets would not be achieved using voluntary measures, President G.W. Bush later championed a system of sanctions and accountability that would shape education in the coming years. Signed into law in 2002, No Child Left Behind (NCLB) represented a significant increase in the federal role in public education (U.S. Department of Education, 2004). Controversial from the start, NCLB passed with strong bipartisan support when concern about the quality of American education was at an apex. The law carried specific performance targets for minority students, special education students, English-language learners, and students in poverty. Tied to Title I funds, states that chose not to comply with the law risked losing money for essential programs. NCLB required schools to make incremental progress toward 100 percent proficiency in reading and math for all students by the end of 2014 (Klein, 2015).

By 2010, it was apparent that the goals set by NCLB would be more challenging to meet than expected. The performance targets set out by NCLB dramatically increased stress levels for teachers of tested subjects. Because of this, the pool of potential qualified teachers in high need areas such as math, reading, science, and special education saw a dramatic decline (Hanushek & Rivkin, 2010). For failing schools already under scrutiny, the impact was more significant as teachers sought positions in higher performing schools (Ryan et al., 2017). As Ingersoll and colleagues (2016) note, sanctions and policies meant to target and support struggling schools often reduced teacher autonomy and increased administrative tasks that can take away from time spent preparing for learning.

Systems to Address Teacher Shortages

To meet the need for highly qualified subject area teachers outlined in "A Nation at Risk," states began exploring options for alternative certification programs outside traditional higher education pathways. The initial intent was to seek individuals with specific content knowledge or experience in scientific or technical fields to teach a new generation of scientists, engineers, and technicians (NCEE, 1983). Alternative certification programs initially started with a handful of states in the early 1980s and expanded to over 40 states in the subsequent 20 years (Feistritzer & Chester, 2003; Zetchner & Schulte, 2001). Alternative certification programs aimed to streamline the requirements for entering the teaching profession by reducing the number of hours and courses needed to serve. In addition to addressing specific content area needs, these programs often focused on recruiting candidates from difficult to staff regions or underserved racial and ethnic backgrounds to increase diversity and quality (Mikulecky et al., 2004).

Non-profit organizations such as Teach for America (TFA) seek to fill high need positions in low-income communities with recent college graduates. Established in 1990, Teach for America has enlisted over 64,000 "corps members" to engage in a two-year teaching commitment (Teach for America, 2020). In addition to salary and benefits from participating schools, corps members are eligible to receive additional funding through AmeriCorps. These funds can be used to pay back student loans or for future educational expenses (Teach for America, 2022). Critics of the program cited concerns over the temporary nature of TFA placements and noted the short five-week training as insufficient to prepare corps members for the classrooms they will fill (Craig, 2022). Attrition in the program is high due to recruits not expected to continue teaching beyond their two-year commitment (Heilig & Jez, 2010).

Other approaches have worked to recruit potential teachers through "Grow Your Own" programs that either engage students before entering college or community members with a stake in the local school system. Student-based programs such as Ready, Set, Teach in Texas or the Teacher Cadet program in North Carolina intend to generate interest in the profession while offering high school credit for participation. The programs include foundational coursework in teaching and learning along with field experiences in local schools (NCFPSC, 2022; TEA, 2019). Programs for adults include variations on alternative certification geared toward current classroom paraprofessionals, teacher substitutes, or other interested community members. These programs partner with local universities or community colleges and offer candidates educational, social, and financial support as they work toward certification (Motamedi et al., 2017). Proponents of these programs have indicated the potential to recruit culturally diverse and multilingual candidates that may find traditional routes to certification more challenging (Connally et al., 2017).

SHIFTING INPUT: RESPONSE TO THE POST-PANDEMIC TEACHER SHORTAGE

COVID-19 has strained a vulnerable K-12 workforce and overwrought schools struggling to find substitute teachers to cover classrooms. Schools have increased pay for teachers and substitutes, asked community members to sub, or returned to remote or hybrid learning to address these shortages (Wright, 2022). Other relief strategies to keep schools open include state education agencies that eased requirements for hiring substitute teachers. With states now having the power to address labor challenges, some have signed legislation allowing secretaries, paraprofessionals, and other support staff to work as substitute teachers. In Oklahoma, the governor signed a directive mandating state agencies to allow employees to serve as substitutes (Wright, 2022). Similarly, New Mexico is asking state workers and its National Guard to join their education workforce by becoming licensed substitute teachers and childcare employees (Wright, 2022).

Well before reports of the novel coronavirus (COVID-19) appeared in the news prompting the World Health Organization (WHO) to declare it a global pandemic on March 11, 2020 (Cucinotta & Vanelli, 2020), a growing teacher shortage crisis was already underway in the United States. Pre-pandemic findings published by Diliberti, Schwartz, and Grant (2021) estimated that approximately 8 percent of public-school teachers leave annually through retirement or attrition. The COVID-19 pandemic worsened the teacher shortage, particularly in historically hard-to-fill teaching positions (Latterman & Steffes, 2017), including math, science, bilingual education, and special education (Carver-Thomas, 2022). Moreover, the pandemic exacerbated the unequaled distribution of highly qualified teachers serving low-income students and high-poverty schools (Carver-Thomas, 2022; U.S. Department of Education, 2022). Additionally, a survey conducted nationally of 4,665 teachers suggests teachers have an increased workload with less time planning due to staffing shortages (Karbowski, 2022).

Though it is hard to directly measure the causes of the teacher labor market shortage in the years before the pandemic, the imbalance between the number of available positions and qualified teachers willing to fill those positions is significant. Part of this may be due in part to insufficient recruitment and retention of teacher practices, the high number of retiring baby boomers, and fewer college students majoring in education (Garcia & Weiss, 2019). Given the risk factors for severe illness, it is little wonder why the teacher shortage has intensified. Considering that many schools struggle to fill vacancies, even a slight decrease in the teacher supply can result in severe disruption (Carver-Thomas, 2022).

Data from the U.S. Department of Education for the 2021-22 school year identified 48 states with shortages of special education teachers, 43 states with shortages of math teachers, and 41 states with shortages of science teachers (U.S. Department of Education, 2022). In Texas, teacher shortages by subject area for the same school year included career and technology education, English as a second language, mathematics, and special education (National Conference of State Legislatures, 2022). Data for the 2022-23 school year showed that the number of subject area shortages in Texas grew from four to seven. The most significant shortages are in career and technical education, computer science, English as a Second Language, language arts, mathematics, science, and special education (National Conference of State Legislatures, 2022). The Texas pre-COVID-19 teacher shortage by subject area data from 2018-19 found no difference compared to the 2021-22 school year with the same four subject areas listed during that span (National Conference of State Legislatures, 2022).

Measures to Increase Input

Increasing teacher pay and easing eligibility requirements for substitute teachers are standard practices that many states have adopted. Mississippi passed a bill increasing teacher salaries by an average of $4,700 over the next several years (Wright, 2022). Legislators in Nebraska used federal funding to give teachers a one-time stipend (Wright, 2022). With all the focus on salary increases, stress is often overlooked as one of the reasons teachers quit the profession (Sutcher et al., 2016; Wright, 2022). Results from a 2020 survey of 1,000 former public-school teachers identified stress as a leading reason for leaving—nearly twice as often as low salaries (Diliberti et al., 2021).

It is reasonable to assume that these excessive teacher attrition rates are not typical and attributable to the stress caused by the COVID-19 pandemic (Sutcher et al., 2016). Amid the increasing teacher shortage, higher than-normal attrition rates matter because of their detrimental effect on student achievement, an impediment to the development and implementation of curricula, and increased expenditures for districts on hiring and training replacements (Sorenson and Ladd, 2019). Considering the federal government has historically maintained organizational influence within the nation's schools to ensure consistency for academic outcomes, its involvement within and contributions toward responsive efforts to mitigate the teacher shortage crisis was inevitable.

Before the COVID-19 pandemic, teachers exiting the profession either through retirement or attrition was approximately 8 percent annually (Goldring et al., 2014). By the beginning of 2020, teacher attrition rates had only increased by a slight margin (Diliberti et al., 2021). However, in a national survey conducted in October 2020, just months after the pandemic began, approximately one-quarter of the respondents indicated that they were likely to leave the profession before the end of the 2020-2021 school year (Diliberti et al., 2021). This shows significant changes in teacher attrition attitudes compared to pre-COVID-19 pandemic attitudes. There is little doubt that the higher rates of teacher burnout and lower morale levels during the 2020-2021 school year foreshadowed higher teacher attrition levels (Diliberti & Kaufman, 2020; Diliberti et al., 2021; Kurtz, 2020).

The National Conference of State Legislatures (NCSL) reported in 2017 that a shortage of teachers had intensified in recent years and disproportionately affected rural communities (Latterman & Steffes, 2017). With more than one in four schools being rural, serving approximately 8.9 million students, there was growing cause for concern, and is a more significant worry for rural school district leaders today. Coupled with declining enrollment in education programs, estimated from 719,081 to 464,250 students between the 2008-2009 and 2013-2014 school years (Podolsky et al., 2016), the projected teacher shortage was expected to reach 112,000 by 2018 (Sutcher et al., 2016). Again, these dire projections were before the global pandemic declaration in 2020.

NAVIGATING EXTERNAL ENVIRONMENTAL INFLUENCES WITH SHIFTING INPUT

In response to the growing national teacher shortage, the United States Department of Education (2022, August 16), released guidance on allowable use for the American Rescue Plan Elementary and Secondary School Emergency Relief Funds (ARP ESSER), in which $122 billion was provided to state and local education agencies in an effort to address needs arising from the pandemic. The guidance document outlines five strategies for states and local education agencies to leverage funding to strategically increase teacher recruitment and retention. However, in applying the lens of open systems theory, there are considerations when inserting the suggested external environmental strategies or "inputs" and the subsequent reactions to the sub-systems. Here, the authors provide a platform for a sample of practitioners—principals, instructional coaches, and teacher leaders—to contextualize their lived experiences of teacher shortages within post-pandemic schools and the resulting influence of the shifting inputs.

Voices From the Field: Collective Case Study Methodology and Connection

The practitioner perspectives presented are derived from a larger, collective case study that aimed to examine school leadership decision-making and response to teacher mental/emotional health challenges exposed from the COVID-19 pandemic over the 2020-2022 school years. The research questions that guided this study included: 1) What pandemic experiences have challenged the mental/emotional health of teachers?; 2) What supports have administrators provided to teachers specific to their mental/emotional health?; 3) Who, other than administrators, has provided guidance to teachers in navigating the mental/emotional health needs for students?; and, 4) How has the pandemic experience informed future crises responses for campus leaders? Situated among school sites in Texas, study participants were selected through purposeful and snowball sampling measures and included eight campus principals and 24 teacher leaders. Data was collected from open-ended survey probes followed by semi-structured interviews with participants. Grounded within the construct of open systems theory, data was analyzed utilizing a phronetic iterative approach and first-level coded through the constant comparative method followed by axial coding until saturation. Three overarching themes, ethic of care—intentionality, reframing communication—responsiveness, and evolving roles—adaptability, provided insight to the school context and framed the environmental and situational conditions influencing educational leadership during extended crises.

Given the teacher and substitute shortages, extended time period of the interruption to schooling, and the lack of resources, it is critical to provide school officials, policy makers, and support agencies with relevant tools for quickly assessing areas of needed improvement with practical steps for principals. As such, this study contributed to the research surrounding education leadership during crises and the conclusions inform organizational practices for future responsive efforts. For the purpose of this chapter, specific discussion related to the teacher and staffing shortages have been categorized within the data strands from this study to provide context to the five federal recommendations previously mentioned. While these perspectives may not be generalized, the sample of voices from the field bring insight into how the evolving measures to increase educator personnel, or inputs, influence and impact schools as open systems.

Input: Preparation Program Investment

The United States Department of Education (2022, August 16) leads the first two suggestions for combatting teacher shortages through investing in a variety of preparation programs, such as: partnering with university educator preparation programs to develop "Grow Your Own" through residency and apprenticeship models; offering stipends, scholarships, course and testing fee coverage for high-need areas; and, supporting alternative measures for para-professional and substitute staff to become teacher trained and certified. As practitioners have embraced these new inputs, they share insight to navigating the realities. One school leader reflected:

Since the pandemic, there has been a challenge with finding certified teachers. The lowering of standards to be a teacher has helped get passionate educators in the classroom sooner. On the flip side, those that don't know much about teaching but need a job have been coming in and not performing.

Sharing a similar sentiment, another school leader noted:

Less stringent hiring guidelines have helped to recruit teachers. However, the challenges lie in the gamble of bringing on a complete novice and the reality that they honestly aren't prepared to have significant impact on drastically increasing student achievement because it is much involved.

Resonating the challenges of leadership to maintain sub-system health, a school leader shared, "The needs of new, non-certified teachers are expansive. When more than half of your staff is new and non-certified, it can be overwhelming to build capacity and meet the needs of every teacher." Here, practitioners situate the impact of the lack of preexisting mapping and the consequential influence on the feedback cycle that ultimately disrupts the sub-systems within their schools. In further reflecting upon the open system lens, it could be argued that peer systems are equally unprepared to navigate pushing back or filling gaps to reach homeostasis, consequently grappling with negative feedback loops that suppress progress.

Input: Recruitment and Hiring Strategies

The third recommendation from the United States Department of Education (2022, August 16) revolves around recruitment and hiring strategies. These strategies include increased compensation, reaching out to retired educators to return to the field, and creating partnerships among other schools and higher education institutions for collaborative opportunities to increase mentorship for new teachers and expand diverse representation. Relatedly, Lindsay, Gnedko-Berry and Wan (2021) conducted a study investigating 61,000 certified teachers in Michigan who were no longer practicing in the field. The results of the study concluded that a leading contributor to returning to teaching would be increased salary, among other financial incentives. As previously discussed, the flexibility in certification requirements, partnered with a wider, broader net of hiring strategies provided through funding incentives, has supported the efforts, as one school leader shared, "We actually had an increase in substitutes as we were able to increase pay." Additionally, substitutes are being utilized creatively to address needs, as a principal explained that he is spending part of his day teaching 3rd grade math due to a lack of qualified teachers. Following his lesson, "several substitutes assist during instruction, and then facilitate small group learning and technology reinforcement," allowing him to return to his leadership duties for the campus. However, another school leader has not had the same positive experience with an increase in staff, disclosing, "my entire leadership team has had to sub because of the inability to find subs. Also, I have had to hire uncertified teachers that have no experience." Similarly, other school leaders shared, "the district has increased incentives to attract applicants but it has not helped with vacancies" and "we focus on certified but accept non-certified. Getting dual credit teachers has been a challenge because of this." Another consideration, one principal pointed out is that "HR is more flexible . . . and our vetting process has relaxed." However, another school leader pointed out these new practices have created difficulties within the balance of sub-systems, stating:

Experienced teachers are vacating the profession and new teachers are entering enthusiastic but ill equipped. Additionally, new teachers do not always have the stamina to educate struggling students, which leads to their exodus and the hunt for new teachers again. Lastly, it is difficult to gain footing when we don't have consistent delivery of instruction or consistent staff to establish a productive culture.

What the practitioners describe creates an image in which the transformation process is heavily burdened with uncertainty in mapping and responsive feedback, negatively impacting and further disrupting sub-systems. This leads to points of reflection toward innovative inputs that are designed to address needs. However, the organizational actors are disconnected from the environment and do not recognize the implications of new stimuli within the residual environment of post-pandemic schools.

Input: Improved Retention Strategies

Scholars indicate that teacher retention and instructional consistency is the leading contributor to student achievement and teacher turnover is disruptive to school progress (Holme et al., 2018; Sorenson and Ladd, 2019). Accordingly, the United States Department of Education (2022, August 16) further recommends improved retention strategies through a) increased support systems, b) quality substitutes for campus stability and teacher absences, c) financial incentives such as retention bonuses, d) scholarships and loan forgiveness, and, e) innovative scheduling to support staff and student needs. Utilizing one of the strategies, one school leader affirmed that her district provided retention bonuses to influence teachers to stay. However, she mused, "the only challenge I see is that districts must continue to provide bonus or increase salaries in order for them to come back." Additionally, there are situations in which the incentives and efforts do not yield results, as one principal shared strategic measures to maintain stability:

I've addressed current and recent shortages by filling in the gap with support staff, paraprofessionals and administrators. I send out an email at 5:30 each morning to outline classroom and duty coverage when staff members are out. Since we do not have a pool of substitutes to pull from, everyone contributes to ensure classrooms are covered and students receive some level of instruction.

Resonating similar challenges, another school leader shared:

We split classes to try and ensure the students are in front of a teacher. We have also utilized our Instructional Facilitator, Academic Interventionist, Counselor, AP and Principal to fill in classrooms. It creates a difficulty in the increased load on other teachers with larger class sizes.

Another principal explained other difficulties related to a lack of regular staff and substitutes, for some new hires have been onboarded as late as October. As a result, a co-curricular class without a teacher "has been directed to report to the gym where coaches take on an additional burden of overseeing and ensuring student safety beyond their normal scope and duties." Similarly, a principal expanded upon the challenges a lack of staff has created, which may negatively impact retention, stating "it has led to an increase in student behavior, a decrease in student achievement, and has caused setbacks regarding operations and systems for success." This point reflects how negative feedback loops present within the mapping process may arguably generate diminished teacher retention and student performance outcomes.

Input: Systemic Support for Mental, Emotional, and Physical Health

The final recommendation from the United States Department of Education (2022, August 16) includes building systems of support to address mental, emotional, and physical health for teachers and staff. Suggestions include increased communication methods, strategic scheduling for planning times, mentorship to prevent burnout, and increased access to human resource support. When the pandemic prompted schools to pivot to remote/virtual instruction for responsive health measures, a new set of concerns emerged once restrictions were lifted and schools were repopulated with in-person learning. One school leader shared:

One major challenge the pandemic presented was instilling the fear of worth. Many teachers, substitutes, and staff in general now face a decision that they perceive if going back into the schools will increase exposure to COVID and other health related illnesses. The pandemic also introduced the possibility of working remotely. This has caused many to leave the field of education all together.

Beyond the concerns related to physical health, emotional and mental strains are also present for educators, which further disrupt the sub-systems and potentially prevent regularity. As mentioned previously, sub-system disruption is a domino effect; when educators are strained, there is resulting influence on students. To this end, a principal noted:

High absenteeism negatively impacts the staff in general. Other teachers are needed to support classes where no substitutes are secured. The splitting of classes creates a constant standard of confusion for scholars who are needing to adjust to their schedules on the fly as well as from a campus perspective where conducting discipline investigations become more cumbersome as another layer is added to locate scholars due to the changed schedules for the day. Naturally, this added stress to teachers and support staff decreases morale and may increase higher absenteeism.

As a result of the challenges, school leaders also shared how they are creating new maps or schemata from the feedback loops and have adapted their practices to be responsive to their educators' needs. Implementing some of the suggested strategies provided by the U.S. Department of Education have been used in response. Accordingly, some school leaders shared that "teachers' SEL is a point of emphasis," and "I am more lenient with staff. I don't write people up for tardiness and have relaxed the dress code." Another school leader identified that "we have tried to be flexible with teachers and show as much appreciation for them as possible." Similarly, a principal resonated with similar points, stating "retention efforts have changed with a focus on mental health and how this role will impact the world." While the sample of practitioners shared intentional efforts toward the holistic well-being of their partner educators and respective sub-systems, the transformation process is still a new territory of mapping and feedback to navigate. Speaking to this point, a principal reflected, "being mindful of mental health during this time and trying to provide grace, while also ensuring accountability for all has been a struggle and balancing act."

LOOKING AHEAD: REFRAMING SYSTEMS TO THRIVE

Crises evoke actions and reactions, as a system navigates, interacts with, and engages with new stimuli to make sense of the situation and envision successful outcomes. As established in the history of teacher shortages, the school system has navigated a number of crises. Through these instances, schools, as open systems, have demonstrated how educators have led the construction of new maps and informative feedback to influence transformation efforts.

Going back to the tenet of collective capacity within systems thinking, it could be argued that the strength of an open system lies within the transformation development. Here, interactions potentially enable or constrain the conditions of the sub-systems working together to produce positive outcomes for the organization. In relation to the school system, there are several considerations to reflect upon. Through the unprecedented time of the pandemic, the teacher shortage has been impacted by evolving physical, emotional, mental, and social needs of the broader school community, including educators, students, and families. Yet, the responsive efforts to mitigate the teacher shortage—input—has been directly influenced by external environmental actors. Accordingly, these external actors are disconnected from the school environments as political entities within state and federal jurisdiction. However, the subsequent responsibility for implementation and transformation of these inputs falls upon the local education agencies and its educators, for the output expectations of student outcomes have not changed. As a result, educators can no longer passively accept inputs, and teacher shortage stakeholders should be the collective responsibility of all environmental actors.

Implications and Recommendations

A potential solution to reframing systems to engage teacher shortage stakeholders may be to create intentional spaces within the transformation process to engage in feedback that creates synergy for more authentic transformation efforts in post-pandemic schools. An unprecedented time calls for unprecedented efforts, and a viable measure may be to have meaningful engagement from all stakeholders involved. Here, representatives from federal agencies, federal legislation, state legislation, local agencies, and educators in the trenches may create intentional sub-systems that support the education sector. A possible focus may be to examine new mapping trends, feedback loops, and exemplars of teacher recruitment and retention success across the nation representing all categorical types of schools. By embracing an open systems perspective, decision-makers can come together to surround, lift and support educators and provide intentional spaces to understand their needs in the post-pandemic transformation process. A successful open system is in sync with its environment; therefore, external environmental actors must first be aware the map educators are deconstructing and constructing to engage in transformation. The education system has demonstrated resilience in navigating the unknown. However, as a collective unit, a reframed system of the school community must lean into the needs of educators to make progress toward improved outcomes in the national teacher crisis. As one school leader concluded:

Post pandemic staffing challenges have created a difficult burden and challenge in striving to create continuous systems that promote effective classroom management and school safety. The only thing consistent about staffing is that it is consistently inconsistent. High absenteeism, unfilled positions, and lack of quality teacher retention results in systems that are mediocre at best. Stresses and burdens are exponentially increased for those of us who strive to effectively and consistently fulfill our duties and expectations as we labor to fill in the gaps by splitting and/or covering class on a regular basis. We are exhausted.

REFERENCES

Carver-Thomas, D. (2022). *Teacher Shortages Take Center Stage*. Learning Policy Institute. Retrieved from: https://learningpolicyinstitute.org/blog/teacher-shortages-take-center-stage

Carver-Thomas, D., & Darling-Hammond, L. (2017). *Teacher turnover: Why it matters and what we can do about it.* Learning Policy Institute. https://learningpolicyinstitute.org/product/teacher-turnover

Connally, K., Garcia, A., Cook, S., & Williams, C. P. (2017). *Teacher talent untapped: Multilingual paraprofessionals speak about the barriers to entering the profession.* New America. https://www.newamerica.org/education-policy/policy-papers/teacher-talent-untapped

Craig, S. (2022, September 11). *If you want to do right by students, reconsider teach for America.* The Georgetown Voice. https://georgetownvoice.com/2022/09/11/reconsider-teach-for-america/

Cucinotta, D., & Vanelli, M. (2020, March 19). WHO Declares COVID-19 a Pandemic. *Acta Biomedica, 91*(1), 157–160. https://pubmed.ncbi.nlm.nih.gov/32191675/ PMID:32191675

Diliberti, M., & Kaufman, J. (2020). *Will This School Year Be Another Casualty of the Pandemic? Key Findings from the American Educator Panels Fall 2020 COVID-19 Surveys.* RAND Corporation. doi:10.7249/RRA168-4

Diliberti, M., Schwartz, H., & Grant, D. (2021). *Stress Topped the reasons Why Public School Teachers Quit, Even Before COVID-19.* RAND Corporation. Retrieved from https://www.rand.org/pubs/research_reports/RRA1121-2.html

Feistritzer, C. E., & Chester, D. T. (2003). *Alternative teacher certification: A state-by-state analysis, 2003.* National Center for Education Information.

Fritzberg, G. (2012). *A brief history of education reform: The federal government's efforts to improve our schools.* Seattle Pacific University Response. https://spu.edu/depts/uc/response/new/2012-spring/features/history-of-reform.asp

Futrell, M. H. (1993). National education reforms. *National Forum, 73*(4), 30.

Garcia, E., & Weiss, E. (2019). *The teacher shortage is real, large and growing, and worse than we thought.* Economic Policy Institute. Retrieved from: https://www.epi.org/publication/the-teacher-shortage-is-real-large-and-growing-and-worse-than-we-thought-the-first-report-in-the-perfect-storm-in-the-teacher-labor-market-series/

Goldring, R., Taie, S., & Riddles, M. (2014). *Teacher Attrition and Mobility: Results From the 2012-13 Teacher Follow-up Survey.* National Center for Education Statistics, U.S. Department of Education. Retrieved from: https://eric.ed.gov/?id=ED546773

Gurr, D., Drysdale, L., & Goode, H. (2022). An open systems model of successful school leadership. *Journal of Educational Administration, 60*(1), 21–40. doi:10.1108/JEA-04-2021-0091

Hanushek, E. A., & Rivkin, S. G. (2010). The quality and distribution of teachers under the no child left behind act. *The Journal of Economic Perspectives, 24*(3), 133–150. doi:10.1257/jep.24.3.133

Heilig, J. V., & Jez, S. J. (2010). *Teach for America: A review of the evidence.* National Education Policy Center. https://nepc.colorado.edu/publication/teach-for-america

Holme, J. J., Jabbar, H., Germain, E., & Dinning, J. (2018). Rethinking teacher turnover: Longitudinal measures of instability in schools. *Educational Researcher, 47*(1), 62–75. doi:10.3102/0013189X17735813

Ingersoll, R., Merrill, L., & May, H. (2016). Do accountability policies push teachers out? *Educational Leadership, 73*(8), 44–49. https://repository.upenn.edu/gse_pubs/551

Karbowski, D. (2022). State of teaching statistics 2022. *Adopt a Classroom.* Retrieved from: https://www.adoptaclassroom.org/2022/04/12/state-of-teaching-statistics-2022/

Kessinger, T. A. (2007). Efforts toward national educational reform: An essentialist political agenda. *Mid-Western Educational Researcher, 20*(2), 16–23.

Kober, N., & Rentner, D. S. (2020). *History and evolution of public education in the US.* Center on Education Policy. https://files.eric.ed.gov/fulltext/ED606970.pdf

Kurtz, H. (2020). Educator morale, school job applicants declining, survey shows. *Education Week.* Retrieved from: https://www.edweek.org/teaching-learning/educator-morale-school-job-applicants-declining-survey-shows/2020/11

Latterman, K., & Steffes, S. (October 2017). Tackling teacher and principal shortages in rural areas. *Legis Brief, 25*(40). Retrieved from: https://www.ncsl.org/research/education/tackling-teacher-and-principal-shortages-in-rural-areas.aspx

Lindsay, J., Gnedko-Berry, N., & Wan, C. (2021). *Michigan teachers who are not teaching: Who are they, and what would motivate them to teach?* (REL 2021–076). U.S. Department of Education, Institute of Education Sciences, National Center for Education Evaluation and Regional Assistance, Regional Educational Laboratory Midwest. Retrieved from https://ies.ed.gov/ncee/edlabs

Marion, R., & Gonzales, L. D. (2014). *Leadership in education: Organizational theory for the practitioner* (2nd ed.). Waveland Press.

Mikulecky, M., Shkodriani, G., & Wilner, A. (2004). *A growing trend to address the teacher shortage. Alternative Certification.* Policy Brief. Education Commission of the States. https://eric.ed.gov/?id=ED484845

National Commission on Excellence in Education. (1983). A nation at risk: The imperative for educational reform. *The Elementary School Journal, 84*(2), 113–130. doi:10.1086/461348

National Conference of State Legislatures. (2022). *Teacher shortage areas by state*. Retrieved from: https://www.ncsl.org/research/education/teacher-shortage-areas-by-state.aspx

NC Foundation for Public School Children. (2022, December 8). *About NC teacher cadet*. https://www.ncfpsc.org/about/teacher-cadet

Podolsky, A., Kini, T., Bishop, J., & Darling-Hammond, L. (2016). *Solving the teacher shortage: How to attract and retain excellent educators*. Learning Policy Institute. doi:10.54300/262.960

Ryan, S. V., von der Embse, N. P., Pendergast, L. L., Saeki, E., Segool, N., & Schwing, S. (2017). Leaving the teaching profession: The role of teacher stress and educational accountability policies on turnover intent. *Teaching and Teacher Education, 66*, 1–11. doi:10.1016/j.tate.2017.03.016

Schwartz, R. B., Robinson, M. A., Kirst, M. W., & Kirp, D. L. (2000). Goals 2000 and the standards movement. *Brookings Papers on Education Policy,* (3), 173-214. https://www.jstor.org/stable/20067222

Sutcher, L., Darling-Hammond, L., & Carver-Thomas, D. (2016). *Coming crisis in teaching? Teacher supply, demand, and shortages in the U.S.* (research brief). Palo Alto, CA: Learning Policy Institute. Retrieved from: https://learningpolicyinstitute.org/product/coming-crisis-teaching

Teach for America. (2020). *Our work*. https://www.teachforamerica.org/what-we-do/our-work

Teach for America. (2022). *Life as a Corps member.* Retrieved December 14, 2022, from https://www.teachforamerica.org/faqs/life-as-a-corps-member#salary-benefits--aid

Texas Education Agency. (2019). *Ready, set, teach!* https://readysetteach.org/index.html

Tirozzi, G. N., & Uro, G. (1997). Education reform in the United States: National policy in support of local efforts for school improvement. *The American Psychologist, 52*(3), 241–249. doi:10.1037/0003-066X.52.3.241

Toch, T. (2011). Who rules? The national drive for education reform has touched off many power struggles, but one has emerged as fundamental. *The Wilson Quarterly, 43*(4), •••. http://archive.wilsonquarterly.com/essays/who-rules

Tozer, S., Senese, G., & Violas, P. (2015). *School and society: Historical and contemporary perspectives* (7th ed.). McGraw-Hill College.

U.S. Department of Education. (2004). *Executive summary of the No Child Left Behind Act of 2001.* https://www2.ed.gov/nclb/overview/intro/execsumm.html

U.S. Department of Education. (2010). *An overview of the U.S. Department of Education.* https://www2.ed.gov/about/overview/focus/what.html

U.S. Department of Education. (2022). *Teacher shortage areas*. Retrieved from: https://tsa.ed.gov/#/home/

U.S. Department of Education. (2022, August 16). *Addressing the teacher shortage with American rescue plan funds*. Retrieved from: https://www2.ed.gov/documents/coronavirus/teacher-shortage.pdf

von Bertalanffy, L. (1968). General systems—A critical review. In W. Buckley (Ed.), *Modern systems research for the behavioral scientist*. Aldine.

Wright, A. (2022). *Teachers 'beaten down' by staff shortages, COVID*. Stateline. Retrieved from: https://stateline.org/2022/01/31/teachers-beaten-down-by-staff-shortages-covid/

Zetchner, K. M., & Schulte, A. K. (2001). What we know and don't know from peer-reviewed research about alternative teacher certification programs. *Journal of Teacher Education, 52*(14), 266–282. doi:10.1177/0022487101052004002

KEY TERMS AND DEFINITIONS

Instructional Coach: The term instructional coach encompasses a school leader or teacher who provides coaching to teachers for curriculum implementation and instructional improvement.

Post-Pandemic Schools: The term post-pandemic schools refers to the generalized time period following the initial 2020 school closures when schools returned to in-person learning and re-established traditional routines.

Practitioners: The term practitioners refers to practicing educators in the field.

School Leaders: The term school leader refers to individuals within a school setting who serve in a leadership capacity for teachers, such as administrators, principals, assistant principals, or supervisors.

Systems Thinking: Systems thinking is a theoretical lens that centers around how a complex system operates and promotes change through the relationships and interactions of its parts.

Teacher Recruitment: The term teacher recruitment encompasses the practices and strategies school officials engage in to obtain new teachers.

Teacher Retention: The term teacher retention encompasses the practices and strategies school officials employ to retain their current teachers.

Teacher Shortage: The teacher shortage refers to research studies utilizing trends in national data indicating teacher-student ratio imbalance, teacher vacancies, and projections for the career field through higher education and certification agency statistics.

Compilation of References

Aagaard, J. (2019). Multitasking as distraction: A conceptual analysis of media multitasking research. *Theory & Psychology, 29*(1), 87–99. doi:10.1177/0959354318815766

Aalioui, L., Gouzi, F., & Tricot, A. (2022). Reducing cognitive load during video lectures in physiology with eye movement modeling and pauses: A randomized controlled study. *Advances in Physiology Education, 46*(2), 288–296. doi:10.1152/advan.00185.2021 PMID:35297692

Abrashera, D., Escribens, M., Sabzalieva, E., Viero do Nascimento, D., & Yerovi, C. (2022). *Resuming or reforming: Tracking the global impact of higher education after two years of disruption.* UNESCO International Institute for Higher Education in Latin America and the Caribbean. https://unesdoc.unesco.org/ark:/48223/pf0000381749

Ackoff, R. L. (1974). *Redesigning the future: A systems approach to societal problems.* John Wiley & Sons.

Aduhene, D. T., & Osei-Assibey, E. (2021). Socio-economic impact of COVID-19 on Ghana's economy: Challenges and prospects. *International Journal of Social Economics, 48*(4), 543–556. doi:10.1108/IJSE-08-2020-0582

Adžić, S., & Al-Mansour, J. (2021). The negative impact of COVID-19 on firms: Insights from Serbia. *Eastern European Economics, 59*(5), 472–486. doi:10.1080/00128775.2021.1953387

Agostinho, S. (2011). The use of a visual learning design representation to support the design process of teaching in higher education. *Australasian Journal of Educational Technology, 27*(6), 961–978. doi:10.14742/ajet.923

Agostinho, S., Lockyer, L., & Bennett, S. (2018). Identifying the characteristics of support Australian university teachers use in their design work: Implications for the learning design field. *Australasian Journal of Educational Technology, 34*(2), 1–15. doi:10.14742/ajet.3776

Aguiar-Castillo, L., Hernandez-Lopez, L., De Saá-Pérez, P., & Perez-Jimenez, R. (2020). Gamification as a motivation strategy for higher education students in tourism face-to-face learning. *Journal of Hospitality, Leisure, Sport and Tourism Education, 27*, 100267. doi:10.1016/j.jhlste.2020.100267

Alahäivälä, T., & Oinas-Kukkonen, H. (2016). Understanding persuasion contexts in health gamification: A systematic analysis of gamified health behavior change support systems literature. *International Journal of Medical Informatics, 96*, 62–70. doi:10.1016/j.ijmedinf.2016.02.006 PMID:26944611

Alfaro, L., Chari, A., Greenland, A. N., & Schott, P. K. (2020). *Aggregate and Firm-level Stock Returns During Pandemics, in Real Time* (National Bureau of Economic Research Working Paper No. 26950). https://www.nber.org/system/files/working_papers/w26950/w26950.pdf

Alon, T., Doepke, M., Olmstead-Rumsey, J., & Tertilt, M. (2020). *The impact of COVID-19 on gender equality* (Issue w26947). National Bureau of economic research.

AlSaad, F. M., & Durugbo, C. M. (2021). Gamification-as-innovation: A review. *International Journal on Management of Innovation & Technology, 18*(05), 2130002. doi:10.1142/S0219877021300020

Alsmairat, A. K. M. (2021). Transformative supply chain drivers during COVID-19: A customer perspective. *Polish Journal of Management Studies, 24*(2), 9–23. doi:10.17512/pjms.2021.24.2.01

Alston, J. A. (2005). Tempered radicals and servant leaders: Black females preserving in the superintendency. *Educational Administration Quarterly, 41*(4), 675–688. doi:10.1177/0013161X04274275

Altinpulluk, H., Kilinc, H., Firat, M., & Yumurtaci, O. (2019). The influence of segmented and complete educational videos on the cognitive load, satisfaction, engagement, and academic achievement levels of learners. *Journal of Computers in Education, 7*(2), 155-182. doi:10.1007/s40692-019-00151-7

Alvarez, J., & Djaouti, D. (2011). An introduction to serious game definitions and concepts. *Serious games & simulation for risks management, 11*(1), 11-15. http//hayka-kultura.com/larsen.html

Ambrosino, P., Fuschillo, S., Papa, A., Di Minno, M. N. D., & Maniscalco, M. (2020). Exergaming as a supportive tool for home-based rehabilitation in the COVID-19 pandemic era. *Games for Health Journal, 9*(5), 311–313. doi:10.1089/g4h.2020.0095 PMID:32876489

AmCham Serbia, Economics Institute & Faculty of Medicine of Belgrade University. (2020). *Studija o mogućim pravcima adekvatnog odgovora zdravstvenog sektora izazovima epidemije COVID-19 u periodu 2020-2021* [A Study on the Possible Directions of an Adequate Response of the Health Sector to the Challenges of the COVID-19 Epidemic in the Period 2020-2021]. https://www.amcham.rs/upload/documents/AmCham%20COVID%20studija%20Final.pdf

American Speech-Language-Hearing Association. (n.d.-a). *Evidence-based practice (EBP).* American Speech-Language-Hearing Association. https://www.asha.org/Research/EBP/

American Speech-Language-Hearing Association. (n.d.-b). *Telepractice.* American Speech-Language-Hearing Association. https://www.asha.org/Practice-Portal/Professional-Issues/Telepractice/

American Speech-Language-Hearing Association. (n.d.-c). *Who are speech-language pathologists, and what do they do?* American Speech-Language-Hearing Association. https://www.asha.org/public/Who-Are-Speech-Language-Pathologists/

Anderson, E. M., Blitz, L. V., & Saastamoinen, M. (2015). Exploring a school-university model for professional development with classroom staff: Teaching trauma-informed approaches. *School Community Journal, 25*(2), 113–134.

Anderson, K. M., Haynes, J. D., Ilesanmi, I., & Conner, N. E. (2022). Teacher professional development on trauma-informed care: Tapping into students' inner emotional worlds. *Journal of Education for Students Placed at Risk, 27*(1), 59–79. doi:10.1080/10824669.2021.1977132

Anderson, T., Rourke, L., Garrison, D. R., & Archer, W. (2001). Assessing teaching presence in a computer conferencing context. *Journal of Asynchronous Learning Networks, 5*(2), 1–17.

Angrist, J. D., & Pischke, J.-S. (2009). *Mostly harmless econometrics: an empiricist's companion.* Princeton University Press. doi:10.1515/9781400829828

Ansmann, M., & Seyfried, M. (2022). Isomorphism and organizational performance: Evidence from quality management in higher education. *Quality Assurance in Education, 30*(1), 135–149. doi:10.1108/QAE-07-2021-0114

Apple, M. (1979). *Ideology and curriculum.* Routledge. doi:10.4324/9780203241219

Aragon, R. (2016). What has an Impact on Grades? Instructor-Made Videos, Communication, and Timing in an Online Statistics Course. *Journal of Humanistic Mathematics, 6*(2), 84–95. doi:10.5642/jhummath.201602.07

Archambault, L. M., & Barnett, J. H. (2010). Revisiting technological pedagogical content knowledge: Exploring the TPACK framework. *Computers & Education, 55*(4), 1656–1662. doi:10.1016/j.compedu.2010.07.009

Argyris, C. (1976). Single-loop and double-loop models in research on decision making. *Administrative Science Quarterly, 21*(3), 363–375. doi:10.2307/2391848

Argyris, C. (1991). Teaching smart people how to learn. *Harvard Business Review, 4*(2). https://hbr.org/1991/05/teaching-smart-people-how-to-learn

Arnold, R. D., & Wade, J. (2015). Notably, the least obvious part of the system, its function or purpose, is often the most crucial determinant of the system's behavior. *Procedia Computer Science, 00*, 2–10.

Arnold, R. D., & Wade, J. P. (2015). A definition of systems thinking: A systems approach. *Procedia Computer Science, 44*, 669–678. doi:10.1016/j.procs.2015.03.050

Artz, B., Johnson, M., Robson, D., & Siemers, S. (2022). Live or lecture capture: Evidence from a classroom random control trial. *International Review of Economics Education, 40*, 100240. doi:10.1016/j.iree.2022.100240

Asensio-Pérez, J. I., Dimitriadis, Y., Pozzi, F., Hernández-Leo, D., Prieto, L. P., Persico, D., & Villagrá-Sobrino, S. L. (2017). Towards teaching as design: Exploring the interplay between full-lifecycle learning design tooling and teacher professional development. *Computers & Education, 114*, 92–116. doi:10.1016/j.compedu.2017.06.011

Ashby, W. R. (1957). *An Introduction to Cybernetics*. Chapman & Hall.

Ashraf, B. N. (2020). Stock markets' reaction to COVID-19: Cases or fatalities*? Research in International Business and Finance, 54*, 101249. doi:10.1016/j.ribaf.2020.101249 PMID:34170989

Astaíza-Martínez, A. F., Mazorco-Salas, J. E., & Castillo-Bohórquez, M. I. (2021). Teacher-researcher training in higher education: *A systems thinking approach. Systemic Practice and Action Research, 34*(2), 187–201. doi:10.100711213-020-09532-x

Balgova, M. (2021). Job Search during a Pandemic Recession: Survey Evidence from the Netherlands". *IZA Discussion Papers, 14180.*

Balingit, M. (2022). Shortages of staff and equipment continue to plague schools, new data shows. *Washington Post.* https://www.washingtonpost.com/education/2022/12/06/teacher-supply-shortage-free-lunch/

Ballé, M. (1996). *Managing with systems thinking: Making dynamics work for you in business decision making*. McGraw Hill.

Banathy, B. H. (1999). Systems thinking in higher education: Learning comes to focus. *Systems Research and Behavioral Science, 16*(2), 133–145. doi:10.1002/(SICI)1099-1743(199903/04)16:2<133::AID-SRES281>3.0.CO;2-T

Barr, M. (2017). Video games can develop graduate skills in higher education students: A randomised trial. *Computers & Education, 113*, 86–97. doi:10.1016/j.compedu.2017.05.016

Barr, M., & Copeland-Stewart, A. (2022). Playing video games during the COVID-19 pandemic and effects on players' well-being. *Games and Culture, 17*(1), 122–139. doi:10.1177/15554120211017036

Bawden, T. (2023, April 12). New Covid variant 'Arcturus' is even more infectious than Kraken and could push UK cases higher. *The Independent.* https://inews.co.uk/news/health/new-arcturus-covid-variant-kraken-uk-cases-higher-2270721

Beer, S. (1972). *The brain of the firm.* John Wiley & Sons.

Beer, S. (1979). *The heart of enterprise: The managerial cybernetics of organization.* John Wiley & Sons.

Beer, S. (2007). *Diagnosing the system for organizations*. John Wiley & Sons.

Belt, E. S., & Lowenthal, P. R. (2021). Video use in online and blended courses: A qualitative synthesis. *Distance Education, 42*(3), 410–440. doi:10.1080/01587919.2021.1954882

Bennett, S., Agostinho, S., & Lockyer, L. (2015). Technology tools to support learning design: Implications derived from an investigation of university teachers' design practices. *Computers & Education, 81*, 211–220. doi:10.1016/j.compedu.2014.10.016

Bennett, S., Agostinho, S., & Lockyer, L. (2017). The process of designing for learning: Understanding university teachers' design work. *Educational Technology Research and Development, 65*(1), 125–145. doi:10.100711423-016-9469-y

Bennett, S., Thomas, L., Agostinho, S., Lockyer, L., Jones, J., & Harper, B. (2011). Understanding the design context for Australian university teachers: Implications for the future of learning design. *Learning, Media and Technology, 36*(2), 151–167. doi:10.1080/17439884.2011.553622

Beraha, I., & Đuričin, S. (2020). The impact of COVID-19 crisis on medium-sized enterprises in Serbia. *Economic Analysis, 53*(1), 14–27. https://www.library.ien.bg.ac.rs/index.php/ea/article/view/1264/1083

Binizi, M., Centorinao, S., Cevik, S., & Gwon, G. (2022). *Here comes the change: The role of global and domestic factors in post-pandemic inflation in Europe.* International Monetary Fund. https://www.imf.org/en/Publications/WP/Issues/2022/12/09/Here-Comes-the-Change-The-Role-of-Global-and-Domestic-Factors-in-Post-Pandemic-Inflation-in-525754#:~:text=Global%20inflation%20has%20surged%20to,to%20become%20an%20entrenched%20phenomenon

Bird, D. (2019). Derivation of a conceptual framework to assess and mitigate identified customer cybersecurity risks by utilizing the public cloud. In *Proceedings Fourth International Congress on Information and Communications Technology* (vol. 2). Springer.

Bird, D. (2020a). Distributed test, track and trace system as a contributor to epidemic containment management. In *Proceedings of the Fourth World Conference on Smart Trends in Systems, Security and Sustainability.* Institute of Electrical and Electronic Engineers. 10.1109/WorldS450073.2020.9210295

Bird, D. (2020b). *Real-time and retrospective analyses of cyber security.* IGI Global.

Bird, D. (2021). A concept of operations to embody the utilization of a distributed test, track and trace system for epidemic containment management. In A. Joshi, A. Nagar, & G. Marin-Raventos (Eds.), *Advances in sustainability, science and technology* (pp. 253–266). Springer.

Blake, R., & Moulton, J. (1968). *Corporate excellence through grid organization development.* Gulf Publishing Company.

Bodroža, D., & Lazić, M. (2021). Economic impact of COVID-19 pandemic on Western Balkan countries. *Economic Analysis, 54*(2), 30–40. doi:10.28934/ea.21.54.2.pp30-40

Boes, S., Sabariego, C., Bickenbach, J., & Stucki, G. (2021). How to capture the individual and societal impacts of syndemics: The lived experience of COVID-19. *BMJ Global Health, 6*(10), 006735. doi:10.1136/bmjgh-2021-006735 PMID:34706878

Bolman, L. G., & Deal, T. E. (2021). *Reframing organizations:Artistry, choice, and leadership.* Jossey-Bass.

Borntrager, C., Caringi, J. C., van den Pol, R., Crosby, L., O'Connell, K., Trautman, A., & McDonald, M. (2012). Secondary traumatic stress in school personnel. *Advances in School Mental Health Promotion, 5*(1), 38–50. doi:10.1080/1754730X.2012.664862

Bourguignon, F., & Spadaro, A. (2006). Microsimulation as a tool for evaluating redistribution policies. *The Journal of Economic Inequality, 4*(1), 77–106. doi:10.100710888-005-9012-6

Britanu, C., & Bejinaru, R. (2021). COVID-19 induced emergent knowledge strategies. *Knowledge and Process Management, 28*(1), 11–17. doi:10.1002/kpm.1656

Bronfenbrenner, U. (1994). Ecological models of human development. In International Encyclopedia of Education (2nd ed., Vol. 3). Elsevier.

Brown, G., & Irby, B. (2003). The synergistic leadership theory: contextualizing multiple realities of female leaders. *Journal of Women in Educational Leadership*, 75. Accessed online: https://digitalcommons.unl.edu/jwel/75

Bruner, J. (1991). The narrative construction of reality. *Critical Inquiry, 18*(1), 1–21. doi:10.1086/448619

Bryant, J., Child, F., Espinosa, J., Hall, S., Kola-Oyeneyin, T., Lim, C., Panier, F., Sarakatsannis, J., Schmautzer, D., Unger, S., & Woord, B. (2022). *How COVID-19 caused a global learning crisis.* McKinsey & Company. https://www.mckinsey.com/industries/education/our-insights/how-covid-19-caused-a-global-learning-crisis

Bryk, A. (2020). *Improvement in action: advancing quality in america's schools.* Harvard University Press.

Bryk, A., Gomez, L., Grunow, A., & LeMahieu, P. (2015). *Learning to improve:how America's schools can get better at getting better.* Harvard University Press.

Buck, J., Loversidge, J., Chipps, E., Gallagher-Frod, L., Genter, L., & Yen, P.-Y. (2018). Top of the license nursing practice: Describing common nursing activities and nurses' experiences that hinder top-of license practice Part 1. *The Journal of Nursing Administration, 48*, 266–271. doi:10.1097/NNA.0000000000000611 PMID:29672373

Buckley, P., & Doyle, E. (2016). Gamification and student motivation. *Interactive Learning Environments, 24*(6), 1162–1175. doi:10.1080/10494820.2014.964263

Buonsenso, D., Roland, D., De Rose, C., Vásquez-Hoyos, P., Ramly, B., Chakakala-Chaziya, J. N., Munro, A., & González-Dambrauskas, S. (2021). Schools closures during the COVID-19 pandemic: A catastrophic global situation. *The Pediatric Infectious Disease Journal, 40*(4), e146–e150. doi:10.1097/INF.0000000000003052

Burgess, M. (2020). *Why the NHS Covid-19 contact tracing app failed.* Wired UK. https://www.wired.co.uk/article/nhs-tracing-app-scrapped-apple-google-uk

Burton, L., Cyr, D., & Weiner, J. (2020). "Unbroken, but bent": Gendered racism in school leadership. *Frontiers in Education, 5*, 52. doi:10.3389/feduc.2020.00052

Butler, S. M. (2020). After COVID-19—Thinking differently about running the health care system. *JAMA Health Forum, 1*(4), 200478–200478. doi:10.1001/jamahealthforum.2020.0478 PMID:36218622

Buus, L., & Georgsen, M. (2018). A learning design methodology for developing short learning programmes in further and continuing education. *Journal of Interactive Media in Education, 1*(8), 1–10. doi:10.5334/jime.469

Byrnes, K. G., Kiely, P. A., Dunne, C. P., McDermott, K. W., & Coffey, J. C. (2021). Communication, collaboration and contagion: "Virtualisation" of anatomy during COVID-19. *Clinical Anatomy (New York, N.Y.), 34*(1), 82–89. doi:10.1002/ca.23649 PMID:32648289

Cabrera, D. (2006). *Systems thinking* [Doctoral dissertation]. Cornell University.

Cabrera, D., Cabrera, L., & Midgley, G. (2023) .The Four Waves of Systems Thinking. In Routledge Handbook of Systems Thinking. London: Routledge.

Cabrera, D., & Cabrera, L. (2015). *Systems thinking made simple: New hope for solving wicked problems in a complex world.* Odyssean Press.

Cabrera, D., & Cabrera, L. (2019). What is systems thinking? In J. M. Spector, B. B. Lockee, & M. D. Childress (Eds.), *Learning, Design, and Technology: An International Compendium of Theory, Research, Practice, and Policy.* SpringerLink. doi:10.1007/978-3-319-17727-4_100-1

Capper, C. A. (2019). *Organizational theory for equity and diversity: Leading integrated, socially just education.* Routledge.

Cardona Mejía, L. M., Pardo del Val, M., & Dasí Coscollar, M. S. À. (2020). The institutional isomorphism in the context of organizational changes in higher education institutions. *International Journal of Research in Education and Science, 6*(1), 61–73. doi:10.46328/ijres.v6i1.639

Carnegie Mellon University. (2020). *Higher Education Was Already Ripe for Disruption. Then, COVID-19 Happened.* Carnegie Mellon University. https://www.cmu.edu/news/stories/archives/2020/september/higher-education-covid-disruption.html

Carras, M. C., Kalbarczyk, A., Wells, K., Banks, J., Kowert, R., Gillespie, C., & Latkin, C. (2018). Connection, meaning, and distraction: A qualitative study of video game play and mental health recovery in veterans treated for mental and/or behavioral health problems. *Social Science & Medicine, 216*, 124–132. doi:10.1016/j.socscimed.2018.08.044 PMID:30257787

Carver-Thomas, D. (2022). *Teacher Shortages Take Center Stage.* Learning Policy Institute. Retrieved from: https://learningpolicyinstitute.org/blog/teacher-shortages-take-center-stage

Carver-Thomas, D., & Darling-Hammond, L. (2017). *Teacher turnover: Why it matters and what we can do about it.* Learning Policy Institute. https://learningpolicyinstitute.org/product/teacher-turnover

Cellan-Jones, R. (2021, July 14). *NHS Covid app: Should it stay or should it go?* BBC News. https://www.bbc.co.uk/news/technology-57836325

Center for Advanced Economic Studies. (2020). *The COVID-Crisis and Serbia's SMEs: Assessment of Impact and Outline of Future Scenarios.* https://ceves.org.rs/wp-content/uploads/2020/11/WB-Covid-19_-Report-final.pdf

Challenges and opportunities in the post COVID-19 world. (2020). World Economic Forum. https://www.weforum.org/reports/post-covid-19-challenges-and-opportunities

Challenges of global governance amid the COVID-19 pandemic . (2020). Council on Foreign Relations, International Institutions and Global Governance Program. https://www.cfr.org/report/challenges-global-governance-amid-covid-19-pandemic

Channel 4 News. (2022, December 29). *Channel 4 News.* Channel 4. https://www.channel4.com/news/

Chatterjee, R., Juvale, D., He, L., & Lundy Evans, L. (2022). Chapter 10 back to design basics: Reflections, challenges, and essentials of a designer's survival kit during a pandemic. In V. Dennen, C. Dickson-Deane, X. Ge, D. Ifenthaler, S. Murthy, & J. C. Richardson (Eds.), Global Perspectives on Educational Innovations for Emergency Situations (pp. 99–108). Association for Educational Communications and Technology. doi:10.1007/978-3-030-99634-5

Chaudhry, I. S. (2022). Viable System Model: A tool for managing sustainable development holistically. *Management & Sustainability: An Arab Review, 1*(1), 50–65. doi:10.1108/MSAR-01-2022-0008

Checkland, P. (1981). *Systems Thinking, Systems Practice.* John Wiley.

Checkland, P. (1981). *Systems thinking, systems practice.* Wiley.

Checkland, P. (1999). *Systems thinking, systems practice: Includes a 30-year retrospective.* John Wiley & Sons.

Checkland, P. (2000). Soft systems methodology: A thirty year retrospective. *Systems Research and Behavioral Science, 17*(S1), S11–S58. doi:10.1002/1099-1743(200011)17:1+<::AID-SRES374>3.0.CO;2-O

Che, L., Du, H., & Chan, K. W. (2020). Unequal pain: A sketch of the impact of the Covid-19 pandemic on migrants' employment in China. *Eurasian Geography and Economics, 61*(4–5), 448–463. doi:10.1080/15387216.2020.1791726

Chen, S. (2020). During the epidemic period, the phenomenon-level game "Animal Crossing" was born. *Beijing Youth Daily.* https://www.chinanews.com.cn/cul/2020/04-07/9149268.shtml

Chen, X., Chen, S., Wang, X., & Huang, Y. (2021). "I was afraid, but now I enjoy being a streamer!" Understanding the Challenges and Prospects of Using Live Streaming for Online Education. *Proceedings of the ACM on Human-Computer Interaction, 4*(CSCW3), (pp. 1-32). ACM. 10.1145/3432936

Cheng, L., Ritzhaupt, A. D., & Antonenko, P. (2018). Effects of the flipped classroom instructional strategy on students' learning outcomes: A meta-analysis. *Educational Technology Research and Development, 67*(4), 793–824. doi:10.100711423-018-9633-7

Chen, S., Yang, J., Yang, W., Wang, C., & Bärnighausen, T. (2020). COVID-19 control in China during mass population movements at New Year. *Lancet, 395*(10226), 764–766. doi:10.1016/S0140-6736(20)30421-9 PMID:32105609

Chernick, H., Copeland, D., & Reschovsky, A. (2020). The fiscal effects of the COVID-19 pandemic on cities: An initial assessment. *National Tax Journal, 73*(3), 669–732. doi:10.17310/ntj.2020.3.04

Children's Law Center. (2021). *A path forward transforming the public behavioral health system for children, youth, and their families in the District of Columbia.* Author.

Childs, J., Grooms, A., & Mozley, M. P. (2022). Hidden in (virtual) plain sight: A charter district's focus on attendance during COVID-19. *Education and Urban Society.* Advance online publication. doi:10.1177/00131245211065414

Cho, I., Kaplanidou, K., & Sato, S. (2021). Gamified wearable fitness tracker for physical activity: A comprehensive literature review. *Sustainability (Basel), 13*(13), 7017. doi:10.3390u13137017

Chung, H. W., Apio, C., Goo, T., Heo, G., Han, K., Kim, T., Kim, H., Yeonghyeon, K., Doeun, L., Lim, J., Lee, S., & Park, T. (2021). Effects of government policies on the spread of COVID-19 worldwide. *Scientific Reports, 11*(1), 20495. doi:10.103841598-021-99368-9 PMID:34650119

Clark, A. J. (2021). *Dealing with pluralism and marginalization processes in multi-organizational collaboration: Augmenting the Viable System Model with team syntegrity.* [Doctoral dissertation, Victoria University of Wellington]. doi:10.26686/wgtn.16960747.v1

Clark, K., & Hoffman, A. (2019). Educating healthcare students: Strategies to teach systems thinking to prepare new healthcare graduates. *Journal of Professional Nursing, 35*(3), 195–200. doi:10.1016/j.profnurs.2018.12.006 PMID:31126396

Clavet, N. J., Tiberti, L., Vladisavljević, M., Rakić, J. Ž., Anić, A., Krstić, G., & Ranđelović, S. (2019). Reduction of child poverty in Serbia: Benefit or employment strategy? *Economics of Transition and Institutional Change, 27*(3), 615–645. doi:10.1111/ecot.12197

Coggi, P. T., & Coggi, C. J. (2022). *COVID-19 after two years: The failure of pandemic governance.* Council of Councils. https://www.cfr.org/councilofcouncils/global-memos/covid-19-after-two-years-failure-pandemic-governance

Cole, D., Pereira, R., & Spray, J. (2021). Gamification and Virtual Reality for Communicating Ecoliteracy and Climate Science: Carbon Transport in the Essequibo River at Iwokrama Guyana. In *Transforming Society and Organizations through Gamification: From the Sustainable Development Goals to Inclusive Workplaces* (pp. 151–180). Springer International Publishing. doi:10.1007/978-3-030-68207-1_9

Cole, H., & Griffiths, M. D. (2007). Social interactions in massively multiplayer online role-playing gamers. *Cyberpsychology & Behavior, 10*(4), 575–583. doi:10.1089/cpb.2007.9988 PMID:17711367

Comas-Quinn, A., de los Arcos, B., & Mardomingo, R. (2012). Virtual learning environments (VLEs) for distance language learning: Shifting tutor roles in a contested space for interaction. *Computer Assisted Language Learning, 25*(2), 129–143. doi:10.1080/09588221.2011.636055

Connally, K., Garcia, A., Cook, S., & Williams, C. P. (2017). *Teacher talent untapped: Multilingual paraprofessionals speak about the barriers to entering the profession.* New America. https://www. newamerica.org/education-policy/policy-papers/teacher-talent-untapped

Constantine. (2020). Behind the explosion of Animal Crossing during the epidemic, is it another escape from the real predicament? *Tencent News.* https://xw.qq.com/cmsid/20200402A07QY100

Cook, J., Threadgold, S., Farrugia, D., & Coffey, J. (2021). Youth, precarious work and the pandemic. *Young, 29*(4), 331–348. doi:10.1177/11033088211018964

Costley, J., Fanguy, M., Lange, C., & Baldwin, M. (2021). The effects of video lecture viewing strategies on cognitive load. *Journal of Computing in Higher Education, 33*(1), 19–38. doi:10.100712528-020-09254-y

Craig, S. (2022, September 11). *If you want to do right by students, reconsider teach for America.* The Georgetown Voice. https://georgetownvoice.com/2022/09/11/reconsider-teach-for-america/

Crawford, J., Butler-Henderson, K., Rudolph, J., Malkawi, B., Glowatz, M., Burton, R., Magni, P. A., & Lam, S. (2020). COVID-19: 20 countries' higher education intra-period digital pedagogy responses. *Journal of Applied Learning & Teaching, 3*(1), 1–20. doi:10.37074/jalt.2020.3.1.7

Creswell, J. W., & Poth, C. N. (2018). *Research design: Qualitative, quantitative, and mixed methods approaches* (4th ed.). Sage Publications.

Csikszentmihalyi, M., & Csikzentmihaly, M. (1990). *Flow: The psychology of optimal experience* (Vol. 1990). Harper & Row.

Cucinotta, D., & Vanelli, M. (2020). WHO declares COVID-19 a pandemic. *Acta Biomedica, 91*(1), 157. doi:10.23750/abm.v91i1.9397 PMID:32191675

Cucinotta, D., & Vanelli, M. (2020, March 19). WHO Declares COVID-19 a Pandemic. *Acta Biomedica, 91*(1), 157–160. https://pubmed.ncbi.nlm.nih.gov/32191675/ PMID:32191675

Cuevas, J., & Dawson, B. L. (2018). A test of two alternative cognitive processing models: Learning styles and dual coding. *Theory and Research in Education, 16*(1), 40–64. doi:10.1177/1477878517731450

Da Silveira, G., Borenstein, D., & Fogliatto, F. S. (2001). Mass customization: Literature review and research directions. *International Journal of Production Economics, 72*(1), 1–13. doi:10.1016/S0925-5273(00)00079-7

Dacin, M. T. (1997). Isomorphism in context: The power and prescription of institutional norms. *Academy of Management Journal, 40*(1), 46–81. doi:10.2307/257020

Davidová, M., Sharma, S., McMeel, D., & Loizides, F. (2022). Co-De|GT: The gamification and tokenisation of more-than-human qualities and values. *Sustainability (Basel), 14*(7), 3787. doi:10.3390u14073787

Davies, A., & Mao, F. (2022, December 29). *Covid in China: US imposes Covid testing for visitors from China.* The BBC. https://www.bbc.co.uk/news/world-asia-china-64111492

Davies, R. (2023, April 17). Spring Covid booster 2023: Who is eligible and can you book them now? *The Evening Standard.* https://www.standard.co.uk/news/health/spring-covid-booster-vaccine-eligibility-booking-b1074507.html

De Savigny, D., & Adam, T. (Eds.). (2009). *Systems Thinking for Health Systems Strengthening.* World Health Organization. https://apps.who.int/iris/bitstream/handle/10665/44204/9789241563895_eng.pdf?sequence=1&isAllowed=y

Deb, P., Furceri, D., Ostry, J. D., Tawk, N., & Yang, N. (2021). *The Effects of Fiscal Measures during COVID-19* (International Monetary Fund Working Paper No. 262). https://www.imf.org/-/media/Files/Publications/WP/2021/English/wpiea2021262-print-pdf.ashx

Del Boca, D., Oggero, N., Profeta, P., & Rossi, M. (2020). Women's and men's work, housework and childcare, before and during COVID-19. *Review of Economics of the Household, 18*(4), 1001–1017. doi:10.100711150-020-09502-1 PMID:32922242

Delahunty, J. (2018). Connecting to learn, learning to connect: Thinking together in asynchronous forum discussion. *Linguistics and Education, 46,* 12–22. doi:10.1016/j.linged.2018.05.003

Deming, W. E. (1982). *Quality, productivity, and competitive position.* Massachusetts Inst Technology.

Deming, W. E. (1988). *Out of the crisis.* Massachusetts Institute of Technology.

Department of Health and Social Care, Office for National Statistics, Government Actuary's Department and Home Office, (UK). (2021). *Direct and Indirect Impacts of COVID-19 on Excess Deaths and Morbidity.* https://assets.publishing.service.gov.uk/government/uploads/system/uploads/attachment_data/file/907616/s0650-direct-indirect-impacts-covid-19-excess-deaths-morbidity-sage-48.pdf

Department of Health and Social Care. (n.d.). *Our governance.* GOV.UK. https://www.gov.uk/government/organisations/department-of-health-and-social-care/about/our-governance

Department of Labor Logo United Statesdepartment of Labor. (n.d.). *1910.134 App A - Fit Testing Procedures (Mandatory). | Occupational Safety and Health Administration.* OSHA. https://www.osha.gov/laws-regs/regulations/standardnumber/1910/1910.134AppA

Desimone, L. M. (2009). Improving impact studies of teachers' professional development: Toward better conceptualizations and measures. *Educational Researcher, 38*(3), 181–199. doi:10.3102/0013189X08331140

Deslatte, A., Hatch, M. E., & Stokan, E. (2020). How can local governments address pandemic inequities? *Public Administration Review, 80*(5), 827–831. doi:10.1111/puar.13257 PMID:32836452

Desson, Z., Lambertz, L., Peters, J. W., Falkenbach, M., & Kauer, L. (2020). Europe's COVID-19 outliers: German, Austrian and Swiss policy responses during the early stages of the 2020 pandemic. *Health Policy and Technology, 9*(4), 405–418. doi:10.1016/j.hlpt.2020.09.003

Deterding, S., Dixon, D., Khaled, R., & Nacke, L. (2011). From game design elements to gamefulness: defining "gamification". *Proceedings of the 15th international academic MindTrek conference: Envisioning future media environments,* (pp. 9-15). ACM. 10.1145/2181037.2181040

Devos, B. (2020, March 20). *Key policy letters signed by the Education Secretary or Deputy Secretary.* U.S. Department of Education. https://www2.ed.gov/policy/gen/guid/secletter/200320.html

Dewi, F. (2014). Edmodo: A social learning platform for blended learning classes in higher education. *Research in Education Technology: Pedagogy and Technology Journal, 11*(2), 1-11. https://www.academia.edu/20300729/EDMODO_ASocial_Learning_Platform_for_Blended_Learning_Class_in_Higher_Education

Dhukaram, A. V., Sgouropoulou, C., Feldman, G., & Amini, A. (2016). Higher education provision using systems thinking approach – case studies. *European Journal of Engineering Education, 43*(1), 3–25. doi:10.1080/03043797.2016.1210569

Dichev, C., & Dicheva, D. (2017). Gamifying education: what is known, what is believed and what remains uncertain: a critical review. *International Journal of Educational Technology in Higher Education, 14*(1), 1–36. doi:10.118641239-017-0042-5

Diliberti, M., & Kaufman, J. (2020). *Will This School Year Be Another Casualty of the Pandemic? Key Findings from the American Educator Panels Fall 2020 COVID-19 Surveys.* RAND Corporation. doi:10.7249/RRA168-4

Diliberti, M., Schwartz, H., & Grant, D. (2021). *Stress Topped the reasons Why Public School Teachers Quit, Even Before COVID-19.* RAND Corporation. Retrieved from https://www.rand.org/pubs/research_reports/RRA1121-2.html

DiMaggio, P. J., & Powell, W. W. (1983). The iron cage revisited: Institutional isomorphism and collective rationality in organizational fields. *American Sociological Review, 48*(2), 147–160.

DiMaggio, P. J., & Powell, W. W. (1983). The iron cage revisited: Institutional isomorphism and collective rationality in organizational fields. *American Sociological Review, 48*(2), 147–160. doi:10.2307/2095101

Donthu, N., & Gustafsson, A. (2020). Effects of COVID-19 on business and research. *Journal of Business Research, 117*, 284–289. doi:10.1016/j.jbusres.2020.06.008 PMID:32536736

Dorado, J. S., Martinez, M., McArthur, L. E., & Leibovitz, T. (2016). Healthy Environments and Response to Trauma in Schools (HEARTS): A whole-school, multi-level, prevention and intervention program for creating trauma-informed, safe and supportive schools. *School Mental Health, 8*(1), 163–176. doi:10.100712310-016-9177-0

Downey, A. (2020, September 22). *Total cost of NHS contact-tracing app set to top £35 million.* DigitalHealth. https://www.digitalhealth.net/2020/09/total-cost-of-nhs-contact-tracing-app-set-to-top-35-million/

Downey, A. (2021, June 2). *Developer of NHS Covid-19 app awarded £10m contract extension.* DigitalHealth. https://www.digitalhealth.net/2021/06/developer-of-nhs-covid-19-app-awarded-10m-contract-extension/

Downs, L. R. (2020a). *New Regulations Review #2: Credit Hour, Title IV Eligibility, & What Exactly is a Week?* https://wcet.wiche.edu/frontiers/2020/04/06/new-regulations-review-2/

Downs, L. R. (2020b). *Surprise! Newly Released Final Regulations.* https://wcet.wiche.edu/frontiers/2020/08/26/newly-released-final-regulations/

Duan, Q. Q. (2020). The fitness ring adventure has become "hard currency" during the epidemic. *Sina Tech.* https://tech.sina.com.cn/it/2020-02-27/doc-iimxxstf4649888.shtml

Dunnion, J., & O'Donovan, B. (2014). Systems thinking and higher education: The Vanguard method. *Systemic Practice and Action Research, 27*(1), 23–37. doi:10.100711213-012-9258-4

Đurović, G. (2020). *Analysis of Economic and Social Consequences of the COVID-19 Pandemic in Western Balkans Administrations.* ReSPA. https://www.respaweb.eu/download/doc/Analysis+of+economic+and+social+consequences+of+the+Covid-19+pandemic+in+Western+Balkans+administrations.pdf/e97586d30e7b07fa0dd3f00ebf84dcc8.pdf

Dwiyogo, W. D. 2018. Developing a blended learning-based method for problem-solving in capability learning. *Turkish Online Journal of Educational Technology-TOJET, 17*(1),51-61. https://files.eric.ed.gov/fulltext/EJ1165753.pdf

Dzemydienė, D., & Tankelevičienė, L. (2010). Multi-layered architecture of the adaptable architectural distance learning system. *Technological and Economic Development of Economy, 15*(2), 229–244. doi:10.3846/1392-8619.2009.15.229-244

Edwards, S. (2013). Forum focus-All the lonely people. *Nursing Older People, 25*(10), 10. doi:10.7748/nop2013.12.25.10.10.s12

Egashira, M., Son, D., & Ema, A. (2022). Serious Game for Change in Behavioral Intention Toward Lifestyle-Related Diseases: Experimental Study with Structural Equation Modeling Using the Theory of Planned Behavior. *JMIR Serious Games, 10*(1), e28982. doi:10.2196/28982 PMID:35188465

Engeström, Y. (1999). Activity theory and individual and social transformation. In Y. Engeström, R. Miettinen, & R. L. Punamaki (Eds.), *Perspectives on activity theory* (pp. 19–38). Cambridge University Press. doi:10.1017/CBO9780511812774.003

Engeström, Y. (2001). Expansive learning at work: Toward an activity theoretical reconceptualization. *Journal of Education and Work, 14*(1), 133–156. doi:10.1080/13639080020028747

Environment. (2022). In *Oxford Lexico*. Lexico.com.

Enz, S., Hall, A. C. G., & Williams, K. K. (2021). The Myth of Multitasking and What It Means for Future Pharmacists. *American Journal of Pharmaceutical Education, 85*(10), 8267–1020. doi:10.5688/ajpe8267 PMID:34965913

Eri, T. (2013). The best way to conduct intervention research: Methodological considerations. *Quality & Quantity, 47*(5), 2459–2472. doi:10.100711135-012-9664-9

Espinosa, A., & Walker, J. (2013). Complexity management in practice: A Viable System Model intervention in an Irish eco-community. *European Journal of Operational Research, 225*(1), 118–129. doi:10.1016/j.ejor.2012.09.015

Eurostat. (2020). *Flash estimates of income inequalities and poverty indicators for 2019*. https://ec.europa.eu/eurostat/documents/7894008/8256843/Flash-estimate-of-income-inequalities-and-poverty-indicators-experimental-results-2019.pdf

Eurostat. (2021). *Nowcasting income indicators during the pandemic (FE 2020) Methodological developments*. https://ec.europa.eu/eurostat/documents/7894008/8256843/Methodological-note-2020.pdf/9a70fb55-ceb7-d25a-1b31-ab0c030095d2?t=1625228771763

Evans, M. G. (1970). The effects of supervisory behavior on the path-goal relationship. *Organizational Behavior and Human Performance, 5*(3), 277–298. doi:10.1016/0030-5073(70)90021-8

Fadda, M., Galimberti, E., Fiordelli, M., Romanò, L., Zanetti, A., & Schulz, P. J. (2017). Effectiveness of a smartphone app to increase parents' knowledge and empowerment in the MMR vaccination decision: A randomized controlled trial. *Human Vaccines & Immunotherapeutics, 13*(11), 2512–2521. doi:10.1080/21645515.2017.1360456 PMID:29125783

Farooq, M. U., Al Asmari, A., & Javid, C. Z. (2012). A study of online English language teacher education programmes in distance education context in Pakistan. *English Language Teaching, 5*(11), 91–103. doi:10.5539/elt.v5n11p91

Farré, L., Fawaz, Y., González, L., & Graves, J. (2020). *How the COVID-19 lockdown affected gender inequality in paid and unpaid work in Spain.*

Feistritzer, C. E., & Chester, D. T. (2003). *Alternative teacher certification: A state-by-state analysis, 2003*. National Center for Education Information.

Ferris, N. (2023, April 16). Why the world is unprepared if new Covid variant Arcturus proves deadly – in graphs. *The Independent.* https://www.independent.co.uk/news/health/arcturus-variant-nhs-covid-vaccine-b2321575.html

Feygin, A., Nolan, L., Hickling, A., & Friedman, L. (2020). *Evidence for networked improvement communities: A systematic review of the literature.* American Institute for Research. https://www.air.org/sites/default/files/NIC-Systematic-Review-Report-123019-Jan-2020.pdf

Fielder, F. E. (1967). *A theory of leadership effectiveness*. McGraw-Hill.

Figari, F., Salvatori, A., & Sutherland, H. (2011). Economic downturn and stress testing European welfare systems. *Research in Labor Economics*, *32*, 257–286. doi:10.1108/S0147-9121(2011)0000032012

Fiscal Council of the Republic of Serbia. (2020). *Ocena antikriznog programa ekonomskih mera* [Evaluation of the Anti-Crisis Program of Economic Measures]. https://www.fiskalnisavet.rs/doc/ocene-i-misljenja/2020/FS-Ocena_anti-kriznog_programa_ekonomskih_mera.pdf

Fiscal Council of the Republic of Serbia. (2022a). *Antikrizne budžetske mere tokom pandemije COVID-19: troškovi, rezultati i pouke* [Anti-crisis Budgetary Measures During the COVID-19 Pandemic: Costs, Results and Lessons Learned]. https://www.fiskalnisavet.rs/doc/analize-stavovi-predlozi/2022/FS_Antikrizne_budzetske_mere_tokom_pandemije.pdf

Fiscal Council of the Republic of Serbia. (2022b). *Predlog mera socijalne i poreske politike za smanjenje nejednakosti i rizika od siromaštva u Republici Srbiji* [Proposed Social and Tax Policy Measures for Reducing Inequality and Poverty in the Republic of Serbia]. https://www.fiskalnisavet.rs/doc/analize-stavovi-predlozi/2022/FS_Predlog%20mera%20socijalne%20i%20poreske%20politike.pdf

Fitch, D., Parker-Barua, L., & Watt, J. W. (2014). Envisioning public child welfare agencies as learning organizations: Applying Beer's Viable System Model to Title IV-E program evaluation. *Journal of Public Child Welfare*, *8*(2), 119–142. doi:10.1080/15548732.2013.879089

Fleming, T. M., Bavin, L., Stasiak, K., Hermansson-Webb, E., Merry, S. N., Cheek, C., Lucassen, M., Lau, H. M., Pollmuller, B., & Hetrick, S. (2017). Serious games and gamification for mental health: Current status and promising directions. *Frontiers in Psychiatry*, *7*, 215. doi:10.3389/fpsyt.2016.00215 PMID:28119636

Flood, R. L., & Romm, N. R. A. (Eds.). (1997). *Critical systems thinking: Current research and practice*. Plenum Press.

Fontana, M. T. (2020). Gamification of ChemDraw during the COVID-19 pandemic: Investigating how a serious, educational-game tournament (molecule madness) impacts student wellness and organic chemistry skills while distance learning. *Journal of Chemical Education*, *97*(9), 3358–3368. doi:10.1021/acs.jchemed.0c00722

Forrester, J. W. (1961). *Industrial dynamics*. MIT Press.

Forsythe, E., Kahn, L. B., Lange, F., & Wiczer, D. (2020). Labor demand in the time of COVID-19: Evidence from vacancy postings and UI claims. *Journal of Public Economics*, *189*, 104238. doi:10.1016/j.jpubeco.2020.104238 PMID:32834178

Fouda, A., Mahmoudi, N., Moy, N., & Paolucci, F. (2020). The COVID-19 pandemic in Greece, Iceland, New Zealand, and Singapore: Health policies and lessons learned. *Health Policy and Technology, 9*(4), 510–524. doi:10.1016/j.hlpt.2020.08.015

Fox, D. (2009). An Overarching Framework for Sustainability. *Built Environment*, *35*(3), 302–307. doi:10.2148/benv.35.3.302

Freeman, D. L., & Cameron, L. (2008). Research methodology on language development from a complex systems perspective. *Modern Language Journal*, *92*(2), 200–213. doi:10.1111/j.1540-4781.2008.00714.x

Frick, T. W. (2020). Education systems and technology in 1990, 2020, and beyond. *TechTrends, 64*(5), 693–703. doi:10.100711528-020-00527-y PMID:32838405

Friedman, B. D., & Neuman Allen, K. (2014). Systems Theory. In J. R. Brandell (Ed.), *Essentials of clinical social work* (1st ed., pp. 3–20). SAGE. doi:10.4135/9781483398266.n2

Fritzberg, G. (2012). *A brief history of education reform: The federal government's efforts to improve our schools.* Seattle Pacific University Response. https://spu.edu/depts/uc/response/new/2012-spring/features/history-of-reform.asp

Futrell, M. H. (1993). National education reforms. *National Forum, 73*(4), 30.

Gajić, M. (2020). *The Impact of the COVID-19 on the Serbian Economy in 2020.* Friedrich Naumann Foundation for Freedom. https://shop.freiheit.org/download/P2@1040/398508/Final____LIFE_WITH_THE_Covid19_Serbia.pdf

Galikyan, I., & Admiraal, W. (2019). Students' engagement in asynchronous online discussion: The relationship between cognitive presence, learner prominence, and academic performance. *The Internet and Higher Education, 43*, 100692. doi:10.1016/j.iheduc.2019.100692

Gallup. (2014). *State of American Schools.* Retrieved from http://www.gallup.com/ services/178709/state-america-schools-report.aspx

Garcia, E., & Weiss, E. (2019). *The teacher shortage is real, large and growing, and worse than we thought.* Economic Policy Institute. Retrieved from: https://www.epi.org/publication/the-teacher-shortage-is-real-large-and-growing-and-worse-than-we-thought-the-first-report-in-the-perfect-storm-in-the-teacher-labor-market-series/

Garrison, D. R. (2016). E-Learning in the 21st Century: A Community of Inquiry Framework for Research and Practice (3rd ed.). Routledge.

Garrison, D. R. (2011). A Framework for Research and Practice (2nd ed.). Routledge.

Garrison, D. R., Anderson, T., & Archer, W. (1999). Critical Inquiry in a Text-Based Environment: Computer Conferencing in Higher Education. *The Internet and Higher Education, 2*(2), 87–105. doi:10.1016/S1096-7516(00)00016-6

GarrisonD. R.Cleveland-InnesM.VaughanN. (2019). *CoI Publications.* https://coi.athabascau.ca/publications/

Gates, B. (2022). *How to prevent the next pandemic.* Allen Lane.

Gates, E. F. (2016). Making sense of the emerging conversation in evaluation about systems thinking and complexity science. *Evaluation and Program Planning, 59*, 62–73. doi:10.1016/j.evalprogplan.2016.08.004 PMID:27591941

Georgia Department of Public Health. (n.d.). *Function and Form of the Command Center: Advanced Healthcare Emergency Management Course.* Department of Public Health.

Getzels, J., & Guba, E. (1955). Role conflict and personality. *Journal of Personality, 24*(1), 74–85. doi:10.1111/j.1467-6494.1955.tb01175.x PMID:13296016

Gewertz, C. (2020). *The Coronavirus Spring: The Historic Closing of U.S. Schools.* https://www.edweek.org/leadership/the-coronavirus-spring-the-historic-closing-of-u-s-schools-a-timeline/2020/07

Ghiga, I., Richardson, S., Álvarez, A. M. R., Kato, M., Naidoo, D., Otsu, S., Nguyen, P. T., Nguyen, P. N., & Nguyen, T. (2021). PIPDeploy: Development and implementation of a gamified table top simulation exercise to strengthen national pandemic vaccine preparedness and readiness. *Vaccine, 39*(2), 364–371. doi:10.1016/j.vaccine.2020.11.047 PMID:33293160

Gierdowski, D. C. (2019). *2019 Study of Undergraduate Students and Information Technology*. https://library.educause.edu/resources/2019/10/2019-study-of-undergraduate-students-and-information-technology

Gilligan, C. (1982). *In a different voice*. Polity Press.

Goldring, R., Taie, S., & Riddles, M. (2014). *Teacher Attrition and Mobility: Results From the 2012-13 Teacher Follow-up Survey*. National Center for Education Statistics, U.S. Department of Education. Retrieved from: https://eric.ed.gov/?id=ED546773

Gopalan, C., & Klann, M. C. (2017). The effect of flipped teaching combined with modified team-based learning on student performance in physiology. *Advances in Physiology Education*, *41*(3), 363–367. doi:10.1152/advan.00179.2016 PMID:28679573

Government Operational Research Service. (2012). *Introduction to systems thinking*. Government Office for Science.

Greenberg, M. T., Brown, J. L., & Abenavoli, R. M. (2016). Teacher stress and health effects on teachers, students, and schools. Edna Bennett Pierce Prevention Research Center, Pennsylvania State University.

Green, J. K., Burrow, M. S., & Carvalho, L. (2020). Designing for transition: Supporting teachers and students cope with emergency remote education. *Postdigital Science and Education*, *2*(3), 906–922. doi:10.100742438-020-00185-6

Greer, S. L., Jarman, H., Falkenbach, M., Massard de Fonseca, E., Raj, M., & King, E. J. (2021). Social policy as an integral component of pandemic response: Learning from COVID-19 in Brazil, Germany, India, and the United States. *Global Public Health: An International Journal for Research, Policy and Practice*, *16*(8-9), 1209–1222. doi:10.1080/17441692.2021.1916831 PMID:33876715

Gregory, A. (2023, April 14). Mapped: Worst areas in UK for Covid cases amid fears over new Arcturus variant. *The Independent*. https://www.independent.co.uk/news/health/covid-cases-uk-map-arcturus-b2323268.html

Grogan, M. (1996). *Voices of women aspiring to the superintendency*. State University Press of New York.

Grogan, M. (2000). Laying the groundwork for reconception of the superintendency from feminist postmodern perspectives. *Educational Administration Quarterly*, *36*(1), 117–142. doi:10.1177/0013161X00361005

Grogan, M., & Shakeshaft, C. (2011). *Women and educational leadership*. Jossey-Bass.

Grogan, M., & Smith, F. (1998). A feminist perspective of women superintendents' approaches to moral dilemmas. *Journal for a Just and Caring Education*, *4*(2), 176–192.

Guggenheim, D., Kimball, B., Chilcott, L., Strickland, B., Canada, G., Rhee, M., Weingarten, R., Roland, E., Richman, B., Finton, G., Cassidy, J., Roberts, K., Beck, C., & Legend, J. (2011). *Waiting for "Superman*. Paramount Home Entertainment.

Gurr, D., Drysdale, L., & Goode, H. (2022). An open systems model of successful school leadership. *Journal of Educational Administration*, *60*(1), 21–40. doi:10.1108/JEA-04-2021-0091

Gutmore, D., & Galloway, A. (2021). Case study of a higher education's department's responses to the Covid 19 pandemic implications for policy and practice. *Journal of Education & Social Policy*, *8*(3), 60–66. doi:10.30845/jesp.v8n3p8

H., S., & Figari, F. (2013). EUROMOD: the European Union tax-benefit microsimulation model. *International Journal of Microsimulation*, *6*(1), 4–26.

Haase, M. (2023, April 15). *What Experts Want You to Know About Arcturus, the New COVID Variant*. Prevention. https://www.prevention.com/health/health-conditions/a43592554/arcturus-xbb16-new-covid-variant-what-to-know/

Haileamlak, A. (2021). The impact of COVID-19 on health and health systems. *Ethiopian Journal of Health, 31*(6), 1073–1074. doi:10.4314/ejhs.v31i6.1 PMID:35392335

Haim, D., Ravanilla, N., & Sexton, R. (2021). Sustained government engagement improves subsequent pandemic risk reporting in conflict zones. *The American Political Science Review, 115*(2), 717–724. doi:10.1017/S0003055420001148

Hall, G., Laddu, D. R., Phillips, S. A., Lavie, C. J., & Arena, R. (2021). A tale of two pandemics: How will COVID-19 and global trends in physical inactivity and sedentary behavior affect one another? *Progress in cardiovascular diseases, 64*, 108. 10. https://doi.org/ doi:1016/j.pcad.2020.04.005

Halpin, A. W., & Winder, B. J. (1957). A factorial study of leader behavior in descriptions. In R. M. Stodgill & A. E. Coons (Eds.), *Leader behavior: Its description and measurement* (pp. 39–51). The Ohio State University Bureau of Business Research.

Hamari, J., Koivisto, J., & Sarsa, H. (2014). Does gamification work? A literature review of empirical studies on gamification. *2014 47th Hawaii international conference on system sciences*, 3025-3034. 10.1109/HICSS.2014.377

Hansen, B., Sabia, J., & Schaller, J. (2022). *In-person schooling and youth suicide: Evidence from school calendars and pandemic school closures.* National Bureau of Economic Research. Accessed online: https://www.nber.org/papers/w30795

Hansen, C., Steinmetz, H., & Block, J. (2022). How to conduct a meta-analysis in eight steps: A practical guide. *Management Review Quarterly, 72*(1), 1–19. doi:10.100711301-021-00247-4

Hanushek, E. A., & Rivkin, S. G. (2010). The quality and distribution of teachers under the no child left behind act. *The Journal of Economic Perspectives, 24*(3), 133–150. doi:10.1257/jep.24.3.133

Han, X., Li, J., Chen, Y., Li, Y., Xu, Y., Ying, B., & Shang, H. (2022). SARS-CoV-2 nucleic acid testing is China's key pillar of COVID-19 containment. *Lancet, 399*(10336), 1690–1691. doi:10.1016/S0140-6736(22)00577-3 PMID:35397850

Hart, P., Stern, E., & Sundelius, B. (Eds.). (n.d.). Beyond Groupthink: Political group dynamics and foreign policy-making. Michigan. doi:10.3998/mpub.11178

Hart, C. W. L. (1995). Mass customization: Conceptual underpinnings, opportunities and limits. *International Journal of Service Industry Management, 6*(2), 36–45. doi:10.1108/09564239510084932

Hatcher, S. S., Bride, B. E., Oh, H., King, D. M., & Catrett, J. J. F. (2011). An assessment of secondary traumatic stress in juvenile justice education workers. *Journal of Correctional Health Care, 17*(3), 208–217. doi:10.1177/1078345811401509 PMID:21571748

Heggart, K. (2021). Formulated professional identity of learning designers and the role of open education in maintaining that identity. In A. Marcus-Quinn & T. Hourigan (Eds.), *Handbook for online learning contexts: digital, mobile and open* (pp. 21–34). Springer Nature. doi:10.1007/978-3-030-67349-9_3

Heggart, K., & Dickson-Deane, C. (2021). Microcredentials. *Pacific Journal of Technology Enhanced Learning, 3*(1), 13–14. doi:10.24135/pjtel.v3i1.87

Heilig, J. V., & Jez, S. J. (2010). *Teach for America: A review of the evidence.* National Education Policy Center. https://nepc.colorado.edu/publication/teach-for-america

Hemphill, H., Hemphill, L., & Runquist, R. (2019). Improving completion in online education systems: An application of systems thinking. In M. J. Spector, B. B. Lockee, & M. D. Childress (Eds.), Learning, Design, and Technology (pp. 1–36). Springer International Publishing. doi:10.1007/978-3-319-17727-4_104-1 doi:10.1007/978-3-319-17727-4_104-1

Hemphill, J. K., & Coons, A. E. (1957). Development of the leader behavior description questionnaire. In R.M. STodgill & A.E. Coons (Eds.), Leader behavior: its description and measurement (pp. 6-38). Columbus, OH: The Ohio State University Bureau of Business Research.

Hensvik, L., Le Barbanchon, T., & Rathelot, R. (2021). Job search during the COVID-19 crisis. *Journal of Public Economics, 194*, 104349. doi:10.1016/j.jpubeco.2020.104349

Hersey, P., & Blanchard, K. (1969). Life cycle theory of leadership. *Training and Development Journal, 23*(5), 26–34.

Higginson, S., Milovanovic, K., Gillespie, J., Matthews, A., Williams, C., Wall, L., Moy, N., Hinwood, M., Melia, A., & Paolucci, F. (2020). COVID-19: The need for an Australian economic pandemic response plan. *Health Policy and Technology, 9*(4), 488–502. doi:10.1016/j.hlpt.2020.08.017

Hill Collins, P. (1990). *Black feminist thought: Knowledge, consciousness and the politics of empowerment*. Routledge.

Hitch, C. (1953). Sub-optimization in operations problems. *Journal of the Operations Research Society of America, 1*(3), 87–99. doi:10.1287/opre.1.3.87

Hodges, C. B., Moore, S., Lockee, B. B., Trust, T., & Bond, M. A. (2020). *The difference between emergency remote teaching and online learning*. EDUCAUSE Review. https://er.educause.edu/articles/2020/3/the-difference-between-emergency-remote-teaching-and-online-learning

Hodges, C., Moore, S., Lockee, B., Trust, T., & Bond, A. (2020). *The difference between emergency remote teaching and online learning*. Educause Review.

Hoekman, L. M., Smits, M. M. V., & Koolman, X. (2020). The Dutch COVID-19 approach: Regional differences in a small country. *Health Policy and Technology, 9*(4), 613–622. doi:10.1016/j.hlpt.2020.08.008

Hoeskma, J. (2023, January 6). *Palantir gets £11.5m six-month NHS contract extension*. DigitalHealth. https://www.digitalhealth.net/2023/01/palantir-gets-11-5m-six-month-nhs-contract-extension/

Hogan, K., & Weathers, K. C. (2003). Psychological and ecological perspectives on the development of systems thinking. In *Understanding Urban Ecosystems* (pp. 233–260). Springer. doi:10.1007/0-387-22615-X_15

Holgersen, H., Jia, Z., & Svenkerud, S. (2020). *Labor demand during the covid-19 crisis in norway: Evidence from vacancy posting data*.

Holme, J. J., Jabbar, H., Germain, E., & Dinning, J. (2018). Rethinking teacher turnover: Longitudinal measures of instability in schools. *Educational Researcher, 47*(1), 62–75. doi:10.3102/0013189X17735813

Hope, D. L., Grant, G. D., Rogers, G. D., & King, M. A. (2021). PROGRAMME DESCRIPTION: Integration of an extended, immersive, gamified pharmacy simulation as a capstone event. *Pharmacy Education, 21*, 656–669. doi:10.46542/pe.2021.211.656669

Horton, R. (2020). Systems-based approaches to speech-language pathology service delivery for school age children. In D. Scott (Ed.), *Cases on Communication Disorders in Culturally Diverse Populations* (pp. 113–136). IGI Global. doi:10.4018/978-1-7998-2261-5.ch006

Huang, R. H., Liu, D. J., Tlili, A., Yang, J. F., & Wang, H. H. (2020). Handbook on facilitating flexible learning during educational disruption: The Chinese experience in maintaining undisrupted learning in COVID-19 outbreak. *Beijing: Smart Learning Institute of Beijing Normal University, 46*. https://creativecommons.org/licenses/by-sa/3.0/igo/

Huang, A., Makridis, C., Baker, M., Medeiros, M., & Guo, Z. (2020). Understanding the impact of COVID-19 intervention policies on the hospitality labor market. *International Journal of Hospitality Management, 91*, 102660. doi:10.1016/j.ijhm.2020.102660 PMID:32904433

Huang, C., Wang, Y., Li, X., Ren, L., Zhao, J., Hu, Y., Zhang, L., Fan, G., Xu, J., Gu, X., Cheng, Z., Yu, T., Xia, J., Wei, Y., Wu, W., Xie, X., Yin, W., Li, H., Liu, M., ... Cao, B. (2020). Clinical features of patients infected with 2019 novel coronavirus in Wuhan, China. *Lancet, 395*(10223), 497–506. doi:10.1016/S0140-6736(20)30183-5 PMID:31986264

Hymas, C., & Lovett, S. (2022, December 30). China accused of covering up Covid data amid new variant fears. *The Telegraph.* https://www.telegraph.co.uk/politics/2022/12/30/china-withholding-covid-data-new-variant-fears/

I.L.O. (2020). *COVID-19 and the World of Work Rapid Assessment of the Employment Impacts and Policy Responses.* International Labour Organization.

Immergluck, D. (2013). Too little, too late, and too timid: The federal response to the foreclosure crisis at the five-year mark. *Housing Policy Debate, 23*(1), 199–232. doi:10.1080/10511482.2012.749933

Ingersoll, R., Merrill, L., & May, H. (2016). Do accountability policies push teachers out? *Educational Leadership, 73*(8), 44–49. https://repository.upenn.edu/gse_pubs/551

Institute of Economic Sciences. (2022a). *Institutional Response to COVID-19 Pandemic Effects on the Labour Market – Lessons Learned and Future Perspectives.* Policy Brief [Research Project INEQ RS COVID-19 No. 7552225]. https://www.ien.bg.ac.rs/data/images/proj_ineqrs/policy_brief_institutional_response_co_covid19_pandemic_effects_labour_market.pdf

Institute of Economic Sciences. (2022b). *Effects of the COVID-19 Pandemic on Poverty and Income Inequality in Serbia and Cost-Benefit Analysis of the Implemented Income Support Measures.* Policy Brief [Research Project INEQ RS COVID-19 No. 7552225]. https://www.ien.bg.ac.rs/data/images/proj_ineqrs/policy_brief_effects_covid19_pandemic_on_poverty_income_inequality_inserbia.pdf

International Society for Technology in Education. (2023). *ISTE Standards.* ISTE.

Ioannidis, J. P. (2022). The end of the COVID-19 pandemic. *European Journal of Clinical Investigation*, e13782. doi:10.1111/eci.13782

Irawan, V. T., Sutadji, E., & Widiyanti, W. (2017). Blended learning based on schoology: Effort of improvement learning outcome and practicum chance in vocational high school. *Cogent Education, 4*(1), 1282031. doi:10.1080/2331186X.2017.1282031

Irby, B., Brown, G., Duffy, J., & Trautman, D. (2001). The synergistic leadership theory. *Journal of Educational Administration, 40*(4), 304–322. doi:10.1108/09578230210433409

Ison, R. (1999). Applying systems thinking to higher education. Systems Research and Behavioral Science. *The Official Journal of the International Federation for Systems Research, 16*(2), 107–112.

Jaaron, A. A. M., & Backhouse, C. J. (2013). Systems thinking for service delivery design: A real time mass customisation model. *IFAC Proceedings Volumes, 46*(9), 228-233. 10.3182/20130619-3-RU-3018.00053

Jenkins, G. P., Olubamiro, O. C., & Miklyaev, M. (2022). *Cost-Benefit Analysis of Tax Incentives in Serbia* (Development Discussion Papers No. 2022-01, JDI Executive Programs Support to Enhancing the Quality of the Tax System in Serbia: Analysis of Tax Incentives in Serbia). https://cri-world.com/publications/qed_dp_4579.pdf

Jennings, P. A. (2020). *Teacher burnout turnaround: Strategies for empowered educators.* WW Norton & Company.

Jones, L., Palumbo, D., & Brown, D. (2021). *Coronavirus: How the pandemic has changed the world economy.* BBC News. https://www.bbc.com/news/business-51706225

Jordan, K., & Elsden-Clifton, J. (2016). Designing meaningful learning environments in a work-integrated-learning course using a DOCC Design. In B. H. Khan (Ed.), *Revolutionizing modern education through meaningful e-learning implementation* (pp. 1–17). IGI Global. doi:10.4018/978-1-5225-0466-5.ch001

Jordà, Ò. (2005). Estimation and inference of impulse responses by local projections. *The American Economic Review, 95*(1), 161–182. doi:10.1257/0002828053828518

Joshua, Ć. (2016). An analysis of current evidence supporting two alternate learning models: learning styles and dual coding. *Journal of Educational Sciences & Psychology, 6*(68).

Josselson, R. (2010). *Narrative research* (N. Salkind, Ed.).

Joyce, P. G., & Prabowo, A. S. (2020). Government responses to the coronavirus in the United States: Immediate remedial actions, rising debt levels and budgetary hangovers. *Journal of Public Budgeting, Accounting & Financial Management, 32*(5), 745–758. doi:10.1108/JPBAFM-07-2020-0111

Joyce, P., Maron, F., & Reddy, P. S. (2020). Managing global pandemics: Public governance matters. In P. Joyce, F. Maron, & P. S. Reddy (Eds.), *Good public governance in a global pandemic* (Vol. 1, pp. 567–591). The International Institute of Administrative Sciences. https://www.iias-iisa.org/page/ebook doi:10.46996/pgs.v1e1

Ju, Z. (2022). There are many styles of nucleic acid testing vouchers in Suzhou. *Netease News.* https://www.163.com/dy/article/H60QATJT0535B3HJ.html

Junpho, M., & Kaewprapan Rosenkranz, W. (2018). A study of systems thinking process: a review of the literature. *Proceedings of International Conference on Technology and Social Science (ICTSS 2018).*

Kali, Y., Goodyear, P., & Markauskaite, L. (2011). Researching design practices and design cognition: Contexts, experiences and pedagogical knowledge-in-pieces. *Learning, Media and Technology, 36*(2), 129–149. doi:10.1080/17439884.2011.553621

Kamberović, S., Mitrović, S., & Behrens, A. (2020). *GAP Analysis - Opportunities for Economic Empowerment through Sustainable Business Models in the Aftermath of the COVID-19 Pandemic in the Republic of Serbia.* Organization for Security and Co-operation in Europe Mission to Serbia. https://www.osce.org/files/f/documents/9/1/479678_0.pdf

Kamenetz, A. (2020). *6 Ways Universities Are Responding To Coronavirus.* https://www.npr.org/2020/03/06/812462913/6-ways-universities-are-responding-to-coronavirus

Karakose, T. (2021). The impact of the COVID-19 epidemic on higher education: Opportunities and implications for policy and practice. *Educational Process: International Journal, 10*(1), 7–12. doi:10.22521/edupij.2021.101.1

Karbowski, D. (2022). State of teaching statistics 2022. *Adopt a Classroom.* Retrieved from: https://www.adoptaclassroom.org/2022/04/12/state-of-teaching-statistics-2022/

Kaye, L. K., Kowert, R., & Quinn, S. (2017). The role of social identity and online social capital on psychosocial outcomes in MMO players. *Computers in Human Behavior, 74*, 215–223. doi:10.1016/j.chb.2017.04.030

Kay, K., & Boss, S. (2021). *Redefining student success: Building a new vision to transform leading, teaching, and learning.* Corwin Press.

Keane, D. (2023, April 7). London study finds first evidence Covid spreads on household surfaces. *The Evening Standard.* https://www.standard.co.uk/news/health/covid-spreads-household-surfaces-imperial-college-london-study-b1072709.html

Keating, C. B. (2000). A systems-based methodology for structural analysis of health care operations. *Journal of Management in Medicine, 14*(3/4), 179–198. doi:10.1108/02689230010359183 PMID:11142060

Kenber, B. (2022, December 27). Covid data dumped as publishing R value 'no longer necessary'. *The Times.* https://www.thetimes.co.uk/article/covid-data-dumped-as-publishing-r-rate-no-longer-necessary-gzhclctnf

Kennedy, M. M. (2016). How does professional development improve teaching? *Review of Educational Research, 86*(4), 945–980. doi:10.3102/0034654315626800

Kessinger, T. A. (2007). Efforts toward national educational reform: An essentialist political agenda. *Mid-Western Educational Researcher, 20*(2), 16–23.

Kim, B. (2015). Playing with gameful activities and assessments: Avatars and experience points in a graduate course. *Proceedings of the Computer-Supported Collaborative Learning (CSCL) Conference.* https://repository.isls.org//handle/1/450

Kim, D. (1999). Introduction to systems thinking. Pegasus Communications (now Leverage Networks Inc.).

Kirschner, P. A., & De Bruyckere, P. (2017). The myths of the digital native and the multitasker. *Teaching and Teacher Education, 67*, 135–142. doi:10.1016/j.tate.2017.06.001

Knussen, L. J. (2020). *Investigating early career teachers' design of technology-integrated learning in context* [Doctoral dissertation]. University of Technology Sydney. https://opus.lib.uts.edu.au/handle/10453/144102

Kobayashi, Y., Heinrich, T., & Bryant, K. A. (2021). Public support for development aid during the COVID-19 pandemic. *World Development, 138*, 105248. doi:10.1016/j.worlddev.2020.105248 PMID:33106725

Kober, N., & Rentner, D. S. (2020). *History and evolution of public education in the US.* Center on Education Policy. https://files.eric.ed.gov/fulltext/ED606970.pdf

Koivisto, J., & Hamari, J. (2019). The rise of motivational information systems: A review of gamification research. *International Journal of Information Management, 45*, 191–210. doi:10.1016/j.ijinfomgt.2018.10.013

Kotsopoulos, K. I., Chourdaki, P., Tsolis, D., Antoniadis, R., Pavlidis, G., & Assimakopoulos, N. (2019). An authoring platform for developing smart apps which elevate cultural heritage experiences: A system dynamics approach in gamification. *Journal of Ambient Intelligence and Humanized Computing*, 1–17. doi:10.100712652-019-01505-w

Kotter, J. (2014). *Accelerate: building strategic agility for a faster moving world.* Harvard Business Review Press.

Kristanto, A., Mustaji, M., & Mariono, A. (2017). The development of instructional materials e-Learning based onblended learning. *International Education Studies, 10*(7), 10–17. doi:10.5539/ies.v10n7p10

Kumari, S., Gautam, H., Nityadarshini, N., Das, B. K., & Chaudhry, R. (2021). Online classes versus traditional classes? Comparison during COVID-19. *Journal of Education and Health Promotion, 10*(457). Advance online publication. doi:10.4103/jehp.jehp_317_21 PMID:35233404

Kurtz, H. (2020). Educator morale, school job applicants declining, survey shows. *Education Week.* Retrieved from: https://www.edweek.org/teaching-learning/educator-morale-school-job-applicants-declining-survey-shows/2020/11

Kwon, R., Zhang, M. L., & VandenBussche, C. J. (2020). Considerations for remote learning in pathology during COVID-19 social distancing. *Cancer Cytopathology, 128*(9), 642–647. doi:10.1002/cncy.22289 PMID:32497399

Kye, B., & Hwang, S.-J. (2020). Social trust in the midst of pandemic crisis: Implications from COVID-19 of South Korea. *Research in Social Stratification and Mobility, 68*, 100523. doi:10.1016/j.rssm.2020.100523 PMID:32572302

Kyriacou, C. (2000). *Stress-busting for Teachers.* Nelson Thornes.

Lai, J., & Widmar, N. O. (2021). Revisiting the Digital Divide in the COVID-19 Era. *Applied Economic Perspectives and Policy*, *43*(1), 458–464. doi:10.1002/aepp.13104 PMID:33230409

Lakkala, M., Toom, A., Ilomäki, L., & Muukkonen, H. (2015). Re-designing university courses to support collaborative knowledge creation practices. *Australasian Journal of Educational Technology*, *31*(5), 521–536. doi:10.14742/ajet.2526

Lamé, G., Jouini, O., & Stal-Le Cardinal, J. (2017). *Combining the Viable System Model and Kotter's 8 Steps for multi-department integration in hospitals.* 12ème Conférence Internationale de Génie Industriel, Compiègne, France. https://hal.archives-ouvertes.fr/hal-01519593

Laster-Loftus, A., & Cooper, L. (2019). The Role of Instructional Videos in Psychology Students' Course Success. *The Journal of Educators Online*, *16*. Advance online publication. doi:10.9743/JEO.2019.16.2.5

Lather, P. (1991). *Getting smart: Feminist research and pedagogy with/in the postmodern.* Routledge. doi:10.4324/9780203451311

Latterman, K., & Steffes, S. (October 2017). Tackling teacher and principal shortages in rural areas. *Legis Brief, 25*(40). Retrieved from: https://www.ncsl.org/research/education/tackling-teacher-and-principal-shortages-in-rural-areas.aspx

Lazarević-Moravčević, M., & Kamenković, S. (2021). The impact of the COVID-19 crisis on the Serbian economy - consequences and recovery. *Economic Analysis*, *54*(2), 41–54. doi:10.28934/ea.21.54.2.pp41-54

Leeuw, L. F. (2012). Linking theory-based evaluation and contribution analysis: Three problems and a few solutions. *Evaluation*, *18*(3), 348–363. doi:10.1177/1356389012452051

Lei, S. I., & So, A. S. I. (2021). Online teaching and learning experiences during the COVID-19 pandemic - A comparison of teacher and student perceptions. *Journal of Hospitality & Tourism Education*, *33*(3), 148–162. doi:10.1080/10963758.2021.1907196

Leonard, A. (2009). The Viable System Model and its application to complex organizations. *Systemic Practice and Action Research*, *22*(4), 223–233. doi:10.100711213-009-9126-z

Levenson, J. (2017). Trauma-informed social work practice. *Social Work*, *62*(2), 105–113. PMID:28339563

Levy, O. (2022, August 15). Covid booster jab approved. *BBC News*.

Lewin, K. (1997). *Resolving social conflicts; &, Field theory in social science.* American Psychological Association. doi:10.1037/10269-000

Lewin, K., Lippitt, R., & White, R. K. (1939). Patterns of aggressive behavior in experimentally created social climates. *The Journal of Social Psychology*, *10*(2), 271–299. doi:10.1080/00224545.1939.9713366

Lewis, A., & Diamond, J. (2015). *Despite the best intentions: How racial inequality thrives in good schools.* Oxford University Press.

Lieberman, M. (2021). How staff shortages are crushing schools. *Education Week.* Retrieved from https://www.edweek.org/leadership/how-staff-shortages-are-crushing-schools/2021/10

Likert, R. (1967). *New patterns of management.* McGraw-Hill.

Li, M., Wang, Y., Wu, Y., & Liu, H. (2021). Gamification narrative design as a predictor for mobile fitness app user persistent usage intentions: A goal priming perspective. *Enterprise Information Systems*, *15*(10), 1501–1545. doi:10.1080/17517575.2021.1941272

Lindsay, J., Gnedko-Berry, N., & Wan, C. (2021). *Michigan teachers who are not teaching: Who are they, and what would motivate them to teach?* (REL 2021–076). U.S. Department of Education, Institute of Education Sciences, National Center for Education Evaluation and Regional Assistance, Regional Educational Laboratory Midwest. Retrieved from https://ies.ed.gov/ncee/edlabs

Lin, X., Yang, H., Wu, Y., Zheng, X., Xie, L., Shen, Z., & Hu, S. (2021). Research on international cooperative governance of the COVID-19. *Frontiers in Public Health*, *9*, 566499. doi:10.3389/fpubh.2021.566499 PMID:33996704

Liyanagunawardena, T., Adams, A., Rassool, N., & Williams, S. (2014). Blended learning in distance education: Sri Lankan perspective. *International Journal of Education and Development using ICT*, *10*(1). https://www.learntechlib.org/p/147446/

Li, Z., Chen, Q., Feng, L., Rodewald, L., Xia, Y., Yu, H., Zhang, R., An, Z., Yin, W., Chen, W., Qin, Y., Peng, Z., Zhang, T., Ni, D., Cui, J., Wang, Q., Yang, X., Zhang, M., Ren, X., & China, C. D. C. (2020). Active case finding with case management: The key to tackling the COVID-19 pandemic. *Lancet*, *396*(10243), 63–70. doi:10.1016/S0140-6736(20)31278-2 PMID:32505220

Li, Z., Liu, F., Cui, J., Peng, Z., Chang, Z., Lai, S., Chen, Q., Wang, L., Gao, G. F., & Feng, Z. (2021). Comprehensive large-scale nucleic acid-testing strategies support China's sustained containment of COVID-19. *Nature Medicine*, *27*(5), 740–742. doi:10.103841591-021-01308-7 PMID:33859409

Locke, V. N., Patarapichayatham, C., & Lewis, S. (2021). *Learning loss in reading and math in US schools due to the COVID-19 Pandemic*. Istation.

López-Santana, M., & Rocco, P. (2021). Fiscal federalism and economic crises in the United States: Lessons from the COVID-19 pandemic and great recession. *Publius*, *51*(3), 365–395. doi:10.1093/publius/pjab015 PMID:34394701

Loversidge, J., Yen, P. Y., Chipps, E., Gallagher-Ford, L., Genter, L., & Buck, J. (2018). Top-of-License Nursing Practice, Part 2: Differentiating BSN and ADN Perceptions of Top-of-License Activities. *The Journal of Nursing Administration*, *48*(6), 329–334. doi:10.1097/NNA.0000000000000623 PMID:29794597

Lovett, S. (2021, August 4). Revealed: Britain's 'lost' blueprint for fighting a coronavirus outbreak – written 16 years ago. *The Independent*. https://apple.news/AswZ6ocEwSnaZJ0NoNv3zIQ

Lowenthal, P., Borup, J., West, R., & Archambault, L. (2020). Thinking Beyond Zoom: Using Asynchronous Video to Maintain Connection and Engagement During the COVID-19 Pandemic. *Journal of Technology and Teacher Education*, *28*(2), 383.

Loyola University. (n.d.). Course list: Community Engaged Learning, Teaching and scholarship. Loyola. https://engage.loyno.edu/service-learning/courses

Luke, C., & Gore, J. (1992). *Feminist and critical pedagogy*. Routledge.

Lustig, N. (Ed.). (2018). *Commitment to equity handbook: Estimating the impact of fiscal policy on inequality and poverty*. Brookings Institution Press.

Lynch, R. (2021). Third of people delete NHS app as 'pingdemic' spreads. *The Telegraph*. https://www.telegraph.co.uk/business/2021/07/16/third-people-delete-nhs-app-pingdemic-spreads/

Maani, K. E., & Cavana, R. Y. (2007). *Systems thinking, system dynamics: Managing change and complexity* (2nd ed.). Pearson Education Canada.

Macadam, R. D., & Packham, R. G. (1989). A case study in the use of soft systems methodology: Restructuring an academic organisation to facilitate the education of systems agriculturalists. *Agricultural Systems, 30*(4), 351–367. doi:10.1016/0308-521X(89)90004-8

Mackenzie, D. (2020). *COVID-19: The pandemic that never should have happened, and how to stop the next one*. The Bridge Street Press.

Mahboubi, P., & Mokaya, M. (2022, April 29). The skills imperative: Workforce development strategies post-COVID. *C.D. SSRN, 609*. doi:10.2139srn.4096106

Majuri, J., Koivisto, J., & Hamari, J. (2018). Gamification of education and learning: A review of empirical literature. *Proceedings of the 2nd international GamiFIN conference, GamiFIN 2018*. https://ceur-ws.org/Vol-2186/paper2.pdf

Ma, N., Xin, S., & Du, J. Y. (2018). A peer coaching-based professional development approach to improving the learning participation and learning design skills of in-service teachers. *Journal of Educational Technology & Society, 21*(2), 291–304. https://www.jstor.org/stable/26388408

Mandryk, R. L., & Birk, M. V. (2019). The potential of game-based digital biomarkers for modeling mental health. *JMIR Mental Health, 6*(4), e13485. doi:10.2196/13485 PMID:31012857

Mann, B. (2019). Compete, conform, or both? School district responses to statewide cyber charter schools. *Journal of School Choice, 14*(1), 49–74. doi:10.1080/15582159.2019.1566996

Manthorpe, J., Woolham, J., Steils, N., Stevens, M., Marntineau, S., Owen, J., & Tinelli, M. (2022). Experiences of adult social work addressing self-neglect during the COVID-19 pandemic. *Journal of Social Work, 22*(5), 1227–1240. doi:10.1177/14680173221083446

Mao, J., & Shearer, R. L. (2019). Technology affordance in online learning: A systems thinking and system dynamics theoretical framework. In M. Spector, B. Lockee, & M. Childress (Eds.), *Learning, Design, and Technology*. Springer. doi:10.1007/978-3-319-17727-4_106-1

Marion, R., & Gonzales, L. D. (2014). *Leadership in education: Organizational theory for the practitioner* (2nd ed.). Waveland Press.

Marjanović, D., & Đukić, M. (2020). Ekonomske mere za ublažavanje posledica COVID-19 [Economic measures to mitigate the consequences of COVID-19]. In P. Mitić & D. Marjanović (Eds.), *Black swan in the world economy 2020* (pp. 91–105). Institute of Economic Sciences.

Markauskaite, L., & Goodyear, P. (2009). Designing for complex ICT-based learning: Understanding teacher thinking to help improve educational design. *Proceedings of the 26th Annual Conference of the Australasian Society for Computers in Learning in Tertiary Education (ASCILITE) in Same Places, Different Spaces*. http://www.ascilite.org.au/conferences/auckland09/procs/markauskaite.pdf

Marshall, C. (1997). *Feminist critical policy analysis I: Perspectives from K-12*. Falmer Press.

Marshall, C., & Young, M. (2013). Policy inroads undermining women in education. *International Journal of Leadership in Education, 16*(2), 205–219. doi:10.1080/13603124.2012.754056

Marsicano, C., Felten, K., Toledo, L., & Buitendorp, M. (2020). *Tracking campus responses to the COVID-19 Pandemic*. American Political Science Association Preprints. doi:10.33774/apsa-2020-3wvrl

Marsicano, C. R., Barnshaw, J., & Letukas, L. (2020). Crisis and change: How COVID-19 exacerbated institutional inequality and how institutions are responding. *New Directions for Institutional Research, 2020*(187/188), 7–30. doi:10.1002/ir.20344

Martin, V. (2020). The response of the monetary and fiscal policies on COVID-19 in Serbia. *Bankarstvo, 49*(2), 92–114. doi:10.5937/bankarstvo2002070M

Maslow, A. H. (1970). *Motivation and personality*. Harper & Row.

Maslow, A. H. (2019). *A Theory of Human Motivation*. HN Publishing.

McGee, P., & Carmean, C. (2012). Accidental pedagogy redux. *Distance Learning, 9*(4), 5. https://www.proquest.com/scholarly-journals/accidental-pedagogy-redux/docview/1353085461/se-2

McGlacken, B., Larkan, F., Mannan, H., & Kodate, N. (2023). Introduction to systems thinking. In F. Larkan, H. Mannan, N. Kodate, & F. Vallières (Eds.), *Systems Thinking for Global Health* (pp. 1–17). Oxford University Press.

McGregor, D. (1957). *The human side of enterprise*. McGraw-Hill.

McNeilly, L. (2018). Why we need to practice at the top of the license. *ASHA Leader, 23*(2), 10–11. doi:10.1044/leader.FMP.23022018.10

Meador, M. (2021). *Learning organization in a time of crisis: A case study of logistics and freight, llc and covid-19* [Dissertation, The University of Memphis]. ProQuest Dissertations Publishing.

Meadows, D. H. (2008). *Thinking in systems: A primer*. Chelsea Green Publishing.

Members of the National Council for Online Education. (2022). *Emergency Remote Instruction Is Not Quality Online Learning*. https://www.insidehighered.com/views/2022/02/03/remote-instruction-and-online-learning-arent-same-thing-opinion

Michaelsen, L. K., & Sweet, M. (2008). The essential elements of team-based learning. *New Directions for Teaching and Learning, 2008*(116), 7–27. doi:10.1002/tl.330

Mikulecky, M., Shkodriani, G., & Wilner, A. (2004). *A growing trend to address the teacher shortage. Alternative Certification.* Policy Brief. Education Commission of the States. https://eric.ed.gov/?id=ED484845

Millar, G. (2009). *The viable governance model: A theoretical model of IT governance with a corporate setting* [Doctoral dissertation, The University of New South Wales]. doi:10.26190/unsworks/22565

Millard, E. (2012). 5 Reasons FLIPPED Classrooms Work. *University Business, 15*(11), 26–29.

Mindeguia, M., Ruiz, M., Igartua, J. I., & Ganzarain, J. (2019). Gamification: An Approach in The Master of Business Innovation and Project Management. In *INTED2019 Proceedings* (pp. 1415-1422). IATED. 10.21125/inted.2019.0440

Ministry of Education, Government of India. (2020). *Students to continue their learning by making full use of the available digital e-Learning platforms – Shri Ramesh Pokhriyal "Nishank" Union HRD Minister appeals to students, teachers, and parents to follow Janata curfew on 22nd March 2020.* Press Information Bureau, Government of India.

Mishra, P., & Koehler, M. J. (2006). Technological pedagogical content knowledge: A framework for teacher knowledge. *Teachers College Record, 108*(6), 1017–1054. doi:10.1111/j.1467-9620.2006.00684.x

Mladovsky, P., Srivastava, D., Cylus, J., Karanikolos, M., Evetovits, T., Thomson, S., & McKee, M. (2012). *Health Policy Responses to the Financial Crisis in Europe* (World Health Organization Policy Summary No. 5). https://www.euro.who.int/__data/assets/pdf_file/0009/170865/e96643.pdf

Mollenkopf, D. A., Ozanne, L. K., & Stolze, H. W. (2021). A transformative supply chain response to COVID-19. *Journal of Service Management, 32*(2), 190–202. doi:10.1108/JOSM-05-2020-0143

Moll, L. C., Amanti, C., Neff, D., & Gonzalez, N. (1992). Funds of knowledge for teaching: Using a qualitative approach to connect homes and classrooms. *Theory into Practice*, *31*(2), 132–141. doi:10.1080/00405849209543534

Monat, J. P., & Gannon, T. F. (2015). What is systems thinking? A review of selected literature plus recommendations. *American Journal of Systems Science*, *4*(1), 11–26. http://article.sapub.org/10.5923.j.ajss.20150401.02.html

Monat, J., & Gannon, T. (2015). What is systems thinking: A review of selected literature plus recommendations. *American Journal of Systems Science*, *4*(1), 11–26.

Montagni, I., Mabchour, I., & Tzourio, C. (2020). Digital gamification to enhance vaccine knowledge and uptake: Scoping review. *JMIR Serious Games*, *8*(2), e16983. doi:10.2196/16983 PMID:32348271

Moon, S. K., Shu, J., Simpson, T. W., & Kumara, S. R. (2010). A module-based service model for mass customization: Service family design. *IIE Transactions*, *43*(3), 153–163. doi:10.1080/07408171003705383

Moreno, C., Wykes, T., Galderisi, S., Nordentoft, M., Crossley, N., Jones, N., Cannon, M., Correll, C. U., Byrne, L., Carr, S., Chen, E. Y. H., Gorwood, P., Johnson, S., Kärkkäinen, H., Krystal, J. H., Lee, J., Lieberman, J., López-Jaramillo, C., Männikkö, M., ... Arango, C. (2020). How mental health care should change as a consequence of the COVID-19 pandemic. *The Lancet. Psychiatry*, *7*(9), 813–824. doi:10.1016/S2215-0366(20)30307-2 PMID:32682460

Moreno-López, I., Ramos-Sellman, A., Miranda-Aldaco, C., & Gomis Quinto, M. T. (2017). Transforming ways of enhancing foreign language acquisition in the Spanish classroom: Experiential learning approaches. *Foreign Language Annals*, *50*(2), 398–409. doi:10.1111/flan.12267

Morgan, G. (2006). *Images of organizations*. Sage.

Mullins, J. K., & Sabherwal, R. (2020). Gamification: A cognitive-emotional view. *Journal of Business Research*, *106*, 304–314. doi:10.1016/j.jbusres.2018.09.023

Mulyadi, D., Arifani, Y., Wijayantingsih, T. D., & Budiastuti, R. E. (2020). Blended learning in English for specific purposes (ESP) instruction: Lecturers' perspectives. *Computer-Assisted Language Learning Electronic Journal, 21*(2), 204-219. http://callej.org/journal/21-2/Mulyadi-Arifani-Wijayantingsih-Budiastuti2020.pdf

Mundasad, S., & Roxby, P. (2023, May 5). *Covid global health emergency is over, WHO says*. BBC. https://www.bbc.co.uk/news/health-65499929

Murphy, M. P. (2020). COVID-19 and emergency eLearning: Consequences of the securitization of higher education for post-pandemic pedagogy. *Contemporary Security Policy*, *41*(3), 492–505. doi:10.1080/13523260.2020.1761749

Mustikasari, D., & Heggart, K. (2022). *Conceptualising Teachers as Designers in the Higher Education Context: Case Studies from Indonesian English University Teachers* [Pecha Kucha Presentation]. *39th International Conference on Innovation, Practice and Research in the Use of Educational Technologies in Tertiary Education, ASCILITE 2022*, Sydney, NSW, Australia. 10.14742/apubs.2022.49

Myers, L. C., & Liu, V. X. (2022). The COVID-19 pandemic strikes again and again and again. *JAMA Network Open*, *5*(3), e221760. Advance online publication. doi:10.1001/jamanetworkopen.2022.1760 PMID:35262720

Nair, A., & Sadasivan, R. (2019). Winning the talent game: HR gamification experience for Generation Z. *International Journal on Leadership*, *7*(1), 44–49. http://publishingindia.com/ijl/

Nair, B. B. (2022). Endorsing gamification pedagogy as a helpful strategy to offset the COVID-19 induced disruptions in tourism education. *Journal of Hospitality, Leisure, Sport and Tourism Education*, *30*, 100362. doi:10.1016/j.jhlste.2021.100362 PMID:34849104

Nash, A. M., & Peters-Hawkins, A. L. (2020). For us: Towards an intersectional leadership conceptualization by Black women for Black girls. *Journal of Educational Administration and History*. Advance online publication. doi:10.1080/00220620.1785403

Natalia, F. (2017). Electronic means of foreign language learning in the system of higher education. *Educational Research Review, 12*(3), 116–119. doi:10.5897/ERR2016.3062

National Alliance for Local Economic Development. (2021). *Efekti ekonomskih mera za ublažavanje negativnih posledica COVID-19 na privredu* [Effects of Economic Measures to Mitigate the Negative Consequences of COVID-19 on the Economy]. https://naled.rs/htdocs/Files/06839/Analiza_efekata_ekonomskih_mera_za_ublazavanje_negativnih_posledica_COVID-19_na_privredu.pdf

National Commission on Excellence in Education. (1983). A nation at risk: The imperative for educational reform. *The Elementary School Journal, 84*(2), 113–130. doi:10.1086/461348

National Conference of State Legislatures. (2022). *Teacher shortage areas by state.* Retrieved from: https://www.ncsl.org/research/education/teacher-shortage-areas-by-state.aspx

National Cyber Security Centre. (2021). *High level privacy and security design for NHS Covid-19 Contract Tracing App*. Crown Copyright.

National Health Service Test and Trace. (n.d.). *How to do a PCR test for coronavirus (COVID-19) at home.* NHS UK.

National Health Service. (2023, March 21). *How to use an NHS COVID-19 rapid lateral flow test.* NHS UK. https://www.nhs.uk/nhs-services/covid-19-services/testing-for-covid-19/how-to-use-an-nhs-covid-19-rapid-lateral-flow-test/

Nayak, J., Mishra, M., Naik, B., Swapnarekha, H., Cengiz, K., & Shanmuganathan, V. (2021). An impact study of COVID-19 on six different industries: Automobile, energy and power, agriculture, education, travel and tourism and consumer electronics. *Expert Systems, 39*(3), 1–32. *Wiley Online Library Database Model, 2023.* Advance online publication. doi:10.1111/exsy.12677 PMID:33821074

NC Foundation for Public School Children. (2022, December 8). *About NC teacher cadet.* https://www.ncfpsc.org/about/teacher-cadet

Nelson, C. E. (2022). *9 skills that make great online school teachers.* Connections Academy; Pearson.

Nelson, G. (2020). *WCAG 2.1 Guidelines Explained with Examples.* https://medium.com/c2-group/wcag-2-1-guidelines-explained-with-examples-5c7c5d8b69eb

Nelson, H. G. (2020). The promise of systemic designing: Giving form to water. *Learning, design, and technology: An international compendium of theory, research, practice, and policy*, 1-49.

Nicolaides, N., & Gaynor, A. K. (1992). The knowledge base informing the teaching administration and organizational theory in UCEA universities: A descriptive and interpretive survey. *Educational Administration Quarterly, 28*(2), 237–265. doi:10.1177/0013161X92028002005

Nieto-Escamez, F. A., & Roldán-Tapia, M. D. (2021). Gamification as online teaching strategy during COVID-19: A Mini-review. *Frontiers in Psychology, 12*, 648552. doi:10.3389/fpsyg.2021.648552 PMID:34093334

Nikolić, I., & Filipović, S. (2021). Comparative analysis of the economic policy response to COVID-19 in the EU-27 and Republic of Serbia. *Ekonomika preduzeća, 7-8*, 460-475. doi:10.5937/EKOPRE2108460N

Nitsche, A. M., Franczyk, B., & Schumann, C. A. (2022). Gamification for Smart and Collaborative Last Mile Networks Knowledge Transfer. In *2022 IEEE 28th International Conference on Engineering, Technology and Innovation (ICE/ITMC) & 31st International Association for Management of Technology (IAMOT) Joint Conference* (pp. 1-9). IEEE. 10.1109/ICE/ITMC-IAMOT55089.2022.10033189

November, A., & Mull, B. (2012). *Flipped Learning: A response to 5 Criticisms -What's Flipped Learning?* https://novemberlearning.com/article/flipped-learning-a-response-to-five-common-criticisms-article/

O.E.C.D. (2013). *Framework for integrated analysis", in OECD Framework for Statistics on the Distribution of Household Income, Consumption and Wealth.* OECD Publishing. https://doi.org/ doi:10.1787/9789264194830-11-en

Oberle, E., & Schonert-Reichl, K. A. (2016). Stress contagion in the classroom? The link between classroom teacher burnout and morning cortisol in elementary school students. *Social Science & Medicine, 159*, 30–37. doi:10.1016/j.socscimed.2016.04.031 PMID:27156042

Oe, H., Takemoto, T., & Ridwan, M. (2020). Is gamification a magic tool? Illusion, remedy, and future opportunities in enhancing learning outcomes during and beyond the COVID-19. *Budapest International Research and Critics in Linguistics and Education (BirLE). Journal, 3*(3), 1401–1414. doi:10.33258/birle.v3i3

Ogan, A., Aleven, V., & Jones, C. (2009). Advancing development of intercultural competence through supporting predictions in narrative video. *International Journal of Artificial Intelligence in Education, 19*(3), 267–288. https://www.learntechlib.org/p/69607/

Ognjenović, K. (2022). Multiple job-holding and informality: may pandemic change their course? In S. Redžepagić, A. Portugal Durate, M. Siničáková & D. Bodroža (Eds.), Economic and financial implications of COVID-19 crisis (pp. 200-218). Balkan Institute of Science and Innovation, Université Côte d'Azur, Nice.

Ognjenović, K., & Đukić, M. (2022). A review into evaluation methods applied in assessing economic and social policies' responses to COVID-19 in Serbia: 2020-2021 [Paper presentation]. PROFEEDBACK Evaluation of Public Policy Responses to 'Black Swans' Conference, Budapest, Hungary.

Ognjenović, K. (2007). The use of propensity score-matching methods in evaluation of active labour market programs in Serbia. *Economic Annals, 52*(172), 21–53. doi:10.2298/EKA0772021O

Ognjenović, K. (2021a). Gender wage gap in Serbia: Inheritance and sources of the wage gap. *Stanovnistvo, 59*(2), 123–141. doi:10.2298/STNV2102123O

Ognjenović, K. (2021b). Prilagođavanje tržišta rada u Srbiji novim uslovima izazvanim COVID-19 krizom [Adjustment of the labour market in Serbia in the conditions of the crisis caused by the COVID-19 pandemic]. In A. Bradić-Martinović, I. Beraha, & S. Đuričin (Eds.), *Poslovanje u periodu pandemije – izazovi i šanse* [Pursuing business during the pandemic – challenges and chances] (pp. 18–37). Institute of Economic Sciences.

Ognjenović, K., & Pavlović, D. (2021). *SILC u Republici Srbiji: metodološki okvir i analiza izabranih pokazatelja siromaštva i nejednakosti* [SILC in the Republic of Serbia: methodological framework and analysis of poverty and inequality indicators]. Institute of Economic Sciences.

Oklahoma State Department of Education (OSDE). (2019). *Student enrollment in Oklahoma public schools tops 700,000.* Oklahoma State Department of Education.

Omukuti, J., Barlow, M., Giraudo, M. E., Lines, T., & Grugel, J. (2021). Systems thinking in COVID-19 recovery is urgently needed to deliver sustainable development for women and girls. *The Lancet. Planetary Health, 5*(12), 921–928. doi:10.1016/S2542-5196(21)00232-1 PMID:34895499

Online Learning Consortium (OLC). (2023). *OLC Quality Scorecard Suite*. Online Learning Consortium.

Organisation for Economic Cooperation and Development. (2001). *Evaluation feedback for effective learning and accountability*. OECD.

Ozemek, C., Lavie, C. J., & Rognmo, Ø. (2019). Global physical activity levels-Need for intervention. *Progress in Cardiovascular Diseases*, *62*(2), 102–107. doi:10.1016/j.pcad.2019.02.004 PMID:30802461

Paivio, A. (1978). *A Dual Coding Approach to Perception and Cognition*. Psychology Press.

Pajvančić, M., Petrušić, N., Nikolin, S., Vladisavljević, A., & Baćanović, V. (2020). *Rodna analiza odgovora na COVID-19 u Republici Srbiji* [Gender Analysis of the Response to COVID-19 in the Republic of Serbia]. Organization for Security and Co-operation in Europe Mission to Serbia & Women's Platform for the Development of Serbia 2014-2020. https://www.osce.org/files/f/documents/3/9/459391.pdf

Parsons, P., & Shelton, K. (2019). Organizational sustainability in online higher education: Reframing through the Viable System Model. *Online Journal of Distance Learning Administration*, *22*(3), 1–14. https://ojdla.com/archive/fall223/parsons_shelton223.pdf

Paterson, P., Chantler, T., & Larson, H. J. (2018). Reasons for non-vaccination: Parental vaccine hesitancy and the childhood influenza vaccination school pilot programme in England. *Vaccine*, *36*(36), 5397–5401. doi:10.1016/j.vaccine.2017.08.016 PMID:28818568

Patton, D., Durand, A., Whipple, K., & Albright, D. L. (2022). Local autonomy and response to the COVID-19 pandemic. *State & Local Government Review*, *54*(2), 165–173. doi:10.1177/0160323X221089661

Paunović, B., & Aničić, Z. (2021). Impact of the COVID-19 crisis on SMEs and possible innovation responses. *Ekonomika preduzeća*, *3-4*, 169-184. doi:10.5937/EKOPRE2103169P

Pedryez, W., & Kwak, K. C. (2006). Linguistic models as a framework of user-centric system modeling. *IEEE Transactions on Systems, Man, and Cybernetics. Part A, Systems and Humans*, *36*(4), 727–745. doi:10.1109/TSMCA.2005.855755

Pedryez, W., & Vasilakos, A. V. (1999). Linguistic models and linguistic modeling. *IEEE Transactions on Systems, Man, and Cybernetics. Part B, Cybernetics*, *29*(6), 745–757. doi:10.1109/3477.809029 PMID:18252354

Pejin Stokić, Lj. (2020). *COVID-19: Impact on the Serbian Economy and the Government Recovery Package* (ESPN Flash Report No. 45). https://ec.europa.eu/social/BlobServlet?docId=22945&langId=en

Pekdağ, B. (2010). Alternative methods in learning chemistry: Learning with animation, simulation, video and multimedia. *Journal of Turkish Science Education*, *7*(2), 79–110.

People. (2022). In *Merriam-Webster*.

Perrow, C. (1986). *Complex organizations a critical essay* (3rd ed.). Scott Foresman.

Perry, D. L., & Daniels, M. L. (2016). Implementing trauma—informed practices in the school setting: A pilot study. *School Mental Health*, *8*(1), 177–188. doi:10.100712310-016-9182-3

Perugini, C., & Vladisavljević, M. (2021). Social stability challenged by Covid-19: Pandemics, inequality and policy responses. *Journal of Policy Modeling*, *43*(1), 146–160. doi:10.1016/j.jpolmod.2020.10.004 PMID:33311816

Petrosino, S., Loria, E., & Pirker, J. (2021). Stay Home Playing LoL-Analyzing Players' Activity and Social Bonds in League of Legends During Covid-19 Lockdowns. *The 16th International Conference on the Foundations of Digital Games (FDG)*, 1-12. 10.1145/3472538.3472551

Pfeiffer, J. (1997). *New directions for organization theory: Problems and prospects*. Oxford University Press.

Picchio, M., & Santolini, R. (2022). The COVID-19 pandemic's effects on voter turnout. *European Journal of Political Economy*, *73*, 102161. doi:10.1016/j.ejpoleco.2021.102161 PMID:34975184

Plack, M. M., Goldman, E. F., Scott, A. R., Pintz, C., Herrmann, D., Kline, K., & Brundage, S. B. (2018). Systems thinking and systems-based practice across the health professions: An inquiry into definitions, teaching practices, and assessment. *Teaching and Learning in Medicine*, *30*(3), 242–254. doi:10.1080/10401334.2017.1398654 PMID:29283669

Podolsky, A., Kini, T., Bishop, J., & Darling-Hammond, L. (2016). *Solving the teacher shortage: How to attract and retain excellent educators*. Learning Policy Institute. doi:10.54300/262.960

Podolsky, M. I., Present, I., Neumann, P. J., & Kim, D. D. (2022). A systematic review of economic evaluations of COVID-19 interventions: Considerations of non-health impacts and distributional issues. *Value in Health*, *25*(8), 1298–1306. doi:10.1016/j.jval.2022.02.003 PMID:35398012

Pokhrel, S., & Chhetri, R. (2021). A Literature Review on Impact of COVID-19 Pandemic on Teaching and Learning. *Higher Education for the Future*, *8*(1), 133–141. doi:10.1177/2347631120983481

Policy brief: Education during COVID-19 and beyond. (2020). United Nations. https://www.un.org/development/desa/dspd/wp-content/uploads/sites/22/2020/08/sg_policy_brief_covid-19_and_education_august_2020.pdf

Porter, L. (1964). Attitudes in management: Perceived deficiencies in need fulfillment as a function of job level. *The Journal of Applied Psychology*, *46*(6), 375–384. doi:10.1037/h0047808

Priyatno, A. (2017). Promoting learner autonomy through Schoology m-learning platform in an EAP class at an Indonesian university. *Teaching English with Technology, 17*(2), 55-76. https://www.ceeol.com/search/article-detail?id=527569

Process. (2022). In *Merriam-Webster*.

Qiu, H., Li, Q., & Li, C. (2021). How technology facilitates tourism education in COVID-19: Case study of nankai University. *Journal of Hospitality, Leisure, Sport and Tourism Education*, *29*, 100288. doi:10.1016/j.jhlste.2020.100288 PMID:34720752

Quality Matters (QM). (2018a). *Home*.

Quality Matters (QM). (2018b). *Standards from the Quality Matters Continuing Education and Professional Development, Sixth Edition*. Quality Matters.

Quality Matters (QM). (2018c). *Standards from the Quality Matters Higher Education Rubric, Sixth Edition*. Quality Matters.

Quality Matters (QM). (2018d). *Standards from the Quality Matters K-12 Rubric, Sixth Edition*.

Quality Matters.

Raiff, B. R., Fortugno, N., Scherlis, D. R., & Rapoza, D. (2018). A mobile game to support smoking cessation: Prototype assessment. *JMIR Serious Games*, *6*(2), e9599. doi:10.2196/games.9599 PMID:29880466

Ramage, M., & Shipp, K. (2009). *Systems thinkers*. Springer. doi:10.1007/978-1-84882-525-3

Ramírez Elizondo, E. (2016). Identification of design patterns for serious games in an educational videogame designed to create awareness on dengue and malaria fever. *IFIP World Information Technology Forum*, 85-95. 10.1007/978-3-319-44447-5_9

Ramírez-Gutiérrez, A. G., Cardoso-Castro, P. P., & Tejeida-Padilla, R. (2021). A methodological proposal for the complementarity of the SSM and the VSM for the analysis of viability in organizations. *Systemic Practice and Action Research, 34*(3), 331–357. doi:10.100711213-020-09536-7

Randjelović, S., & Vladisavljević, M. (2020). Social Welfare Effects of Progressive Income Taxation under Increasing Inequality. *Prague Economic Papers, 2020*(5), 575–599. doi:10.18267/j.pep.750

Rapp, A., Hopfgartner, F., Hamari, J., Linehan, C., & Cena, F. (2019). Strengthening gamification studies: Current trends and future opportunities of gamification research. *International Journal of Human-Computer Studies, 127*, 1–6. doi:10.1016/j.ijhcs.2018.11.007

Rastrigina, O., Leventi, C., Vujackov, S., & Sutherland, H. (2016). *Nowcasting: Estimating developments in median household income and risk of poverty in 2014 and 2015* (Issue EM8/16)). Euromod Working Paper.

Rauhaus, B. M., & Johnson, A. F. (2021). Social inequities highlighted by the prolonged pandemic: Expanding sick leave. *Journal of Public and Nonprofit Affairs, 7*(1), 154–163. doi:10.20899/jpna.7.1.154-163

Rawal, V., Kumar, M., Verma, A., & Pais, J. (2020). COVID-19 lockdown: Impact on agriculture and rural economy. *Society for Social and Economic Research, 13*(2), 34.

Reer, F., & Quandt, T. (2020). Digital games and well-being: An overview. *Video games and well-being*, 1-21. doi:10.1007/978-3-030-32770-5_1

Reimers, F. M., & Chung, C. K. (Eds.). (2019). *Teaching and learning for the twenty-first century: Educational goals, policies, and curricula from six nations*. Harvard Education Press.

Reinecke, L. (2009). Games and recovery: The use of video and computer games to recuperate from stress and strain. *Journal of Media Psychology, 21*(3), 126–142. doi:10.1027/1864-1105.21.3.126

Republic Secretariat for Public Policies. (2018). *Law on the Planning System of the Republic of Serbia*. Official Gazette RS No. 30.

Reyes, R. R., García, S. G., & García, M. G. (2018). Applying the Viable System Model to an organization with CSR goals: The case of a charity organization. *European Journal of Economics and Business Studies, 4*(3), 7–21. https://econpapers.repec.org/article/eurejesjr/267.htm. doi:10.2478/ejes-2018-0055

Reynolds, M., & Holwell, S. (2010). *Introducing systems approaches. Systems approaches to managing change: A practical guide*. Academic Press.

Rezapour, A., Souresrafil, A., Peighambari, M. M., Heidarali, M., & Tashakori-Miyanroudi, M. (2021). Economic evaluation of programs against COVID-19: A systematic review. *International Journal of Surgery, 85*, 10–18. doi:10.1016/j.ijsu.2020.11.015 PMID:33227532

Richmond, B. (1994). Systems dynamics/systems thinking: Let's just get on with it. In *International Systems Dynamics Conference*, Sterling, UK. 10.1002dr.4260100204

Rincon-Flores, E. G., & Santos-Guevara, B. N. (2021). Gamification during Covid-19: Promoting active learning and motivation in higher education. *Australasian Journal of Educational Technology, 37*(5), 43–60. doi:10.14742/ajet.7157

Rivas, C. L. S., Pérez, H. K. F., Rivas, L. J. M., & Almaguer, C. A. G. (2022). Gamification for the Development of Competencies in Tec21 Based on Mixed Reality. In *European Conference on Games Based Learning* (Vol. 16, No. 1, pp. 482-490). 10.34190/ecgbl.16.1.679

Rodriguez, A., & Alltucker, K. (2023, May 05). The COVID public health emergency is over in the US. Here's what that means for you. *USA Today*. https://eu.usatoday.com/story/news/health/2023/05/11/end-covid-public-health-emergency-explained/70182594007/

Rodriguez-Ulloa, R. (2022). Cybernetic governance of the Peruvian State: A proposal. *AI & Society*, *37*(3), 1207–1229. doi:10.100700146-021-01329-3 PMID:35309097

Rose, K. K. (2009). Student perceptions of the use of instructor-made videos in online and face-to-face classes. *Journal of Online Learning and Teaching*, *5*(3), 487–495.

Russell-Babin, K., & Wurmser, T. (2016). Transforming care through top-of-license practice. *Nursing Management*, *24-28*(5), 24–28. doi:10.1097/01.NUMA.0000482527.15743.12

Ryan, S. V., von der Embse, N. P., Pendergast, L. L., Saeki, E., Segool, N., & Schwing, S. (2017). Leaving the teaching profession: The role of teacher stress and educational accountability policies on turnover intent. *Teaching and Teacher Education*, *66*, 1–11. doi:10.1016/j.tate.2017.03.016

S.O.R.S. (2021). *STATISTICAL RELEASE PD10 Survey on income and living conditions* (Issue 282 year LXXI). https://publikacije.stat.gov.rs/G2021/PdfE/G20211282.pdf

Sabah, S. (2017). Interactional competence and SLA: A systems-thinking perspective. [JALLR]. *Journal of Applied Linguistics and Language Research*, *4*(4), 185–191.

Sailer, M., Hense, J. U., Mayr, S. K., & Mandl, H. (2017). How gamification motivates: An experimental study of the effects of specific game design elements on psychological need satisfaction. *Computers in Human Behavior*, *69*, 371–380. doi:10.1016/j.chb.2016.12.033

Salmons, J. (2014). *Qualitative online interviews: Strategies, design, and skills*. Sage Publications. doi:10.4135/9781071878880

Salvador, F., De Holan, P. M., & Piller, F. (2009). Cracking the code of mass customization. *MIT Sloan Management Review*, *50*(3), 71–78.

Sanchez, D. R., Langer, M., & Kaur, R. (2020). Gamification in the classroom: Examining the impact of gamified quizzes on student learning. *Computers & Education*, *144*, 103666. doi:10.1016/j.compedu.2019.103666

Sardi, L., Idri, A., & Fernández-Alemán, J. L. (2017). A systematic review of gamification in e-Health. *Journal of Biomedical Informatics*, *71*, 31–48. doi:10.1016/j.jbi.2017.05.011 PMID:28536062

Sato, T., Shimizu, K., Shiko, Y., Kawasaki, Y., Orita, S., Inage, K., Shiga, Y., Suzuki, M., Sato, M., Enomoto, K., Takaoka, H., Mizuki, N., Kim, G., Hozumi, T., Tsuchiya, R., Otagiri, T., Mukaihata, T., Furuya, T., Maki, S., ... Eguchi, Y. (2021). Effects of nintendo ring fit adventure exergame on pain and psychological factors in patients with chronic low back pain. *Games for Health Journal*, *10*(3), 158–164. doi:10.1089/g4h.2020.0180 PMID:33891508

Sawer, B. J. (1992). *Evaluating for accountability: a practical guide for the inexperienced evaluator*. Oregon State University Extension Service.

Scagnoli, N. I., Choo, J., & Tian, J. (2019). Students' insights on the use of video lectures in online classes: Students' insights on video lectures. *British Journal of Educational Technology*, *50*(1), 399–414. doi:10.1111/bjet.12572

Schmidt-Kraepelin, M., Warsinsky, S., Thiebes, S., & Sunyaev, A. (2020). The role of gamification in health behavior change: a review of theory-driven studies. *Proceedings of the 53rd Hawaii international conference on system sciences*, 1256-1265. http://hdl.handle.net/10125/63894

Schwaninger, M. (2006). Design for viable organizations: The diagnostic power of the Viable System Model. *Kybernetes, 35*(7/8), 955–966. doi:10.1108/03684920610675012

Schwaninger, M., & Scheef, C. (2016). A test of the Viable System Model: Theoretical claim vs. empirical evidence. *Cybernetics and Systems, 47*(7), 1–26. doi:10.1080/01969722.2016.1209375

Schwartz, R. B., Robinson, M. A., Kirst, M. W., & Kirp, D. L. (2000). Goals 2000 and the standards movement. *Brookings Papers on Education Policy,* (3), 173-214. https://www.jstor.org/stable/20067222

SeCons. (2020a). *Impact of COVID-19 Pandemic and Measures for its Prevention on Employment and Working Conditions of Women and Men in Serbia.* https://secons.net/wp-content/uploads/2022/01/133-publication.pdf

SeCons. (2020b). *Consequences of COVID-19 on Women's and Men's Economic Empowerment.* https://secons.net/wp-content/uploads/2022/01/134-publication.pdf

SeCons. (2021). *Ocena efikasnosti institucionalnog odgovora na pandemiju COVID-19 u multietničkim zajednicama u Sandžaku/jugozapadnoj Srbiji i na jugu Srbije* [Evaluation of the Effectiveness of the Institutional Response to the COVID-19 Pandemic in Multi-Ethnic Communities in Sandžak/Southwest Serbia and in the South of Serbia]. https://secons.net/wp-content/uploads/2021/12/131-publication.pdf

Seddon, J. (2008). *Systems thinking in the public sector.* Triarchy Press.

Seddon, J., & Brand, C. (2008). Debate: Systems thinking and public sector performance. *Public Money & Management, 28*(1), 7–9. doi:10.1111/j.1467-9302.2008.00611.x

Senge, P. (1990). *The fifth discipline, the art and practice of the learning organization.* Doubleday/Currency.

Senge, P. (1990). *The Fifth Discipline: The art and practice of the learning organization.* Crown.

Senge, P. (1990). *The Fifth Discipline: The Art and Practice of The Learning Organization.* Currency Doubleday.

Seo, S., & Kim, H. J. (2021). How COVID-19 Influences Hospitality and Tourism Education: Challenges, Opportunities, and New Directions. *Journal of Hospitality & Tourism Education, 33*(3), 147–147. doi:10.1080/10963758.2021.1929531

Serbian Association of Employers. (2020). *Uticaj pandemije COVID-19 na aktivnosti preduzeća u Srbiji: izazovi i očekivanja* [The Impact of the COVID-19 Pandemic on the Activities of Companies in Serbia: Challenges and Expectations]. https://drive.google.com/file/d/1oxElSqLaUB8mibQV06xMZW77L5hrTurA/view

Shaked, H., & Schechter, C. (2017). *Systems Thinking for School Leaders: Holistic Leadership for Excellence in Education.* Springer International Publishing. doi:10.1007/978-3-319-53571-5

Shakeshaft, C. (1987). *Women in educational administration.* Sage.

Shakeshaft, C. (2011). Wild patience: The glass ceiling for women in school administration. In *Sage handbook of educational leadership.* Sage. doi:10.4135/9781412980036.n15

Shapovalova, A. O., Ivanov, Y. B., Tyschenko, V. F., & Karpova, V. V. (2021). Assessment of the effectiveness of anti-COVID tax support for innovation activities of small and medium-sized enterprises in OECD countries. *Journal of Tax Reform, 7*(1), 68–86. doi:10.15826/jtr.2021.7.1.091

Shearing, H., & Lee, J. (2021, July 19). England Covid restrictions ease as PM urges caution. *The BBC.* https://www.bbc.co.uk/news/uk-57882029

Sherman Newcomb, W. (2014). *Continuing to disrupt the status quo? New and young women professors of educational leadership.* Information Age Publishing.

Sherman Newcomb, W., & Niemeyer, A. (2015). African American women principals: Heeding the call to serve as conduits for transforming urban school communities. *International Journal of Qualitative Studies in Education : QSE, 28*(7), 786–799. doi:10.1080/09518398.2015.1036948

Sherman, W. H. (2005). Preserving the status quo in renegotiating leadership: Women's experiences with a district-based aspiring leaders program. *Educational Administration Quarterly, 41*(5), 707–740. doi:10.1177/0013161X05279548

Sherman, W., & Wrushen, B. (2009). Intersecting leadership knowledge from the field: Diverse women secondary principals. *Journal of School Leadership, 19*(2), 172–199. doi:10.1177/105268460901900203

Shmis, T., Sava, A., Nunes Teixeira, J. E., & Patrinos, H. A. (2020). *Response to COVID-19 in Europe and Central Asia.* World Bank Group Education.

Sibley, C. G., Greaves, L. M., Satherley, N., Wilson, M. S., Overall, N. C., Lee, C. H. J., Milojev, P., Bulbulia, J., Osborne, D., Milfont, T. L., Houkamau, C. A., Duck, I. M., Vickers-Jones, R., & Barlow, F. K. (2020). Effects of the COVID-19 pandemic and nationwide lockdown on trust, attitudes toward government, and well-being. *The American Psychologist, 75*(5), 618–630. doi:10.1037/amp0000662 PMID:32496074

Siddik, M. N. A. (2020). Economic stimulus for COVID-19 pandemic and its determinants: Evidence from cross-country analysis. *Heliyon, 6*(12), e05634. doi:10.1016/j.heliyon.2020.e05634 PMID:33319096

Sims, S., & Fletcher-Wood, H. (2021). Identifying the characteristics of effective teacher professional development: A critical review. *School Effectiveness and School Improvement, 32*(1), 47–63. doi:10.1080/09243453.2020.1772841

Singh, V., & Thurman, A. (2019). How many ways can we define online learning? A systematic literature review of definitions of online learning (1988-2018). *American Journal of Distance Education, 33*(4), 289–306. doi:10.1080/08923647.2019.1663082

Singh, J., Steele, K., & Singh, L. (2021). Combining the best of online and face-to-face learning: Hybrid and blended learning approach for COVID-19, post vaccine, & post-pandemic world. *Journal of Educational Technology Systems, 50*(2), 140–171. Sage Journals. doi:10.1177/00472395211047865

Singh, S., & Arya, A. (2020). A hybrid flipped-classroom approach for online teaching of biochemistry in developing countries during Covid-19 crisis. *Biochemistry and Molecular Biology Education, 48*(5), 502–503. doi:10.1002/bmb.21418 PMID:32790218

Sky News. (2022a, December 29). *Half of passengers on China flight to Italy have COVID - as US becomes fifth country to tighten restrictions on Chinese travellers.* Sky. https://news.sky.com/story/covid-patients-no-longer-have-to-quarantine-in-hong-kong-as-restrictions-are-lifted-12775498

Sky News. (2022b, December 31). *Travellers arriving in England from China will need to show negative COVID test.* Sky. https://news.sky.com/story/uk-expected-to-require-travellers-from-china-to-provide-negative-pre-departure-covid-test-12777027

Smith, H. C. (2015). Student Clinicians Self-Evaluating Performance by Reviewing Session Videos and Using a Rubric. *Perspectives on Administration and Supervision, 25*(2), 78–84. doi:10.1044/aas25.2.78

Smuts, J. C. (1936). *Holism and Evolution* (3rd ed.). MacMillan & Co.

Social Inclusion and Poverty Reduction Unit. (2020). *Impact of the COVID-19 on Vulnerable Groups and Groups at Risk – Causes, Outcomes and Recommendations.* https://socijalnoukljucivanje.gov.rs/wp-content/uploads/2020/12/Posledice_Kovid_19_na_polozaj_osetljivih_grupa_i_grupa_u_riziku_ENG.pdf

Sockman, B. R., Clossey, L., Carducci, O. M., Batson-Magnuson, L., Muzure, D., White, G., Wehmeyer, A., Green, B. A., & Wells, H. (2019). Systems thinking as a heuristic for the implementation of service learning in a university. In M. Spector, B. B. Lockee, & M. Childress (Eds.), *Learning, Design, and Technology* (pp. 1–26). Springer., doi:10.1007/978-3-319-17727-4_91-1

Sovilj, R., & Stojković Zlatanović, S. (2021). Tackling the impact of the COVID-19 pandemic in economy and labour – a case study of Serbia regulation. *Medicine. Law & Society*, *14*(2), 301–320. doi:10.18690/mls.14.2.301-320.2021

Spencer, B. (2023, March 19). Disease X is coming. Are we ready for the next pandemic? *The Times.* https://www.thetimes.co.uk/article/disease-x-is-coming-are-we-ready-for-the-next-pandemic-ck29hmxx7

Squire, L. R., & Dede, A. J. O. (2015). Conscious and unconscious memory systems. *Cold Spring Harbor Perspectives in Biology*, *7*(3), a021667–a021667. doi:10.1101/cshperspect.a021667 PMID:25731765

Srdić, M. (2021). *Podrška preduzetnicima tokom pandemije COVID-19* [Support for Entrepreneurs during the COVID-19 Pandemic]. Organization for Security and Co-operation in Europe Mission to Serbia. https://www.osce.org/files/f/documents/2/7/513133.pdf

Stacey, R. D., Griffin, D., & Shaw, P. (2000). *Complexity and management: Fad or radical challenge to systems thinking?* Routledge.

State University of New York. (2013). *OSCQR*. Online Learning Consortium; State University of New York.

Stave, K. A., & Hopper, M. (2007). What constitutes systems thinking? A proposed taxonomy. *25th International Conference of the System Dynamics Society.*

Steam. (2022). China's War Epidemic. *Steam.* https://store.steampowered.com/app/1731360/_/

Steiner, E. D., & Woo, A. (2021). *Job-related stress threatens the teacher supply: Key findings from the 2021 State of the US Teacher Survey.* RAND.

Stouten, J., Rousseau, D. M., & De Cremer, D. (2018). Successful organizational change: Integrating the management practice and scholarly literatures. *The Academy of Management Annals*, *12*(2), 752–788. doi:10.5465/annals.2016.0095

Sulisworo, D., Agustin, S. P., & Sudarmiyati, E. (2016). Cooperative-blended learning using Moodle as an open source learning platform. *International Journal of Technology Enhanced Learning*, *8*(2), 187–198. doi:10.1504/IJTEL.2016.078089

Sun, S. Y. (2014). Learner perspectives on fully online language learning. *Distance Education*, *35*(1), 18–42. doi:10.1080/01587919.2014.891428

Sun, S. Y. (2021). Using patterns-based learning design for CALL tasks. *Computer Assisted Language Learning*, *34*(8), 1061–1084. doi:10.1080/09588221.2019.1657902

Sun, S. Y., & Goodyear, P. (2020). Social co-configuration in online language learning. *Australasian Journal of Educational Technology*, *36*(2), 13–26. doi:10.14742/ajet.5102

Sutcher, L., Darling-Hammond, L., & Carver-Thomas, D. (2016). *Coming crisis in teaching? Teacher supply, demand, and shortages in the U.S.* (research brief). Palo Alto, CA: Learning Policy Institute. Retrieved from: https://learningpolicyinstitute.org/product/coming-crisis-teaching

Sweller, J. (2020). Cognitive load theory and educational technology. *Educational Technology Research and Development*, *68*(1), 1–16. doi:10.100711423-019-09701-3

Tabnak, M., & Hosseini, S. (2021). The role of gamification and game-based learning in assessment of systems thinking. *Journal of Strategic Management Studies*, *12*(45), 119–138. https://www.smsjournal.ir/article_120683_en.html?lang=en

Tamayo, P. D. (2016). *Report to the legislature: Digital citizenship recommendations*. Office of Superintendent of Public Instruction.

Tamim, S. R. (2020). Analyzing the complexities of online education systems: A systems thinking perspective. *TechTrends, 64*(5), 740–750. doi:10.100711528-020-00538-9

Tan, Y. L. L., Varma, N., & Balakrishnan, N. P. (2020). Teaching systems thinking concepts during a pandemic: 3 strategies forengaging learners. In *Higher Education Campus Conference (e-HECC) 2020*. ScholarBank@NUS Repository. https://scholarbank.nus.edu.sg/handle/10635/202904

Tannenbaum, R., & Schmidt, W. (1973). How to choose a leadership pattern. *Harvard Business Review, 51*, 162–180.

Teach for America. (2020). *Our work*. https://www.teachforamerica.org/what-we-do/our-work

Teach for America. (2022). *Life as a Corps member*. Retrieved December 14, 2022, from https://www.teachforamerica.org/faqs/life-as-a-corps-member#salary-benefits--aid

Tesar, M. (2020). Towards a post-COVID-19 "new normality?": Physical and social distancing, the move to online and higher education. *Policy Futures in Education, 18*(5), 556–559. doi:10.1177/1478210320935671

Texas Education Agency. (2019). *Ready, set, teach!* https://readysetteach.org/index.html

Thang, S. M., Mustaffa, R., Wong, F. F., Noor, N. M., Mahmud, N., Latif, H., Aziz, M., & Abd, S. (2013). A Quantitative Inquiry into the effects of blended learning on English language learning: The case of Malaysian undergraduates. *International Education Studies, 6*(6), 1–7. doi:10.5539/ies.v6n6p1

The Deming Institute. (n.d.). https://deming.org/explore/pdsa/

The Electronic Code of Federal Regulations. (2022). *eCFR: 34 CFR Part 600 -- Institutional Eligibility Under the Higher Education Act of 1965, as Amended*. https://www.ecfr.gov/current/title-34/subtitle-B/chapter-VI/part-600

The European Anti-Poverty Network Serbia. (2020). *COVID Crisis and Poverty: Poverty Watch Serbia 2020*. https://www.eapn.eu/wp-content/uploads/2020/09/EAPN-EAPN-Serbia-Poverty-Watch-2020_ENG-4698.pdf

The global education crisis is even worse than we thought. Here's what we need to do about it. (2022). World Economic Forum. https://www.weforum.org/agenda/2022/01/global-education-crisis-children-students-covid19/

The Mind Tools Content Team. (2022). *Lewin's Change Management Model: Understanding the Three Stages of Change*. https://www.mindtools.com/ajm9l1e/lewins-change-management-model

The Open University. (1993a). *T301 Block4: The soft systems approach*. Open University Press.

The Open University. (1993b). *T301 Block2: A systems approach to failures*. Open University Press.

The Open University. (2007a). *T889 Block5: The soft systems approach*. Open University Press.

The Open University. (2007b). *T889 Block 1 problem solving and improvement: Quality and other approaches – Introduction*. Open University Press.

The Open University. (2022, March 23). *Systems explained by Peter Checkland*. Open Learn. https://www.open.edu/openlearn/money-business/leadership-management/systems-explained-peter-checkland

The Open University. (2023). *5.1 Definition of a system*. Open Learn. https://www.open.edu/openlearn/digital-computing/systems-thinking-and-practice/content-section-5.1

The territorial impact of COVID-19: Managing the crisis across levels of government. (n.d.). OECD Policy Responses to Coronavirus (COVID-19). https://www.oecd.org/coronavirus/policy-responses/the-territorial-impact-of-COVID-19-managing-the-crisis-across-levels-of-government-d3e314e1/

The W. Edwards Deming Institute. (2021). *Enriching society through the Deming philosophy| PDSA cycle.*

The World Bank Group. (2023). *GNI per capita, Atlas method (current US$) - India.*

Thomas, M. S., Crosby, S., & Vanderhaar, J. (2019). Trauma-Informed Practices in Schools Across Two Decades: An Interdisciplinary Review of Research. *Review of Research in Education, 43*(1), 422–452. doi:10.3102/0091732X18821123

Tian, F., Li, H., Tian, S., Yang, J., Shao, J., & Tian, C. (2020). Psychological symptoms of ordinary Chinese citizens based on SCL-90 during the level I emergency response to COVID-19. *Psychiatry Research, 288*, 112992. doi:10.1016/j.psychres.2020.112992 PMID:32302816

Tikkanen, A. (2023, April 11). *National Health Service.* Encyclopaedia Britannica. https://www.britannica.com/topic/National-Health-Service

Tirivayi, N., Richardson, D., Gavrilovic, M., Groppo, V., Kajula, L., Valli, E., & Viola, F. (2020). *A Rapid Review of Economic Policy and Social Protection Responses to Health and Economic Crises and Their Effects on Children: Lessons for the COVID-19 Pandemic Response* (UNICEF Office of Research – Innocenti Working Paper No. 02). https://www.unicef-irc.org/publications/pdf/WP2020-02.pdf

Tirozzi, G. N., & Uro, G. (1997). Education reform in the United States: National policy in support of local efforts for school improvement. *The American Psychologist, 52*(3), 241–249. doi:10.1037/0003-066X.52.3.241

Toch, T. (2011). Who rules? The national drive for education reform has touched off many power struggles, but one has emerged as fundamental. *The Wilson Quarterly, 43*(4), •••. http://archive.wilsonquarterly.com/essays/who-rules

Tomei, L. A. (2008). *Online and distance learning concepts, methodologies, tools, and applications.* Information Science Reference. doi:10.4018/978-1-59904-935-9

Topic, M., & Mitchell, B. (2019). Generation Z & Consumer Trends in Environmental Packaging. Project Report. The Retail Institute, Leeds.

Tozer, S., Senese, G., & Violas, P. (2015). *School and society: Historical and contemporary perspectives* (7th ed.). McGraw-Hill College.

Trbovich, P. (2014). Five ways to incorporate systems thinking into healthcare organizations. *Biomedical Instrumentation & Technology, 48*(s2, Suppl), 31–36. doi:10.2345/0899-8205-48.s2.31 PMID:25244335

Trendall, S. (2021). Sensitivity of contact-tracing app will not be adjusted. *PublicTechnology.net.* https://publictechnology.net/articles/news/sensitivity-contact-tracing-app-will-not-be-adjusted

Troiano, G., & Nardi, A. (2021). Vaccine hesitancy in the era of COVID-19. *Public Health, 194*, 245–251. doi:10.1016/j.puhe.2021.02.025 PMID:33965796

Tsolakis, N., Zissis, D., & Tjahjono, B. (2021). Scrutinising the interplay between governance and resilience in supply chain management: A systems thinking framework. *European Management Journal.*

Tsytsiura, K. V., & Romanova, G. M. (2020). Analysis of the Efficiency of China's Distance Economic Education in Force Majeure Circumstances. *International Journal of Learning. Teaching and Educational Research, 19*(9), 175–195. doi:10.26803/ijlter.19.9.10

Tumwesige, J. (2020). *COVID-19 educational disruption and response: Rethinking e-learning in Uganda*. University of Cambridge.

U.S. Department of Education. (2004). *Executive summary of the No Child Left Behind Act of 2001.* https://www2.ed.gov/nclb/overview/intro/execsumm.html

U.S. Department of Education. (2010). *An overview of the U.S. Department of Education.* https://www2.ed.gov/about/overview/focus/what.html

U.S. Department of Education. (2020a). *Accessibility Statement.* https://www2.ed.gov/notices/accessibility/index.html

U.S. Department of Education. (2020b). *Protecting Students With Disabilities.* https://www2.ed.gov/about/offices/list/ocr/504faq.html

U.S. Department of Education. (2020c). *UPDATED Guidance for interruptions of study related to Coronavirus (COVID-19) | Knowledge Center.* https://fsapartners.ed.gov/knowledge-center/library/electronic-announcements/2020-05-15/updated-guidance-interruptions-study-related-coronavirus-covid-19-updated-june-16-2020

U.S. Department of Education. (2022). *Teacher shortage areas.* Retrieved from: https://tsa.ed.gov/#/home/

U.S. Department of Education. (2022, August 16). *Addressing the teacher shortage with American rescue plan funds.* Retrieved from: https://www2.ed.gov/documents/coronavirus/teacher-shortage.pdf

Uddin, M. S., Routray, J. K., & Warnitchai, P. (2019). Systems thinking approach for resilient critical infrastructures in urban disaster management and sustainable development. In E. N. Farsangi, I. Takewaki, T. Y. Yang, A. Astaneh-Asl, & P. Gardoni (Eds.), *Resilient Structures and Infrastructure* (pp. 379–415). Springer. doi:10.1007/978-981-13-7446-3_15

Udovički, K., & Medić, P. (2021). *Impact of the COVID-19 Crisis on Employment: Focus on Vulnerable Categories.* Social Inclusion and Poverty Reduction Unit & CEVES. https://socijalnoukljucivanje.gov.rs/wp-content/uploads/2021/01/Uticaj_COVID-19_krize_na_zaposlenost.pdf

UK Health Security Agency. (2020, May 27). *NHS test and trace: What to do if you are contacted.* GOV.UK. https://www.gov.uk/guidance/nhs-test-and-trace-how-it-works

UK Health Security Agency. (2022a, February 24). *COVID-19: People with COVID-19 and their contacts.* GOV.UK. https://www.gov.uk/government/publications/covid-19-people-with-covid-19-and-their-contacts

UK Health Security Agency. (2022b, April 1). *People with symptoms of a respiratory infection including COVID-19.* GOV.UK. https://www.gov.uk/guidance/people-with-symptoms-of-a-respiratory-infection-including-covid-19

UK Health Security Agency. (2023, April 27). *NHS COVID-19 app.* GOV.UK. https://www.gov.uk/government/collections/nhs-covid-19-app

UK Health Security Agency. (n.d.). *Featured.* GOV.UK. https://www.gov.uk/government/organisations/uk-health-security-agency

Ulrich, W. (1996). *A primer to Critical Systems Heuristics for action researchers. Independent Research Monograph.* Centre for Systems Studies, University of Hull.

Ulrich, W., & Reynolds, M. (2010). Critical systems heuristics. In *Systems approaches to managing change: A practical guide* (pp. 243–292). Springer. doi:10.1007/978-1-84882-809-4_6

Ultra low interest rates are here to stay : 2021 central bank guide. (2021). Bloomberg News. https://www.bloomberg.com/news/articles/2021-01-05/ultra-low-interest-rates-here-to-stay-2021-central-bank-guide#xj4y7vzkg

UNESCO. (2020). *COVID-19 response toolkit - UNESCO. Globaleducationcoalition.unesco.org*. UNESCO.

United Nations Development Programme Serbia. (2020). *COVID-19: Socio-Economic Impact Assessment*. https://serbia.un.org/sites/default/files/2020-09/seia_report%20%281%29.pdf

United Nations Economic Commission for Europe. (2021). *The Impact of COVID-19 on Trade and Structural Transformation in Serbia – Evidence from UNECE's Survey of Micro, Small and Medium Enterprises*. https://unece.org/sites/default/files/2021-06/Impact_COVID-19_Serbia-Eng.pdf

United States Census Bureau (USCB). (2022). *Income in the United States: 2021*. U.S. Department of Commerce.

University of Cambridge. (2016). *Soft Systems Methodology*. Institute for Manufacturing. https://www.ifm.eng.cam.ac.uk/research/dstools/soft-systems-methodology/

van Bavel, J. J., Baiker, K., Boggio, P. S., Capraro, V., Cichocka, A., Cikara, M., Crockett, M. J., Crum, A. J., Douglas, K. M., Druckman, J. N., Drury, J., Oeindrila, D., Ellemers, N., Finkel, E. J., Fowler, J. H., Gelfand, M., Han, S., Haslam, S. A., Jetten, J., & Willer, R. (2020). Using social and behavioral science to support COVID-19 pandemic response. *Nature Human Behaviour*, *4*(5), 460–471. doi:10.103841562-020-0884-z PMID:32355299

Vangelisti, A. L., & Perlman, D. (Eds.). (2006). *The Cambridge handbook of personal relationships*. Cambridge University Press., doi:10.1017/CBO9780511606632

Verhoef, P. C., Noordhoff, C. S., & Sloot, L. (2022). Reflections and predictions on effects of COVID-19 pandemic on retailing. *Journal of Service Management*, *34*(2), 274–293. doi:10.1108/JOSM-09-2021-0343

Virtual Learning Leadership Alliance, & the Digital Learning Collaborative. (2023). *The Standards*. QM Quality Matters.

Vladisavljević, M., & Lebedinski, L. (2023). *Effects of the COVID-19 pandemic on the labour market and the position of vulnerable groups in Serbia*. Institute of Econonomic Sciences. Belgrade https://www.library.ien.bg.ac.rs/index.php/monog/article/view/1586/1299

Vladisavljević, M., Lebedinski, L., Banović, J., & Vukmirović, V. (2021). *Effects of the COVID-19 Pandemic on the Labour Market and the Position of Vulnerable Groups in Serbia*. Institute of Economic Sciences. (Research Project INEQ RS COVID-19 No. 7552225)

Vladisavljević, M., Zubović, J., Jovanovic, O., & Đukić, M. (2023). Crowding-out effect of tobacco consumption in Serbia. *Tobacco Control*. Advance online publication. doi:10.1136/tc-2022-057727 PMID:36990684

von Bertalanffy, L. (1968). General systems—A critical review. In W. Buckley (Ed.), *Modern systems research for the behavioral scientist*. Aldine.

W3C Web Accessibility Initiative (WAI). (2018). *Web Content Accessibility Guidelines (WCAG) 2.1*. https://www.w3.org/TR/WCAG21/

W3C Web Accessibility Initiative (WAI). (2021). *Planning Audio and Video Media*. https://www.w3.org/WAI/media/av/planning/

Wagner, T. (2010). *The global achievement gap: Why even our best schools don't teach the new survival skills our children need-and what we can do about it*. Academic Press.

Walter, C., & Briggs, J. (2012). *What professional development makes the most difference to teachers*. Oxford University Press. https://clie.org.uk/wp-content/uploads/2011/10/Walter_Briggs_2012.pdf

Wang, S. M., Chen, Y. C., Hou, H. T., Hsu, H. Y., & Li, C. T. (2020). Exploring the effects of card gamebased gamification instructional activity on learners' flow experience, learning anxiety, and performance-A preliminary study. *ICCE 2020-28th International Conference on Computers in Education, 2*, 190-198. https://apsce.net/icce/icce2020/proceedings/W1-13/W4/ICCE2020-Proceedings-Vol2-W4_5.pdf

Wang, F., Wang, Y., & Hu, X. (2017). Gamification Teaching Reform for Higher Vocational Education in China: A Case Study on Layout and Management of Distribution Center. *International Journal of Emerging Technologies in Learning, 12*(9), 130. Advance online publication. doi:10.3991/ijet.v12i09.7493

Wang, T., Olivier, D. F., & Chen, P. (2020). Creating individual and organizational readiness for change: Conceptualization of system readiness for change in school education. *International Journal of Leadership in Education*, 1–25. doi:10.1080/13603124.2020.1818131

Wang, Y., Luangasanatip, N., Pan-Ngum, W., Isaranuwatchai, W., Prawjaeng, J., Saralamba, S., Painter, C., Briones, J. V., & Teerawattananon, Y. (2022). Assessing the cost-effectiveness of COVID-19 vaccines in a low incidence and low mortality setting: The case of Thailand at start of the pandemic. *The European Journal of Health Economics*, 1–14. doi:10.100710198-022-01505-2 PMID:35951243

Watson, S. (2020). *EPIC releases county-by-county enrollment numbers*. Epic Charter Schools.

Watson, T. N. (Ed.). (2020). A Seat at the Table: Examining the impact, ingenuity, and leadership practices of Black women and girls in PK – 20 Contexts. Journal of Educational Administration and History.

Wearmouth, R. (2021). *Major overhaul to NHS Covid app to reduce number of contacts pinged to isolate*. The Mirror. https://apple.news/AtKXj1KuLQ7S4F3pla4vb2Q

Weiner, J., & Burton, L. (2016). The double bind for women: Exploring the gendered nature of turnaround leadership. *Harvard Educational Review, 86*(3), 339–365. doi:10.17763/1943-5045-86.3.339

Weiner, J., Cyr, D., & Burton, L. (2019). *How "identity-blindness" and other microaggressions shape the experiences of black women in principal preparation programs. Journal of Research on Leadership Education.*

Weissert, C. S., Uttermark, M. J., Mackie, K. R., & Artiles, A. (2021). Governors in control: Executive orders, state-local preemption, and the COVID-19 pandemic. *Publius, 51*(3), 396–428. doi:10.1093/publius/pjab013 PMID:34394702

Wheeler, F. P., & Checkland, P. (2000). Review of systems thinking, systems practice: Includes a 30-year retrospective. *The Journal of the Operational Research Society, 51*(5), 647–648. doi:10.2307/254200

Whipple, T. (2022, December 31). Could China's Covid 'catastrophe' be bad for Britain too? *The Times*. https://www.thetimes.co.uk/article/04648ec8-8855-11ed-bb21-8f4d97ec7b02?shareToken=3c0b21689f49dcd47784b8b6ff64d330

White, H. (2009). Theory-based impact evaluation: Principles and practice. *Journal of Development Effectiveness, 1*(3), 271–284. doi:10.1080/19439340903114628

Wiener Institut für Internationale Wirtschaftsvergleiche. (2021). *Getting Stronger after COVID-19: Nearshoring Potential in the Western Balkans* (wiiw Research Report No. 453). https://wiiw.ac.at/getting-stronger-after-covid-19-nearshoring-potential-in-the-western-balkans-dlp-5814.pdf

Wiener, N. (1961). *Cybernetics: or control in the animal and the machine*. The MIT Press. doi:10.1037/13140-000

Willis, N., Hill, S., Kaufman, J., Lewin, S., Kis-Rigo, J., De Castro Freire, S. B., Bosch-Capblanch, X., Glenton, C., Lin, V., Robinson, P., & Wiysonge, C. S. (2013). "Communicate to vaccinate": The development of a taxonomy of communication interventions to improve routine childhood vaccination. *BMC International Health and Human Rights, 13*(1), 1–16. doi:10.1186/1472-698X-13-23 PMID:23663327

Wilson, B. (2008). *Soft systems methodology: Conceptual model building and its contribution*. Wiley.

Wong, G., Apthorpe, H. C., Ruiz, K., & Nanayakkara, S. (2019). An innovative educational approach in using instructional videos to teach dental local anaesthetic skills. *European Journal of Dental Education, 23*(1), 28–34. doi:10.1111/eje.12382 PMID:30069994

World Economic Forum. (2021, January 18). *What 'systems thinking' actually means - and why it matters for innovation today.* World Economic Forum. https://www.weforum.org/agenda/2021/01/what-systems-thinking-actually-means-and-why-it-matters-today/

World Health Organization (WHO). (2020). *Naming the coronavirus disease (COVID-19) and the virus that causes it.*

World Health Organization. (2020). *COVID-19 disease situational reports.* https://www.who.int/emergencies/diseases/novel-coronavirus-2019/situation-reports/

World Health Organization. (2021). *COVID-19 significantly impacts health services for noncommunicable diseases.* https://www.who.int/newsroom/detail/01-06-2020-covid-19-significantly-impacts-health-services-for-noncommunicable-diseases

World Health Organization. (2023). *Coronavirus disease (COVID-10) pandemic.* https://www.who.int/emergencies/diseases/novel-coronavirus-2019

World Health Organization. (2023). *Epidemic and Pandemic Foresight Initiative.* WHO. https://pandemic-foresight.who.int/

Wright, A. (2022). *Teachers 'beaten down' by staff shortages, COVID.* Stateline. Retrieved from: https://stateline.org/2022/01/31/teachers-beaten-down-by-staff-shortages-covid/

Wu, Y. S., Wang, W. Y., Chan, T. C., Chiu, Y. L., Lin, H. C., Chang, Y. T., Wu, H. Y., Liu, T. C., Chuang, Y. C., Wu, J., Chang, W. Y., Sun, C. A., Lin, M. C., Tseng, V. S., Hu, J. M., Li, Y. K., Hsiao, P. J., Chen, C. W., Kao, H. Y., ... Chu, C. M. (2022). Effect of the Nintendo Ring Fit Adventure Exergame on Running Completion Time and Psychological Factors Among University Students Engaging in Distance Learning During the COVID-19 Pandemic: Randomized Controlled Trial. *JMIR Serious Games, 10*(1), e35040. doi:10.2196/35040 PMID:35315780

Xi, N., & Hamari, J. (2019). Does gamification satisfy needs? A study on the relationship between gamification features and intrinsic need satisfaction. *International Journal of Information Management, 46*, 210–221. doi:10.1016/j.ijinfomgt.2018.12.002

Yin, R. K. (2018). *Case study research and applications: Design and methods* (6th ed.). Sage Publications.

Young, D. J. (2008). An empirical investigation of the effects of blended learning on student outcomes in a redesigned intensive Spanish course. *CALICO Journal, 26*(1), 160–181. doi:10.1558/cj.v26i1.160-181

Zain, N. H. M., Johari, S. N., Aziz, S. R. A., Teo, N. H. I., Ishak, N. H., & Othman, Z. (2021). Winning the Needs of the Gen Z: Gamified Health Awareness Campaign in Defeating COVID-19 Pandemic. *Procedia Computer Science, 179*, 974–981. doi:10.1016/j.procs.2021.01.087

Zain, N. H. M., Othman, Z., Noh, N. M., Teo, N. H. I., Zulkipli, N. H. B. N., & Yasin, A. M. (2020). GAMEBC model: Gamification in health awareness campaigns to drive behaviour change in defeating COVID-19 pandemic. *International Journal of Advanced Trends in Computer Science and Engineering, 9*(4), 229–236. Advance online publication. doi:10.30534/ijatcse/2020/3491.42020

Zamarro, G., Camp, A., Fuchsman, D., & McGee, J. B. (2021). *How the pandemic has changed teachers' commitment to remaining in the classroom.* Brookings. Retrieved from https://www.brookings.edu/blog/brown-center-chalkboard/2021/09/08/how-the-pandemic-has-changed-teachers-commitment-to-remaining-in-the-classroom/

Žarković Rakić, J., Randjelović, S., & Vladisavljević, M. (2016). Labour market effects of social security contributions reform in Serbia. *Economic Annals*, *61*(208), 73–91. doi:10.2298/EKA1608073Z

Žarković-Rakić, J., & Vladisavljević, M. (2021). Tax reforms and income inequality in former Yugoslav countries: Escaping the avant-garde neoliberalism in the income tax policy. *Panoeconomicus*, *68*(2), 231–252. doi:10.2298/PAN2102231Z

Zdravković, A., Đukić, M., & Bradić-Martinović, A. (2017). Impact of FDI on unemployment in transition countries: panel cointegration approach. *Industrija*, *45*(1), 161-174. http://ebooks.ien.bg.ac.rs/1093/1/0350-03731701161Z.pdf

Zetchner, K. M., & Schulte, A. K. (2001). What we know and don't know from peer-reviewed research about alternative teacher certification programs. *Journal of Teacher Education*, *52*(14), 266–282. doi:10.1177/0022487101052004002

Zhang, Y., Li, X., Zhu, L., Dong, X., & Hao, Q. (2019). What is a smart classroom? A literature review. *Perspectives on Rethinking and Reforming Education*, 25–40. doi:10.1007/978-981-13-9439-3_2

Zhan, L., Guo, D., Chen, G., & Yang, J. (2018). Effects of Repetition Learning on Associative Recognition Over Time: Role of the Hippocampus and Prefrontal Cortex. *Frontiers in Human Neuroscience*, *12*, 277–277. doi:10.3389/fnhum.2018.00277 PMID:30050418

Zhao, J., & Li, Y. (2020). Experimental Study on Gamification Teaching of Mathematics Classroom Practice Based on Flow Experience. *2020 IEEE 2nd International Conference on Computer Science and Educational Informatization (CSEI)*, 36-40. 10.1109/CSEI50228.2020.9142485

Zhao, S. (2012). Global competency leadership and an empirical study of Chinese managers. In China's reality and global vision: Management research and development in China (pp. 104-124). Academic Press.

Zhong, M. (2021). Vaccine rewards keep escalating. *Netease News*. https://www.163.com/dy/article/GFHC-SMP105328YVR.html

Zhu, L. (2021). The psychology behind video games during COVID-19 pandemic: A case study of Animal Crossing: New Horizons. *Human Behavior and Emerging Technologies*, *3*(1), 157–159. doi:10.1002/hbe2.221

Zieba, K. (2021). How can systems thinking help us in the COVID-19 crisis? *Knowledge and Process Management*, *29*(3), 221–230. doi:10.1002/kpm.1680

Zubović, J., Đukić, M., & Jovanović, O. (2020). *Ekonomski aspekti kontrole duvana i empirijski nalazi u Srbiji* [Economic aspects of tobacco control and empirical findings in Serbia]. Institute of Economic Sciences.

About the Contributors

M. Elizabeth Azukas is currently an Assistant Professor at East Stroudsburg University and has been an educator for more than 25 years. She has held a number of different leadership roles including working as a curriculum supervisor, principal, curriculum director, and as an assistant superintendent. Her career has focused primarily on innovation in education with the goal of increasing equity and access for all students. Most recently, Dr. Azukas's work has focused on personalized learning, online and blended learning, systems thinking, mixed reality, and virtual leadership.

Minkyoung Kim | is an assistant professor at the University of West Florida. For the past 25 years, she has devoted herself to developing scholarship as well as practical professionalism in the field of Instructional Design & Technology. She has earned a doctoral degree in Instructional Systems Technology from Indiana University and bachelor's and master's degrees in the same field at Ewha Womans University in Korea. Previously, she was employed at IBM Korea as a senior business consultant and instructional designer for several years. Her research interests center on instructional design theories related to the learner-centered paradigm, personalized learning, quality of online education, and Her recent research focuses primarily on virtual reality learning environments and the application of artificial intelligence in teaching and learning. She has been on the leadership board of the Systems Thinking and Change (STC) division at the Association for Educational Communications and Technology (AECT) since 2015, and she currently holds the position of president of the STC at AECT.

* * *

Jennifer Bailey currently serves as an assistant professor of practice in educational leadership at The University of Texas at Tyler. A previous school principal and central office administrator, Dr. Bailey's research centers around school leadership learning.

Sarah Beach is a former public school teacher and current PhD candidate in Research, Statistics, and Evaluation at the University of Virginia School of Education and Human Development. Her research focuses on K-12 assessment and accountability as well as K-20 social justice initiatives.

David Bird has worked at a senior level as a UK National Cyber Security Centre certificated cyber security consultant for the UK Government over the past decade and beyond; providing advice and guidance on various topics including cyber security architecture, review of system designs and system level implementations. He holds a Master's Degree in Computing for Commerce and Industry with the Open

University; he has also been awarded a National Qualifications Framework (NQF) Level 7 qualification in Engineering by the City and Guilds Institute as well as NQF Level 6 in Intelligence Management and Leadership and Management by the same institution. He is a Chartered Fellow of the British Computer Society, a Fellow of the Institute of Analysts and Programmers, a Fellow of the Institute of Scientific and Technical Communicators, a Fellow of the Institute of Leadership and Management, and a Member of the Institution of Engineering Technology. He has written many published articles in several reputable magazines over the years comprising topical technical and information security subject matter that includes: British Computer Society ITNoW and Digital Leaders editions, CyberTalk and the then Institute of Information Security Professionals Pulse Magazine. He has authored book chapters for the Institute of Scientific and Technical Communicators and for the Sustainable Intelligent Systems library printed by Springer. David has also published papers with IGI Global, Springer, Institute of Electrical and Electronic Engineers Xplore and Association of Computing Machinery Digital libraries. His work has culminated in a full book publication with IGI Global in 2020 entitled 'Real-time and Retrospective Analyses of Cyber Security'.

Monalisa Dash is a Lecturer in Education in Brajrajnagar College, Sambalpur University, Odisha, India. Her main field of study is that of educational technology, and more specifically how it may be used to foster the growth of students' higher-level cognitive abilities and factors facilitating educators' digital competencies. Dr. Dash is the Associate Editor of EdTechnica and Editor of Abbreviations Page of EdTechnica. In 2022, Dr. Dash won the best research paper award at National Annual Young Researchers' Meet organized by Maharaja Sayajirao University, Gujarat, India. Dr. Dash worked in a team that was awarded the People's Choice Award by the Research and Theory Division at AECT Convention 2022.

Mihajlo Đukić graduated from the Faculty of Economics of the Belgrade University, Serbia (Bachelor and Master studies) and obtained PhD degree in finance at the Faculty of Economics, Technical University of Kosice, Slovakia. He has been employed as a researcher with the Institute of Economic Sciences (IES) since 2010. His research focuses on economic policy, development and evaluation of public policies. Mihajlo has participated in several projects commissioned by the state institutions at the national and local level, as well as by international organizations, which have dealt with the issues related to drafting and evaluation of strategic documents and policy analyses in various fields – local development, youth policy, health economics, etc. As a local coordinator for Serbia, over the period 2012-2017 he was in charge for monitoring over 20 regional research projects with a total value of more than 2 million CHF, funded by SDC and conducted by the research institutions from the Western Balkans. He is a member of the Western Balkan Evaluation Network (WBEN).

Ericka Eppler is an online teacher, instructional designer, and doctoral student at Oklahoma State University studying Learning, Design, and Technology.

Al Galloway, Phd., Assistant Program Administrator, Seton Hall University

Daniel Gutmore, Phd., Senior Faculty Associate, Seton Hall University

Keith Heggart is an early career researcher with a focus on learning and instructional design, educational technology and civics and citizenship education. He is currently exploring the way that online learning platforms can assist in the formation of active citizenship amongst Australian youth. Keith is a former high school teacher, having worked as a school leader in Australia and overseas, in government and non-government sectors. In addition, he has worked as an Organiser for the Independent Education Union of Australia, and as an independent Learning Designer for a range of organisations. Keith has been successful in attaining grants from the Royal Society for Arts, Manufactures and Commerce and also Penrith City Council.

Marvin Huang is an associate professor at the School of Animation and Digital Arts, Communication University of China, Nanjing. He is a PhD student at the School of Communication, Universiti Sains Malaysia under the supervision of Dr. Mohamad Saifudin Mohamad Saleh and Dr. Izzal Asnira Zolkepli. His main research areas include social media and game development.

Audrey Lombardo graduated from Malone College with a Bachelor of Arts in Psychology and Communication Studies and from Cleveland State University with a Master of Arts in Public Administration. She has extensive experience in strategic planning, project management, and business growth and development. She currently works for The Ohio State University Wexner Medical Center in Business Development and Operations.

Timothy Lombardo graduated from The Pennsylvania State University with a Bachelor of Science in Information Sciences and Technology. After graduation, he joined The Pennsylvania State University team as the Help Desk/PC Support Assistance at their Hazleton Campus. During this time, Timothy continued his education, earning a Master of Science in Instructional Technology with a specialization in Instructional Technology Specialist from the Bloomsburg University of Pennsylvania. After completing his master's program, he decided to continue his career, moving west to Ohio. Timothy accepted a position with Ashland University as an Instructional Designer/Course Developer, a position I held for five years. In 2013, he moved to Columbus, Ohio, where he accepted an Instructional Designer position at The Ohio State University and taught online through the Dennis Learning Center at Ohio State. During that time, he earned a Doctorate of Education in Curriculum and Instruction from the University of South Carolina. Timothy then took a Senior Instructional Designer Position at Purdue University. Timothy also began teaching online for the University of the People.

Matt Meador is a Systems Thinking educator, researcher, and scholar. He has been teaching in higher education for ten years, and currently teaching at the University of Tennessee at Martin. Matt consults on the principles of people, process, and environment through project management preparation for those pursuing careers in project management. Matt is supported by his loving wife of 14 years and five children.

Gary J. Miller is an associate professor of Educational Leadership in the School of Education at The University of Texas at Tyler. Dr. Miller's research interests include recruiting and retaining teachers in rural schools, building educational policymaking coalitions, and emerging technologies in language acquisition.

Helen Min is a Ph.D. candidate and Dean's Fellow in Curriculum and Instruction at the University of Virginia School of Education and Human Development with a research focus on evaluating trauma-sensitive pedagogy, understanding the impact of stress on teacher well-being, and assessing the extent of vicarious trauma on teachers. Before starting her Ph.D. program, she taught for six years in Baltimore, MD, Osaka, Japan, and Cairo, Egypt. She received her B.A. from the University of California Davis and her M.S.Ed. from Johns Hopkins University.

Mohamad Saifudin Mohamad Saleh holds a PhD Degree in Environmental Communication from the Faculty of Sustainability, Leuphana University of Lüneburg, Germany (2016). Upon receiving his PhD in September 2016, he began his career as a Senior Lecturer at the School of Communication, USM. His area of specialisation includes environmental communication and sustainability communication. Furthermore, he has a strong interest and passion for sustainability communication-oriented issues relating to religion.

Sheila E. Murphy, Ph.D., has served as a consultant to government, private sector, and nonprofit organizations for 30 years. She served on the faculty of the School of Public Affairs at Arizona State University, where she co-designed and instructed the 10-month Certified Public Manager program, serving federal, tribal, state, county, and municipal government professionals. Her research interests include social systems design, systems-based education, distance education, and organizational communication.

Kosovka Ognjenović is a Research Associate at the Institute of Economic Sciences in Belgrade, Serbia. She received her Ph.D. from the Faculty of Economics, University of Belgrade, where she completed her master's and undergraduate studies. She has published several scientific and professional papers and presented her work at domestic and international conferences. Dr. Ognjenović participates in teaching activities as a teaching assistant and as an instructor in summer schools and courses. She was an instructor at the 'Summer School of Econometrics' organized by the Institute of Economic Sciences. She has participated in several domestic and international projects, including the most recent one on ex-ante analysis of the employment strategy of Serbia. Dr. Ognjenović is a reviewer of scientific papers in several domestic and international journals and a member of Serbia's Scientific Society of Economists and the Association for Comparative Economic Studies (ACES).

Akila T. Rajappa, PhD, CCC-SLP, BCS-S is an Assistant Professor at Department of Communicative Sciences and Disorders (CSD) program at College of Health Sciences, East Stroudsburg University of Pennsylvania. Dr. Rajappa has been a practicing clinician for many years and is a Board-Certified Specialist in Swallowing and Swallowing Disorders (BCS-S). Dr. Rajappa is passionate about teaching and researching scientific topics related to medical speech pathology and in evaluating evidence-based practice (EBP), community out-reach and service-learning initiatives in speech, swallowing and cough rehabilitation for aging, stroke, and neurodegenerative disease population. She is also interested in clinical service delivery in multicultural population and has served in the capacity of President of Asian Indian Caucus (AIC), a multicultural constituency group of American Speech-Language & Hearing Association (ASHA).

Tracey A. Regenold, Ph.D., is an independent scholar and consultant. She develops education for adults in the workplace, primarily to assist them with designing and implementing more effective organizational and social systems. Tracey earned her Ph.D. in education from Old Dominion University. She also holds the Master of Public Administration from Arizona State University. Tracey's research interests include education on systems thinking and social systems design.

Marko Vladisavljević is a Research Associate at the Institute of Economic Sciences, Belgrade and an Assistant professor at the Faculty of Economics, University of Belgrade. He holds a PhD in Economics from the University of Belgrade. He has worked work as a consultant for the World Bank, ILO and UN organisations and has received numerous scholarships for academic and applied research projects. His main research interests include the economics of inequality, gender and demographic economics, health economics, and tax and transfer policy. He has academic papers published in international economics journals such as the Journal of policy modeling, Labor Economics, Cambridge Journal of Economics and the Review of Household Economics, as well as several book chapters for Palgrave and Springer.

Dewi Wahyu Mustikasari is a PhD Candidate of the Faculty of Arts and Social Sciences, University of Technology Sydney, Australia. She works for the English Education Department of UIN Salatiga, a government university in Central Java, Indonesia. Her research interests include technology-enhanced language learning, learning design, and teacher practice.

Rachel Wolf, PhD. CCC-SLP, is an associate professor and chair of the department of Communication Sciences and Disorders at East Stroudsburg University in PA and is an ASHA certified SLP. She is director of the grant funded ESU Autism Education Center. She completed her PhD at New York University in 2009. Her published research examines pragmatic repetition in normal and disordered verbal discourse. Current research has focused on impacts of disorder, specifically ASD, on pragmatics and discourse.

Index

U

V

www.ingramcontent.com/pod-product-compliance
Lightning Source LLC
LaVergne TN
LVHW062141300926
849161LV00010B/2590